Endorsed by
University of Cambridge International Examinations

Complete

Physics

for IGCSE

Stephen Pople

OXFORD

OXFORD
UNIVERSITY PRESS

Great Clarendon Street, Oxford OX2 6DP

Oxford University Press is a department of the University of Oxford. It
furthers the University's objective of excellence in research, scholarship, and
education by publishing worldwide in

Oxford   New York

Auckland   Cape Town   Dar es Salaam   Hong Kong   Karachi
Kuala Lumpur   Madrid   Melbourne   Mexico City   Nairobi
New Delhi   Shanghai   Taipei   Toronto

With offices in

Argentina   Austria   Brazil   Chile   Czech Republic   France   Greece
Guatemala   Hungary   Italy   Japan   Poland   Portugal   Singapore
South Korea   Switzerland   Thailand   Turkey   Ukraine   Vietnam

Oxford is a registered trade mark of Oxford University Press
in the UK and in certain other countriesThe moral rights of the author have
been asserted

Database right Oxford University Press (maker)

First published 1999 as Complete Physics (ISBN 0 19 914734 5)

This edition: Complete Physics for IGCSE first published in 2007

British Library Cataloguing in Publication Data
Data available

ISBN: 978-0-19-915133-2
10 9 8 7 6 5 4 3 2

Printed in China by Printplus

Acknowledgements

The publisher would like to thank the following for their kind permission to reproduce the following photographs:

Cover image courtesy of Photodisc/Don Farrall
Page 9, Tony Stone, J Strachan, **Page 16**, Science Photo Library/J Thomas, **Page 21** (right), Science Photo Library/A-Hart-Davis, (left) TCL, **Page 25**, PA News, **Page 26**, Science Photo Library/K Kent, **Page 30** Tony Stone/A Husmo, **Page 32** Robert Harding Picture Library, **Page 37**, J Allan Cash, **Page 40** Science Photo Library/T Takahara, **Page 41**, Volvo, **Page 48** Allsport, **Page 53** David Simson, **Page 57**, David Simson, **Page 61** David Simson, **Page 64**, Science Photo Library/ R Church, **Page 67**, David Simson, **Page 68** Science Photo Library/Nasa, **Page 70** Science Photo Library/P Plailley, **Page 71** Tony Stone/C Condina, **Page 78** Allsport, **Page 86** (top) Alamy Images, **Page 86** (lower) Tony Stone/A Levenson, **Page 88** Science Photo Library/G Parker, **Page 91** David Simson, **Page 97**, David Simson, **Page 98** TCL, **Page 109** Oxford Scientific Films/M Birjhead, **Page 113** J Allan Cash, **Page 115** David Simson, **Page 119** David Simson, **Page 123** Oxford Scientific Films/J Brown, **Page 128** David Simson, **Page 130** David Simson, **Page 134** Oxford Scientific Films/T Tilford, **Page 135** Science Photo Library/H Schneebeli, **Page 139** David Simson, **Page 140** (top), David Simson, **Page 140** (lower) Science Photo Library/P Menzel, **Page 141** Science Photo Library/Nasa, **Page 142** David Simson, **Page 149** (left) Science Photo Library/D Parker, **Page 149** (right) Science Photo Library/Dr K Schiller, **Page 150** Science Photo Library/A Pasieka, **Page 159**, Science Photo Library, **Page 162** David Simson, **Page 165** (top right), Science Photo Library/P Hayson, Dr J Burgess (left and right), **Page 169** Tony Stone/C Shinn, **Page 170** Science Photo Library/P Menzel, **Page 172** J Allan Cash, **Page 186**, J Allan Cash, **Page 201**, Science Photo Library/Los Alamos National Laboratory, **Page 205** B and C Alexander, **Page 208** David Simson, **Page 217** J Allan Cash, **Page 219** J Allan Cash, **Page 220** Science Photo Library/Salisbury Hospital, **Page 222** David Simson, **Page 225**, Robert Harding, **Page 229** Science Photo Library/Bio Photo Associates, **Page 249**,Tony Stone, **Page 254** Science Photo Library/D Parker, **Page 261** Science Photo Library/M Bond, **Page 262** Science Photo Library/US Dept. of Energy, **Page 264** Science Photo Library/L Mulvehill, **Page 268** Science Photo Library/Fermilab, **Page 273** Science and Society Picture Library, **Page 275** (top and lower), David Simson.

Technical Photography by Peter Gould.

We have tried to trace and contact all copyright holders. If notified the publishers will be pleased to rectify any errors or omissions at the earliest opportunity.

Illustrations by Jeff Bowles, Roger Courthold, Mike Ogden, Jeff Edwards, Russell Walker, Clive Goodyer, Jamie Sneddon, Q2A and Tech Graphics

We are grateful to the following Examination Boards for permission to use GCSE examinations questions and specimen questions (Sp = specimen).

Questions (including all multiple choice questions attributed to L. EXAM.) have been provided by London Examinations, a division of Edexcel Foundation. Edexcel Foundation, **London Examinations** accepts no responsibility whatsoever for the accuracy or method of working in the answers given.
MEG: questions from MEG reproduced by kind permission of the Midland Examining Group. The Midland Examining Group bears no responsibility for the example answers to questions taken from its past question papers which are contained in this publication.
UCLES/CIE: all CIE questions are taken from the IGCSE Physics paper and are reproduced by permission of The University of Cambridge Local Examinations Syndicate. The University of Cambridge Local Examinations Syndicate bears no responsibility for the example answers to questions taken from its past question papers which are contained in this publication. **SEG**: questions from SEG are reproduced with the kind permission of the Southern Examining Group. The Southern Examining Group bears no responsibility for the example answers to questions taken from its past question papers which are contained in this publication.
WJEC: questions from WJEC are reproduced with the kind permission of the Welsh Joint Education Committee. The Welsh Joint Education Committee bears no responsibility for the example answers to questions taken from its past question papers which are contained in this publication.
NEAB: questions from NEAB are reproduced with the kind permission of Northern Examination and Assessment Board. The Northern Examination and Assessment Board bears no responsibility for the example answers to questions taken from its past question papers which are contained in this publication.

All answers are the responsibility of the author.

The author would like to thank Susan Pople for her help in preparing the manuscript and Dr Darren Lewis for providing answers to the questions in sections 1–11.

Introduction

If you are studying physics for IGCSE, then this book is designed for you. It explains the concepts that you will meet, and should help you with your experimental procedures. It is written mostly in double-page units that we have called **spreads**. These are grouped into sections.

Contents This lists the sections, and the spreads in each one.

Syllabus and spreads This gives you guidance about which spreads relate to the particular examination syllabus you are following.

Sections 1 to 11 The main areas of physics are covered here.

Revision checklists At the end of each section (1 to 11), there is a revision checklist giving the main topics covered in each spread.

History of Key Ideas Section 12 describes how scientists have developed their understanding of physics over the years.

Experimental physics Section 13 tells you how to plan and carry out experiments and interpret the results. It includes suggestions for investigations, and guidance on taking practical tests.

Mathematics for physics Section 14 summarizes the mathematical skills you will need when studying physics for IGCSE.

Examination questions There are practice examination questions at the end of each section (1 to 11). In addition, Section 15 contains a collection of questions taken mainly from CIE's IGCSE papers, including some alternative-to-practical questions.

Section 16 is a reference section. It contains the following:

Useful equations A list of the various equations used in the book.

Units and elements A summary of the main units of measurement, and a list of chemical elements.

Electrical symbols and codes The main symbols used in circuit diagrams, and how to interpret the markings on resistors.

Answers Check your numerical (and shorter descriptive) answers here.

Index Use this if you need to look up a particular word or term.

I hope that you will find the book useful as you develop your understanding of physics and prepare for your examinations.

Stephen Pople

Contents

* Watch for this symbol, below and throughout the book. It indicates spreads or parts of spreads that have been included to provide extension material to set physics in a broader context.

For information about the link between spreads and the CIE syllabus, see pages 7–8.

Syllabus and spreads

Below is an outline of CIE's IGCSE syllabus as it stood at the time of publication, along with details of where each topic is covered in the book. Before using this information to construct a teaching or revision programme, please check with the latest version of the syllabus/specification for any changes.

Spread(s)

Units and physical quantities	1.01, p324
General physics	
Length and time	1.02–1.03
Speed, velocity, and acceleration	2.01–2.06
Mass and weight	1.02, 2.09
Density	1.04–1.06
Forces	
a) Effects of forces	2.07–2.08, 2.11, 2.12, 3.04
b) Turning effect	3.01–3.03
c) Conditions for equilibrium	3.01
d) Centre of mass	3.02
e) Scalars and vectors	2.01, 2.11
Energy, work, and power	
a) Energy	4.01–4.03
b) Energy resources	4.05–4.08, 11.07
c) Work	4.01–4.02
d) Power	4.04
Pressure	3.05–3.06, 3.08

Thermal physics	
Simple kinetic molecular model of matter	
a) States of matter	5.01
b) Molecular model	5.01–5.02, 5.05
c) Evaporation	5.09
d) Pressure changes	3.09
Thermal properties	
a) Thermal expansion of solids, liquids, and gases	5.04–5.05
b) Measurement of temperature	5.02–5.03
c) Thermal capacity	5.10
d) Melting and boiling	5.09, 5.11
Transfer of thermal energy	
a) Conduction	5.06
b) Convection	5.07
c) Radiation	5.08
d) Consequences of energy transfer	5.06–5.08

Properties of waves including light and sound

General wave properties	6.01–6.03
Light	
a) Reflection of light	7.02–7.03
b) Refraction of light	7.04–7.06
c) Thin converging lens	7.07–7.08
d) Dispersion of light	7.04
e) Electromagnetic spectrum	7.11–7.12
Sound	6.03–6.06

Electricity and magnetism

Simple phenomena of magnetism	9.01–9.02, 9.04
Electrical quantities	
a) Electric charge	8.01–8.03
b) Current	8.04
c) Electromotive force	8.05
d) Potential difference	8.05
e) Resistance	8.06–8.08
f) Electrical energy	8.11, 8.14
Electric circuits	
a) Circuit diagrams	8.04 and on p325
b) Series and parallel circuits	8.09–8.10
c) Action and use of circuit components	8.06, 9.04, 10.01–10.04
d) Digital electronics	10.01, 10.05–10.06
Dangers of electricity	8.12–8.13
Electromagnetic effects	
a) Electromagnetic induction	9.07–9.08
b) AC generator	9.09
c) Transformer	9.10–9.12
d) The magnetic effect of a current	9.03–9.04
e) Force on a current-carrying conductor	9.05
f) DC motor	9.06
Cathode ray oscilloscopes	
a) Cathode rays	10.07
b) Simple treatment of cathode ray oscilloscope	10.08

Atomic physics

Radioactivity	
a) Detection of radioactivity	11.02–11.03
b) Characteristics of the three kinds of emission	11.02
c) Radioactive decay	11.04
d) Half-life	11.05
e) Safety precautions	11.03, 11.06
The nuclear atom	
a) Atomic model	11.01, 11.09
b) Nucleus	11.01
c) Isotopes	11.01, 11.08

1 Measurements and Units

- PHYSICAL QUANTITIES
- UNITS AND PREFIXES
- SCIENTIFIC NOTATION
- SI UNITS
- MASS
- TIME
- LENGTH
- VOLUME
- DENSITY

Astronomical clock in Prague, in the Czech Republic. As well as giving the time, the clock also shows the positions of the Sun and Moon relative to the constellations of the zodiac. Until about fifty years ago, scientists had to rely on mechanical clocks, such as the one above, to measure time. Today, they have access to atomic clocks whose timekeeping varies by less than a second in a million years.

1.01 Numbers and units

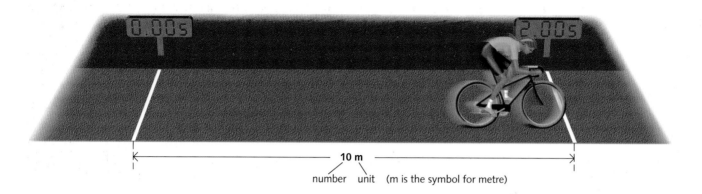

10 m
number unit (m is the symbol for metre)

When you make a measurement, you might get a result like the one above: a distance of 10 m. The complete measurement is called a **physical quantity**. It is made up of two parts: a number and a unit.

10 m really means $10 \times m$ (ten times metre), just as in algebra, $10x$ means $10 \times x$ (ten times x). You can treat the m just like a symbol in an algebraic equation. This is important when combining units.

Combining units

In the diagram above, the girl cycles 10 metres in 2 s. So she travels 5 metres every second. Her *speed* is 5 metres per second. To work out the speed, you divide the distance travelled by the time taken, like this:

$$\text{speed} = \frac{10\,\text{m}}{2\,\text{s}} \qquad \text{(s is the symbol for second)}$$

As m and s can be treated as algebraic symbols:

$$\text{speed} = \frac{10}{2} \cdot \frac{\text{m}}{\text{s}} = 5\,\frac{\text{m}}{\text{s}}$$

To save space, $5\,\dfrac{\text{m}}{\text{s}}$ is usually written as 5 m/s.

So m/s is the unit of speed.

Rights and wrongs

This equation is correct: $\text{speed} = \dfrac{10\,\text{m}}{2\,\text{s}} = 5\,\text{m/s}$

This equation is incorrect: $\text{speed} = \dfrac{10}{2} = 5\,\text{m/s}$

It is incorrect because the m and s have been left out. 10 divided by 2 equals 5, and not 5 m/s.

Strictly speaking, units should be included at *all* stages of a calculation, not just at the end. However, in this book, the 'incorrect' type of equation will sometimes be used so that you can follow the arithmetic without units which make the calculation look more complicated.

Advanced units

5 m/s is a space-saving way of writing $5\,\dfrac{\text{m}}{\text{s}}$.

But $5\,\dfrac{\text{m}}{\text{s}}$ equals $5\,\text{m}\,\dfrac{1}{\text{s}}$.

Also, $\dfrac{1}{\text{s}}$ can be written as s^{-1}.

So the speed can be written as $5\,\text{m}\,\text{s}^{-1}$.

This method of showing units is more common in advanced work.

Bigger and smaller

You can make a unit bigger or smaller by putting an extra symbol, called a prefix, in front. (Below, W stands for watt, a unit of power.)

prefix	meaning		example
G (giga)	1 000 000 000	(10^9)	GW (gigawatt)
M (mega)	1 000 000	(10^6)	MW (megawatt)
k (kilo)	1000	(10^3)	km (kilometre)
d (deci)	$\frac{1}{10}$	(10^{-1})	dm (decimetre)
c (centi)	$\frac{1}{100}$	(10^{-2})	cm (centimetre)
m (milli)	$\frac{1}{1000}$	(10^{-3})	mm (millimetre)
µ (micro)	$\frac{1}{1\,000\,000}$	(10^{-6})	µW (microwatt)
n (nano)	$\frac{1}{1\,000\,000\,000}$	(10^{-9})	nm (nanometre)

Powers of 10
$1000 = 10 \times 10 \times 10 = 10^3$
$100 = 10 \times 10 = 10^2$
$0.1 = \frac{1}{10} = 10^{-1}$
$0.01 = \frac{1}{100} = \frac{1}{10^2} = 10^{-2}$
$0.001 = \frac{1}{1000} = \frac{1}{10^3} = 10^{-3}$

'milli' means 'thousandth', *not* 'millionth'

Scientific notation

An atlas says that the population of Iceland is this:

270 000

There are two problems with giving the number in this form. Writing lots of zeros isn't very convenient. Also, you don't know which zeros are accurate. Most are only there to show you that it is a six-figure number. These problems are avoided if the number is written using powers of ten:

2.7×10^5 ($10^5 = 10 \times 10 \times 10 \times 10 \times 10 = 100\,000$)

'2.7×10^5' tells you that the figures 2 and 7 are important. The number is being given to *two significant figures*. If the population were known more accurately, to three significant figures, it might be written like this:

2.70×10^5

Numbers written using powers of ten are in **scientific notation** or **standard form**. The examples on the right are to one significant figure.

decimal	fraction	scientific notation
500		5×10^2
0.5	$\frac{5}{10}$	5×10^{-1}
0.05	$\frac{5}{100}$	5×10^{-2}
0.005	$\frac{5}{1000}$	5×10^{-3}

1 How many grams are there in 1 kilogram?
2 How many millimetres are there in 1 metre?
3 How many microseconds are there in 1 second?
4 This equation is used to work out the area of a rectangle: area = length × width.
 If a rectangle measures 3 m by 2 m, calculate its area, and include the units in your calculation.

5 Write down the following in km:
 2000 m   200 m 2×10^4 m
6 Write down the following in s:
 5000 ms 5×10^7 µs
7 Using scientific notation, write down the following to two significant figures:
 1500 m   1 500 000 m   0.15 m   0.015 m

1.02 A system of units

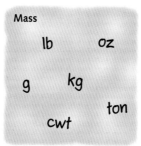

Mass: lb, oz, g, kg, ton, cwt

Length: cm, yd, m, mile, ft, mm, km

Time: s, hour, day, month, year, ms

There are many different units – including those above. But in scientific work, life is much easier if everyone uses a common system of units.

SI units

Most scientists use **SI units** (full name: Le Système International d'Unités). The basic SI units for measuring mass, time, and length are the kilogram, the second, and the metre. From these **base units** come a whole range of units for measuring volume, speed, force, energy, and other quantities.

Other SI base units include the ampere (for measuring electric current) and the kelvin (for measuring temperature).

Mass

Mass is a measure of the amount of substance in an object. It has two effects:
- All objects are attracted to the Earth. The greater the mass of an object, the stronger is the Earth's gravitational pull on it.
- All objects resist attempts to make them go faster, slower, or in a different direction. The greater the mass, the greater is the resistance to change in motion.

The SI base unit of mass is the **kilogram** (symbol **kg**). The standard kilogram is a block of platinum alloy kept at the Office of Weights and Measures in Paris. Other units based on the kilogram are shown below:

The mass of an object can be found using a **balance** like this. The balance really detects the gravitational pull on the object on the pan, but the scale is marked to show the mass.

mass		comparison with base unit	scientific notation	approximate size
1 tonne (t)		1000 kg	10^3 kg	medium-sized car
1 kilogram (kg)		1 kg		bag of sugar
1 gram (g)	1 g	$\frac{1}{1\,000}$ kg	10^{-3} kg	banknote
1 milligram (mg)	$\frac{1}{1\,000}$ g	$\frac{1}{1\,000\,000}$ kg	10^{-6} kg	human hair

Note: the SI base unit of mass is the **kilogram**, not the gram

Time

The SI base unit of time is the **second** (symbol **s**). Here are some shorter units based on the second:

$$1 \text{ millisecond (ms)} = \frac{1}{1000} \text{ s} = 10^{-3} \text{ s}$$

$$1 \text{ microsecond (µs)} = \frac{1}{1\,000\,000} \text{ s} = 10^{-6} \text{ s}$$

$$1 \text{ nanosecond (ns)} = \frac{1}{1\,000\,000\,000} \text{ s} = 10^{-9} \text{ s}$$

To keep time, clocks and watches need something that beats at a steady rate. Some old clocks used the swings of a pendulum. Modern digital watches count the vibrations made by a tiny quartz crystal.

> The second was originally defined as $\frac{1}{60 \times 60 \times 24}$ of a day, one day being the time it takes the Earth to rotate once. But the Earth's rotation is not quite constant. So, for accuracy, the second is now defined in terms of something that never changes: the frequency of an oscillation which can occur in the nucleus of a caesium atom.

Length

The SI base unit of length is the **metre** (symbol **m**). At one time, the standard metre was the distance between two marks on a metal bar kept at the Office of Weights and Measures in Paris. A more accurate standard is now used, based on the speed of light, as on the right.

> By definition, one metre is the distance travelled by light in a vacuum in $\frac{1}{299\,792\,458}$ of a second.

There are larger and smaller units of length based on the metre:

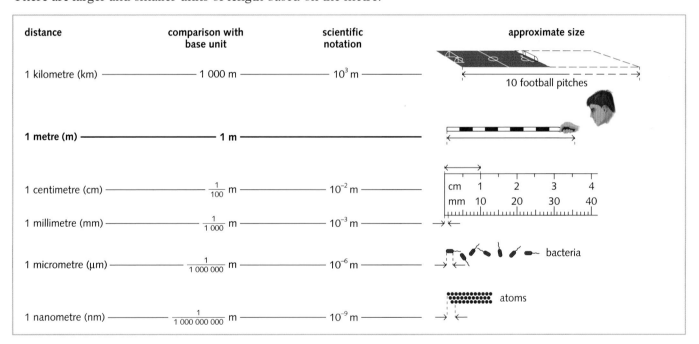

Q

1 What is the SI unit of length?
2 What is the SI unit of mass?
3 What is the SI unit of time?
4 What do the following symbols stand for?
   g   mg   t   µm   ms
5 Write down the value of
  a) 1564 mm in m   **b)** 1750 g in kg
  c) 26 t in kg   **d)** 62 µs in s
  e) 3.65×10^4 g in kg **f)** 6.16×10^{-7} mm in m
6 The 500 pages of a book have a mass of 2.50 kg. What is the mass of each page **a)** in kg **b)** in mg?

7 km   µg   µm   t   nm   kg   m
  ms   s   mg   ns   µs   g   mm

Arrange the above units in three columns as below. The units in each column should be in order, with the largest at the top.

mass	length	time

1.03 Measuring length and time

Measuring length

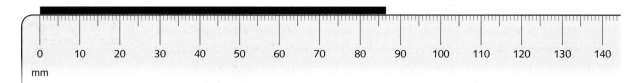

Lengths from a few millmetres up to a metre can be measured using a **rule**, as shown above. When using the rule, the scale should be placed right next to the object being measured. If this is not possible, **calipers** can be used, as shown on the left. The calipers are set so that their points exactly match the ends of the object. Then they are moved across to a rule to make the measurement.

Lengths of several metres can be measured using a **tape** with a scale on it.

With small objects, more accurate length measurements can be made using the methods shown below:

Micrometer (below left) This has a revolving barrel with an extra scale on it. The barrel is connected to a screw thread and, in the example shown, each turn of the barrel closes (or opens) the gap by one millimetre. First, the gap is opened wide. Then it is closed up until the object being measured *just* fits in it (a 'clicking' sound is heard). The diagram shows you how to take the reading.

Vernier calipers (below right) This is an extra sliding scale fitted to some length-measuring instruments. Its divisions are set slightly closer together than normal so that one of them coincides with a division on the fixed scale. The diagram shows you how to take the reading. (The vernier shown is part of a set of calipers used for making external measurements. A second type of caliper has jaws for making internal measurements.)

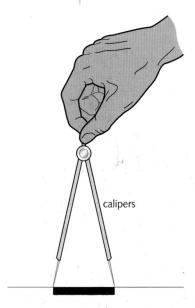

▲ If the rule cannot be placed next to the object being measured, calipers can be used.

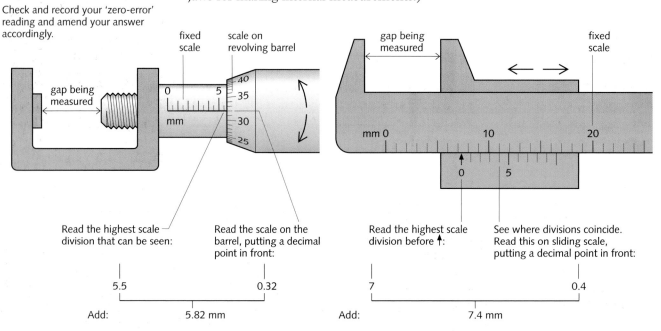

▲ Reading a micrometer

▲ Reading a vernier

Measuring time

Time intervals of many seconds or minutes can be measured using a **stopclock** or a **stopwatch**. Some instruments have an **analogue** display, with a needle ('hand') moving round a circular scale. Others have a **digital** display, which shows a number. There are buttons for starting the timing, stopping it, and resetting the instrument to zero.

With a hand-operated stopclock or stopwatch, making accurate measurements of short time intervals (a few seconds or less) can be difficult. This is because of the time it takes you to react when you have to press the button. Fortunately, in some experiments, there is an simple way of overcoming the problem. Here is an example:

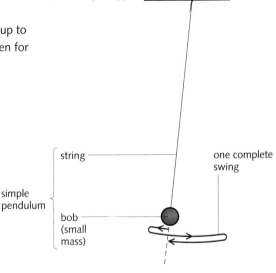

A pendulum can be set up to investigate the time taken for a single swing.

The pendulum above takes about two seconds to make one complete swing. Provided the swings are small, every swing takes the same time. This time is called its **period**. You can find it accurately by measuring the time for 25 swings, and then dividing the result by 25. For example:

Time for 25 swings = 55 seconds

So: time for 1 swing = 55/25 seconds = 2.2 seconds

Another method of improving accuracy is to use automatic timing, as shown in the example on the right. Here, the time taken for a small object to fall a short distance is being measured. The timer is started automatically when the ball cuts one light beam and stopped when it cuts another.

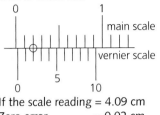

Zero error

Vernier calipers are said to have a **zero error** if the zero marking on the main scale is not in line with the zero marking on the vernier scale when the jaws are fully closed. For example, on the vernier calipers below, the zero error is +0.02 cm.

If the scale reading = 4.09 cm
Zero error = 0.02 cm
Then, the corrected reading
= (4.09 − 0.02) = 4.07 cm

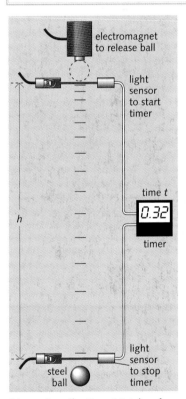

Measuring the time t it takes for a steel ball to fall a distance h

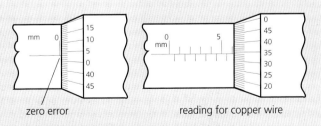

Q

1 A student measures the time taken for 20 swings of a pendulum. He finds that the time taken is 46 seconds.
a) What time does the pendulum take for one swing?
b) How could the student have found the time for one swing more accurately?
2 A student wants to find the thickness of one page of this book. Explain how she might do this accurately.
3 A micrometer is used to measure the diameter of a length of copper wire. The zero error and scale reading are as shown.

zero error

reading for copper wire

a) What is the zero error of the micrometer?
b) What is the correct diameter of the wire?

1.04 Volume and density

Volume

The quantity of space an object takes up is called its volume.

The SI unit of volume is the **cubic metre** (**m³**). However, this is rather large for everyday work, so other units are often used for convenience, as shown in the diagrams below:

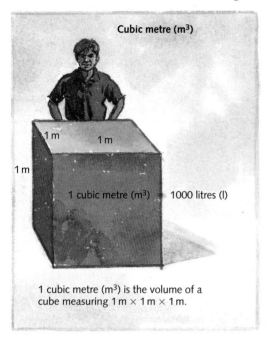

Cubic metre (m³)

1 m
1 m
1 m

1 cubic metre (m³) = 1000 litres (l)

1 cubic metre (m³) is the volume of a cube measuring 1 m × 1 m × 1 m.

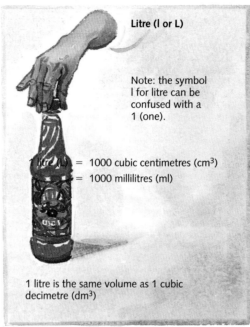

Litre (l or L)

Note: the symbol l for litre can be confused with a 1 (one).

1 litre (L) = 1000 cubic centimetres (cm³)
= 1000 millilitres (ml)

1 litre is the same volume as 1 cubic decimetre (dm³)

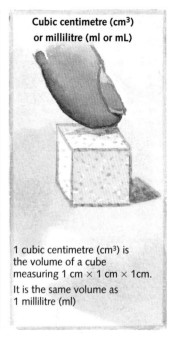

Cubic centimetre (cm³) or millilitre (ml or mL)

1 cubic centimetre (cm³) is the volume of a cube measuring 1 cm × 1 cm × 1cm.
It is the same volume as 1 millilitre (ml)

Density

Is lead heavier than water? Not necessarily. It depends on the volumes of lead and water being compared. However, lead is more dense than water: it has more kilograms packed into every cubic metre.

The **density** of a material is calculated like this:

$$\text{density} = \frac{\text{mass}}{\text{volume}}$$

In the case of water:
a mass of 1000 kg of water has a volume of 1 m³
a mass of 2000 kg of water has a volume of 2 m³
a mass of 3000 kg of water has a volume of 3 m³, and so on.

Using any of these sets of figures in the above equation, the density of water works out to be 1000 kg/m³.

If masses are measured in grams (g) and volumes in cubic centimetres (cm³), it is simpler to calculate densities in g/cm³. Converting to kg/m³ is easy:

$$1 \text{ g/cm}^3 = 1000 \text{ kg/m}^3$$

The density of water is 1 g/cm³. This simple value is no accident. The kilogram (1000 g) was originally supposed to be the mass of 1000 cm³ of water (pure, and at 4 °C). However, a very slight error was made in the early measurement, so this is no longer used as a definition of the kilogram.

The glowing gas in the tail of a comet stretches for millions of kilometres behind the comet's core. The density of the gas is less than a kilogram per cubic *kilo*metre.

substance	density kg/m³	density g/cm³	substance	density kg/m³	density g/cm³
air	1.3	0.0013	granite	2700	2.7
expanded polystyrene	14	0.014	aluminium	2700	2.7
wood (beech)	750	0.75	steel (stainless)	7800	7.8
petrol	800	0.80	copper	8900	8.9
ice (0 °C)	920	0.92	lead	11 400	11.4
polythene	950	0.95	mercury	13 600	13.6
water (4 °C)	1000	1.0	gold	19 300	19.3
concrete	2400	2.4	platinum	21 500	21.5
glass (varies)	2500	2.5	osmium	22 600	22.6

The densities of solids and liquids vary slightly with temperature. Most substances get a little bigger when heated. The increase in volume reduces the density.

The densities of gases can vary enormously depending on how compressed they are.

The rare metal osmium is the densest substance found on Earth. If this book were made of osmium, it would weigh as much as a heavy suitcase.

Density calculations

The equation linking density, mass, and volume can be written in symbols:

$$\rho = \frac{m}{V}$$ where ρ = density, m = mass, and V = volume

This equation can be rearranged to give: $V = \frac{m}{\rho}$ and $m = V\rho$

These are useful if the density is known, but the volume or mass is to be calculated. On the right is a method of finding all three equations.

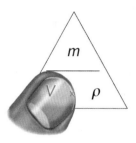

Cover V in the triangle and you can see what V is equal to. It works for m and ρ as well.

Example Using density data from the table above, calculate the mass of steel having the same volume as 5400 kg of aluminium.

First, calculate the volume of 5400 kg of aluminium. In this case, ρ is 2700 kg/m³, m is 5400 kg, and V is to be found. So:

$$V = \frac{m}{\rho} = \frac{5400\,\text{kg}}{2700\,\text{kg/m}^3} = 2\,\text{m}^3$$

This is also the volume of the steel. Therefore, for the steel, ρ is 7800 kg/m³, V is 2 m³, and m is to be found. So:

$$m = V\rho = 7800\,\text{kg/m}^3 \times 2\,\text{m}^3 = 15\,600\,\text{kg}$$

So the mass of steel is 15 600 kg.

In the density equation, the symbol ρ is the Greek letter 'rho'.

1 How many cm³ are there in 1 m³?
2 How many cm³ are there in 1 litre?
3 How many ml are there in 1 m³?
4 A tankful of liquid has a volume of 0.2 m³. What is the volume in **a)** litres **b)** cm³ **c)** ml?
5 Aluminium has a density of 2700 kg/m³.
   a) What is the density in g/cm³?
   b) What is the mass of 20 cm³ of aluminium?
   c) What is the volume of 27 g of aluminium?

Use the information in the table of densities at the top of the page to answer the following:
6 What material, of mass 39 g, has a volume of 5 cm³?
7 What is the mass of air in a room measuring 5 m × 2 m × 3 m?
8 What is the volume of a storage tank which will hold 3200 kg of petrol?
9 What mass of lead has the same volume as 1600 kg of petrol?

1.05 Measuring volume and density

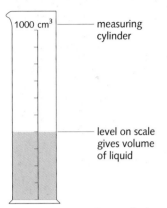

Measuring the volume of a liquid

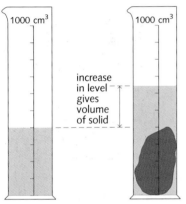

Measuring the volume of a small solid

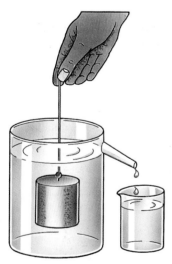

Using a displacement can. Provided the can is filled to the spout at the start, the volume of water collected in the beaker is equal to the volume of the object lowered into the can.

Measuring volume

Liquid A volume of about a litre or so can be measured using a measuring cylinder. When the liquid is poured into the cylinder, the level on the scale gives the volume.

Most measuring cylinders have scales marked in millilitres (ml), or cubic centimetres (cm^3).

Regular solid If an object has a simple shape, its volume can be calculated. For example:

volume of a rectangular block = length × width × height

volume of a cylinder = π × radius² × height

Irregular solid If the shape is too awkward for the volume to be calculated, the solid can be lowered into a partly filled measuring cylinder as shown on the left. The *rise* in level on the volume scale gives the volume of the solid.

If the solid floats, it can be weighed down with a lump of metal. The total volume is found. The volume of the metal is measured in a separate experiment and then subtracted from this total.

Using a displacement can If the solid is too big for a measuring cylinder, its volume can be found using a displacement can, shown below left. First, the can is filled up to the level of the spout (this is done by overfilling it, and then waiting for the surplus water to run out). Then the solid is slowly lowered into the water. The solid is now taking up space once occupied by the water – in other words, it has displaced its own volume of water. The displaced water is collected in a beaker and emptied into a measuring cylinder.

The displacement method, so the story goes, was discovered by accident, by Archimedes. You can find out how on the opposite page.

Measuring density

The density of a material can be found by calculation, once the volume and mass have been measured. The mass of a small solid or of a liquid can be measured using a balance. However, in the case of a liquid, you must remember to allow for the mass of its container.

Here are some readings from an experiment to find the density of a liquid:

volume of liquid in measuring cylinder	= 400 cm³	(A)
mass of measuring cylinder	= 240 g	(B)
mass of measuring cylinder with liquid in	= 560 g	(C)

Therefore: mass of liquid = 560 g − 240 g = 320 g   (C − B)

Therefore density of liquid = $\dfrac{\text{mass}}{\text{volume}} = \dfrac{320 \text{ g}}{400 \text{ cm}^3} = 0.8 \text{ g/cm}^3$

Relative density★

The relative density of a substance tells you how the density compares with that of water. It is calculated like this:

$$\text{relative density} = \frac{\text{density of substance}}{\text{density of water}}$$

For example, lead has a density of 11 300 kg/m³ (11.3 g/cm³) and water has a density of 1000 kg/m³ (1 g/cm³). So:

$$\text{relative density of lead} = \frac{11\,300\,\text{kg/m}^3}{1000\,\text{kg/m}^3} = \frac{11.3\,\text{g/cm}^3}{1\,\text{g/cm}^3} = 11.3$$

Relative density has no units. It is a number whose value is the same as that of the density in g/cm³. It used to be known as 'specific gravity'.

substance	density	relative density
water	1.0 g/cm³	1.0
aluminium	2.7 g/cm³	2.7
copper	8.9 g/cm³	8.9
lead	11.4 g/cm³	11.4
mercury	13.6 g/cm³	13.6
gold	19.3 g/cm³	19.3

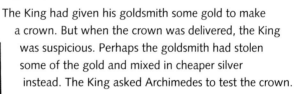

Archimedes and the crown

Archimedes, a Greek mathematician, lived in Syracuse (now in Sicily) around 250 BC. He made important discoveries about levers and liquids, but is probably best remembered for his clever solution to a problem set him by the King of Syracuse.

The King had given his goldsmith some gold to make a crown. But when the crown was delivered, the King was suspicious. Perhaps the goldsmith had stolen some of the gold and mixed in cheaper silver instead. The King asked Archimedes to test the crown.

Archimedes knew that the crown was the correct mass. He also knew that silver was less dense than gold. So a crown with silver in it would have a greater volume than it should have. But how could he measure the volume? Stepping into his bath one day, so the story goes, Archimedes noticed the rise in water level. Here was the answer! He was so excited that he lept from his bath and ran naked through the streets, shouting "Eureka!", which means "I have found it!".

Later, Archimedes put the crown in a container of water and measured the rise in level. Then he did the same with an equal mass of pure gold. The rise in level was different. So the crown could not have been pure gold.

	crown A	crown B	crown C
mass/ g	3750	3750	3750
volume/ cm³	357	194	315

density: gold 19.3 g/cm³; silver 10.5 g/cm³

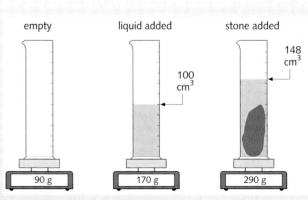

1 Use the information above to decide which crown is gold, which is silver, and which is a mixture.

2 Use the information above to calculate:
a) the mass, volume, and density of the liquid,
b) the mass, volume, and density of the stone.

1.06 More about mass and density

Density essentials

$$\text{density} = \frac{\text{mass}}{\text{volume}}.$$

Comparing masses

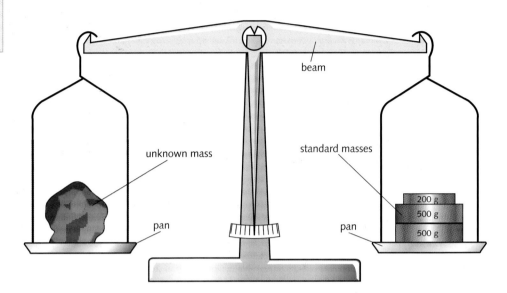

beam

unknown mass

standard masses

200 g

500 g

pan

pan

500 g

▶ A simple beam balance

The device above is called a **beam balance**. It is the simplest, and probably the oldest, way of finding the mass of something. You put the object in one pan, then add standard masses to the other pan until the beam balances in a level position. If you have to add 1.2 kg of standard masses, as in the diagram, then you know that the object also has a mass of 1.2 kg.

The balance is really comparing weights rather than masses. Weight is the downward pull of gravity. The beam balances when the downward pull on one pan is equal to the downward pull on the other. However, masses can be compared because of the way gravity acts on them. If the objects in the two pans have the same weight, they must also have the same mass.

When using a balance like the one above, you might say that you were 'weighing' something. However, 1.2 kg is the mass of the object, not its weight. Weight is a force, measured in force units called **newtons**. For more on this, and the difference between mass and weight, see spreads 2.07 and 2.09.

A more modern type of balance is shown on the left.

A more modern type of balance. It detects the gravitational pull on the object on the pan, but gives its reading in units of mass.

Q

1 On the Moon, the force of gravity on an object is only about one sixth of its value on Earth. Decide whether each of the following would give an accurate measurement of mass if used on the Moon.

a) A beam balance like the one in the diagram at the top of the page.

b) A balance like the one in the photograph above.

2 A balloon like the one on the opposite page contains 2000 m³ of air. When the air is cold, its density is 1.3 kg/m³. When heated, the air expands so that some is pushed out of the hole at the bottom, and the density falls to 1.1 kg/m³. Calculate the following.

a) The mass of air in the balloon when cold.

b) The mass of air in the balloon when hot.

c) The mass of air lost from the balloon during heating.

Planet density

The density of a planet increases towards the centre. However, the average density can be found by dividing the total mass by the total volume. The mass of a planet affects its gravitational pull and, therefore, the orbit of any moon circling it. The mass can be calculated from this. The volume can be calculated once the diameter is known.

The average density gives clues about a planet's structure:

Earth

Average density 5520 kg/m³

This is about double the density of the rocks near the surface, so the Earth must have a high density core – probably mainly iron.

Jupiter

Average density 1330 kg/m³

The low average density is one reason why scientists think that Jupiter is a sphere mostly of hydrogen and helium gas, with a small, rocky core.

not to scale

Float or sink?

You can tell whether a material will float or sink by comparing its density with that of the surrounding liquid (or gas). If it is less dense, it will float; if it is more dense, it will sink. For example, wood is less dense than water, so it floats; steel is more dense, so it sinks.

Density differences are not the *cause* of floating or sinking, just a useful guide for predicting which will occur. Floating is made possible by an upward force produced whenever an object is immersed in a liquid (or gas). To experience this force, try pushing an empty bottle down into water.

Hot air is less dense than cold air, so a hot-air balloon will rise upwards – provided the fabric, gas cylinders, basket, and passengers do not increase the average density by too much.

Ice is less dense than water in its liquid form, so icebergs float.

Related topics: mass **1.02**; volume and density **1.04–105**; force **2.06**; mass and weight **2.09**; convection **5.07**

1 Copy and complete the table shown below:

measurement	unit	symbol
length	?	?
?	kilogram	?
?	?	s

[6]

2 Write down the number of

A mg in 1 g
B g in 1 kg
C mg in 1 kg
D mm in 4 km
E cm in 5 km   [5]

3 Write down the values of

a) 300 cm, in m
b) 500 g, in kg
c) 1500 m, in km
d) 250 ms, in s
e) 0.5 s, in ms
f) 0.75 km, in m
g) 2.5 kg, in g
h) 0.8 m, in mm   [8]

4 The volume of a rectangular block can be calculated using this equation:

volume = length × width × height

Using this information, copy and complete the table below.   [4]

length	width	height	volume of rectangular block
2 cm	3 cm	4 cm	?
5 cm	5 cm	?	100 cm³
6 cm	?	5 cm	300 cm³
?	10 cm	10 cm	50 cm³

5 In each of the following pairs, which quantity is the larger?

a) 2 km or 2500 m?
b) 2 m or 1500 mm?
c) 2 tonnes or 3000 kg?
d) 2 litres or 300 cm³?   [4]

6 Which of the following statements is/are correct?

A One milligram equals one million grams.
B One thousand milligrams equals one gram.
C One million milligrams equals one gram.
D One million milligrams equals one kilogram.   [2]

7

m	g/cm³	m³	km	cm³
kg	ms	ml	kg/m³	s

Which of the above are
a) units of mass?
b) units of length?
c) units of volume?
d) units of time?
e) units of density?   [11]

8 Which block is made of the densest material?

block	mass/g	length/cm	breadth/cm	height/cm
A	480	5	4	4
B	360	10	4	3
C	800	10	5	2
D	600	5	4	3

9 The mass of a measuring cylinder and its contents are measured before and after putting a stone in it.

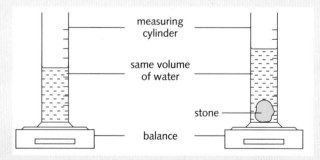

Which of the following could you calculate using measurements taken from the apparatus above?

A the density of the liquid only
B the density of the stone only
C the densities of the liquid and the stone   [2]

10 A plastic bag filled with air has a volume of 0.008 m³. When air in the bag is squeezed into a rigid container, the mass of the container (with air) increases from 0.02 kg to 0.03 kg. Use the formula

$$density = \frac{mass}{volume}$$

to calculate the density of the air in the bag.   [2]

11

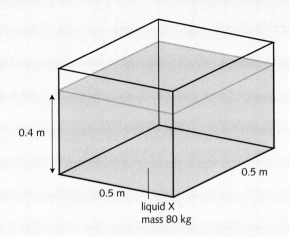

0.4 m

0.5 m

0.5 m

liquid X
mass 80 kg

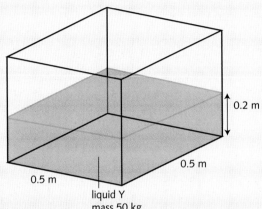

0.2 m

0.5 m

0.5 m

liquid Y
mass 50 kg

In the diagram above, the tanks contain two different liquids, X and Y.

a) What is the volume of each liquid in m³? [2]

b) If you had 1 m³ of the liquid X, what would its mass be? [2]

c) What is the density of liquid X? [2]

d) What is the density of liquid Y? [2]

12 The table shows the density of various substances.

substance	density/ g/cm³
copper	8.9
iron	7.9
kerosene	0.87
mercury	13.6
water	1.0

Consider the following statements:

A 1 cm³ of mercury has a greater mass than 1 cm³ of any other substance in this table – true or false?

B 1 cm³ of water has a smaller mass than 1 cm³ of any other substance in this table – true or false?

C 1 g of iron has a smaller volume than 1 g of copper – true or false?

D 1 g of mercury has a greater mass than 1 g of copper – true or false? [2]

13 A student decides to measure the period of a pendulum (the period is the time taken for one complete swing). Using a stopwatch, he finds that 8 complete swings take 7.4 seconds. With his calculator, he then uses this data to work out the time for one swing. The number shown on his calculator is 0.925.

a) Is it acceptable for the student to claim that the period of the pendulum is 0.925 seconds? Explain your answer. [2]

b) How could the student measure the period more accurately? [2]

c) Later, another student finds that 100 complete swings take 92.8 seconds. From these measurements, what is the period of the pendulum? [2]

Photocopy the list of topics below and tick the boxes of the ones that are included in your examination syllabus. (Your teacher should be able to tell you which they are.) Use your list when you revise. The spread number in brackets tells you where to find more information.

❏ **1** How to use units. (1.01)

❏ **2** Making bigger or smaller units using prefixes. (1.01)

❏ **3** Writing numbers in scientific notation. (1.01)

❏ **4** Significant figures. (1.01)

❏ **5** SI units, including the metre, kilogram, and second. (1.02)

❏ **6** What calipers are used for. (1.03)

❏ **7** How to read a micrometer. (1.03)

❏ **8** How to read a vernier scale. (1.03)

❏ **9** How to measure short intervals of time. (1.03)

❏ **10** How to find the period of a simple pendulum. (1.03)

❏ **11** The meaning of zero error. (1.03)

❏ **12** Units for measuring volume. (1.04)

❏ **13** How density is defined. (1.04)

❏ **14** Using the equation linking mass, volume, and density. (1.04)

❏ **15** Methods of measuring the volume of

– a liquid

– a regular solid

– an irregular solid. (1.05)

❏ **16** How to use a displacement can. (1.05)

❏ **17** The meaning of relative density. (1.05)

❏ **18** How to compare masses with a beam balance. (1.06)

2

Forces and Motion

- SPEED AND VELOCITY
- ACCELERATION
- FREE FALL
- FORCE AND MASS
- FRICTION AND BRAKING
- GRAVITY AND WEIGHT
- ACTION AND REACTION
- VECTORS AND SCALARS
- CIRCULAR MOTION

A bungee jumper leaps more than 180 metres from the top of the Sky Tower in Auckland, New Zealand. With nothing to oppose his fall, he would hit the ground at a speed of 60 metres per second. However, his fall is slowed by the resistance of the air rushing past him, and eventually stopped by the pull of the bungee rope. Side ropes are also being used to stop him crashing into the tower.

2.01 Speed, velocity, and acceleration

▲ *Thrust* supersonic car travelling faster than sound. For speed records, cars are timed over a measured distance (either one kilometre or one mile). The speed is worked out from the average of two runs – down the course and then back again – so that the effects of wind are cancelled out.

Travel times

time taken to travel
1 kilometre (1000 m)

runner   150 s

Grand Prix car   10 s

jet car   3 s

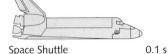
Concorde   1.5 s

Space Shuttle   0.1 s

Speed

If a car travels between two points on a road, its average speed can be calculated like this:

$$\text{average speed} = \frac{\text{distance moved}}{\text{time taken}}$$

If distance is measured in metres (m) and time in seconds (s), speed is measured in metres per second (m/s). For example: if a car moves 90 m in 3 s, its average speed is 30 m/s.

On most journeys, the speed of a car varies, so the actual speed at any moment is usually different from the average speed. To find an actual speed, you need to discover how far the car moves in the shortest time you can measure. For example, if a car moves 0.20 metres in 0.01 s:

$$\text{speed} = \frac{0.20\,\text{m}}{0.01\,\text{s}} = 20\text{ m/s}$$

Velocity

Velocity means the speed of something *and* its direction of travel. For example, a cyclist might have a velocity of 10 m/s due east. On paper, this velocity can be shown using an arrow:

$$\xrightarrow{\hspace{1cm} 10\text{ m/s} \hspace{1cm}}$$

For motion in a straight line you can use a + or − to indicate direction. For example:

+10 m/s (velocity of 10 m/s *to the right*)
−10 m/s (velocity of 10 m/s *to the left*)

Note: +10 m/s may be written without the +, just as 10 m/s.

Quantities, such as velocity, which have a direction as well as a magnitude (size) are called **vectors**.

Acceleration

Something is accelerating if its velocity is *changing*. Acceleration is calculated like this:

> average acceleration = $\dfrac{\text{change in velocity}}{\text{time taken}}$

In symbols:
$$a = \frac{v - u}{t}$$

time	velocity
0 s	0 m/s
1 s	3 m/s
2 s	6 m/s
3 s	9 m/s
4 s	12 m/s

The velocity of this car is increasing by 3 m/s every second. The car has a steady acceleration of 3 m/s².

where u is the initial velocity and v is the final velocity.

For example, if a car increases its velocity from zero to 12 m/s in 4 s:

average acceleration = 12/4 = 3 m/s² (omitting some units for simplicity)

Note that acceleration is measured in metres per second² (m/s²).

Acceleration is a vector. It can be shown using an arrow (usually double-headed). Alternatively, a + or − sign can be used to indicate whether the velocity is increasing or decreasing. For example:

+3 m/s² (velocity *increasing* by 3 m/s every second)
−3 m/s² (velocity *decreasing* by 3 m/s every second)

A *negative* acceleration is called a **deceleration** or a **retardation**.
A *uniform* acceleration means a constant (steady) acceleration.

Solving a problem

> *Example* The car on the right passes post A with a velocity of 12 m/s. If it has a steady acceleration of 3 m/s², what is its velocity 5 s later, at B?

The car is gaining 3 m/s of velocity every second. So in 5 s, it gains an extra 15 m/s on top of its original 12 m/s. Therefore its final velocity is 27 m/s. Note that the result is worked out like this:

final velocity = original velocity + extra velocity

So: final velocity = original velocity + (acceleration × time)

The above equation also works for retardation. If a car has a retardation of 3 m/s², you treat this as an acceleration of −3 m/s².

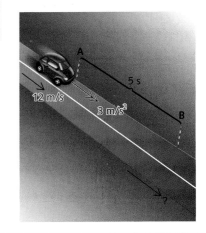

1 A car travels 600 m in 30 s. What is its average speed? Why is its actual speed usually different from its average speed?

2 How is velocity different from speed?

3 A car has a steady speed of 8 m/s.
   a) How far does the car travel in 8 s?
   b) How long does the car take to travel 160 m?

4 Calculate the average speed of each thing in the chart of travel times on the opposite page.

5 A car has an acceleration of +2 m/s². What does this tell you about the velocity of the car? What is meant by an acceleration of −2 m/s²?

6 A car takes 8 s to increase its velocity from 10 m/s to 30 m/s. What is its average acceration?

7 A motor cycle, travelling at 20 m/s, takes 5 s to stop. What is its average retardation?

8 An aircraft on its take-off run has a steady acceleration of 3 m/s².
   a) What velocity does the aircraft gain in 4 s?
   b) If the aircraft passes one post on the runway at a velocity of 20 m/s, what is its velocity 8 s later?

9 A truck travelling at 25 m/s puts its brakes on for 4 s. This produces a retardation of 2 m/s². What does the truck's velocity drop to?

2.02 Motion graphs

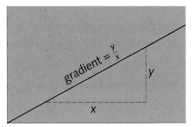

On a straight line graph like this, the gradient has the same value wherever you measure y and x.

Distance–time graphs

Graphs can be useful when studying motion. Below, a car is travelling along a straight road, away from a marker post. The car's distance from the post is measured every second. The charts and graphs show four different examples of what the car's motion might be.

On a graph, the line's rise on the vertical scale divided by its rise on the horizontal scale is called the **gradient**, as shown on the left. With a distance–time graph, the gradient tells you how much extra distance is travelled every second. So:

On a distance–time graph, the gradient of the line is numerically equal to the speed.

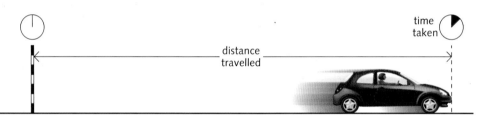

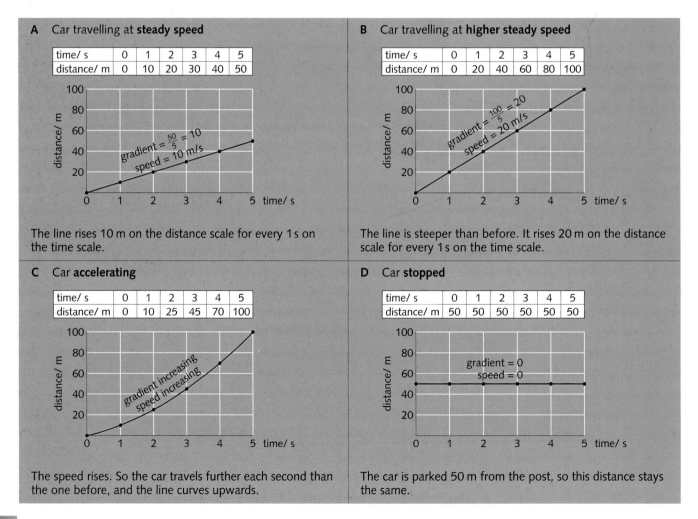

A Car travelling at **steady speed**

time/ s	0	1	2	3	4	5
distance/ m	0	10	20	30	40	50

gradient = $\frac{50}{5}$ = 10
speed = 10 m/s

The line rises 10 m on the distance scale for every 1 s on the time scale.

B Car travelling at **higher steady speed**

time/ s	0	1	2	3	4	5
distance/ m	0	20	40	60	80	100

gradient = $\frac{100}{5}$ = 20
speed = 20 m/s

The line is steeper than before. It rises 20 m on the distance scale for every 1 s on the time scale.

C Car **accelerating**

time/ s	0	1	2	3	4	5
distance/ m	0	10	25	45	70	100

gradient increasing
speed increasing

The speed rises. So the car travels further each second than the one before, and the line curves upwards.

D Car **stopped**

time/ s	0	1	2	3	4	5
distance/ m	50	50	50	50	50	50

gradient = 0
speed = 0

The car is parked 50 m from the post, so this distance stays the same.

Speed–time graphs

Each speed–time graph below is for a car travelling along a straight road. The gradient tells you how much extra speed is gained every second. So:

> On a speed–time graph, the gradient of the line is numerically equal to the acceleration.

In graph E, the car travels at a steady 15 m/s for 5 s, so the distance travelled is 75 m. The area of the shaded rectangle, calculated using the scale numbers, is also 75. This principle works for more complicated graph lines as well. In graph F, the area of the shaded triangle, $^1/_2 \times$ base $\times$ height, equals 50. So the distance travelled is 50 metres.

> On a speed–time graph, the area under the line is numerically equal to the distance travelled.

<table>
<tr><td colspan="2">Velocity–time graphs</td></tr>
</table>

Velocity–time graphs

Velocity is speed in a particular direction.

Where there is no change in the direction of motion, a velocity–time graph looks the same as a speed–time graph.

E   Car travelling at steady speed

time/ s	0	1	2	3	4	5
speed/ m/s	15	15	15	15	15	15

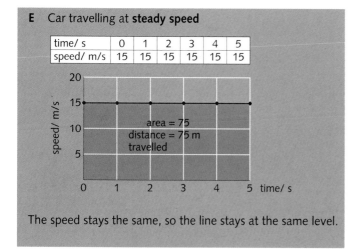

The speed stays the same, so the line stays at the same level.

F   Car with steady acceleration

time/ s	0	1	2	3	4	5
speed/ m/s	0	4	8	12	16	20

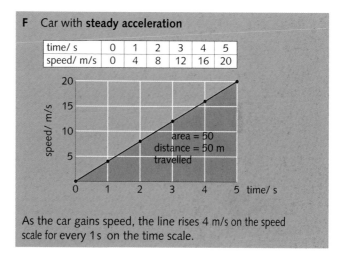

As the car gains speed, the line rises 4 m/s on the speed scale for every 1 s on the time scale.

Q

1

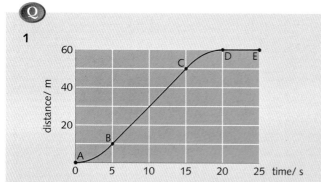

The distance–time graph above is for a motor cycle travelling along a straight road.
a) What is the motor cycle doing between points D and E on the graph?
b) Between which points is it accelerating?
c) Between which points is its speed steady?
d) What is this steady speed?
e) What is the distance travelled between A and D?
f) What is the average speed between A and D?

2

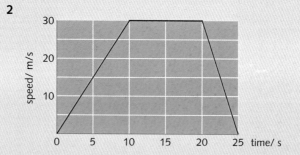

The speed–time graph above is for another motor cycle travelling along the same road.
a) What is the motor cycle's maximum speed?
b) What is the acceleration during the first 10 s?
c) What is its deceleration during the last 5 s?
d) What distance is travelled during the first 10 s?
e) What is the total distance travelled?
f) What is the time taken for the whole journey?
g) What is the average speed for the whole journey?

Recording motion

Using ticker-tape

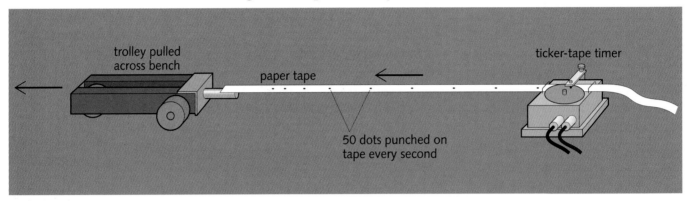

trolley pulled
across bench

ticker-tape timer

paper tape

50 dots punched on
tape every second

Speed, velocity, and acceleration essentials

$$speed* = \frac{distance\ moved}{time\ taken}$$

velocity is speed in a particular direction

$$acceleration* = \frac{change\ in\ velocity}{time\ taken}$$

*average

In the laboratory, motion can be investigated using a trolley like the one above. As the trolley travels along the bench, it pulls a length of paper tape (ticker-tape) behind it. The tape passes through a ticker-tape timer which punches carbon dots on the tape at regular intervals. A typical timer produces 50 dots every second.

Together, the dots on the tape form a complete record of the motion of the trolley. The further apart the dots, the faster the trolley is moving. Here are some examples:

start

steady speed — distance between dots stays the same

higher steady speed — distance between dots greater than before

acceleration — distance between dots increases

acceleration ———————— then ———————— retardation

▼ Motion can also be recorded by taking lots of photographs one after another, like this. These pictures were taken at hour intervals, at midsummer, in Norway. Even at midnight (extreme left) the sun is still above the horizon.

Calculations from tape

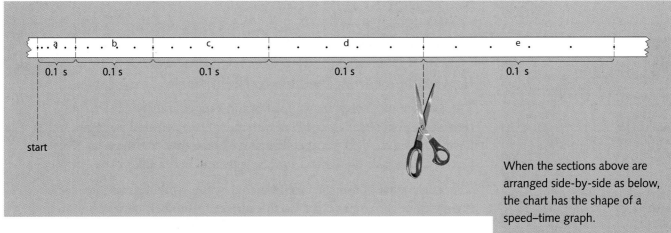

start

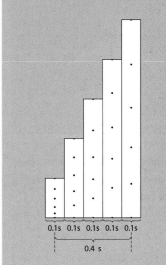

When the sections above are arranged side-by-side as below, the chart has the shape of a speed–time graph.

The ticker-tape record above is for a trolley with steady acceleration.

The tape has been marked off in sections 5 dot-spaces long. One dot-space is the distance travelled by the trolley in 1/50 second (0.02 s). So 5 dot-spaces is the distance travelled in 1/10 second (0.1 s).

If the tape is chopped up into its 5 dot-space sections, and the sections put side-by-side in order, the result is a chart like the one on the right. The chart is the shape of a speed–time graph. The lengths of the sections represent speeds because the trolley travels further in each 0.1 s as its speed increases. Side-by-side, the sections provide a time scale because each section starts 0.1 s after the one before.

The acceleration of the trolley can be found from measurements on the tape. Do questions 2 and 3 below to discover how.

Q

1

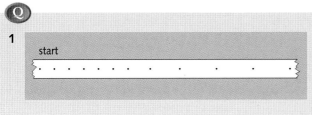

start

Describe the motion of the trolley that produced the ticker-tape record above.

2 The dots on the tape below were made by a ticker-tape timer producing 50 dots per second.
a) Count the number of dot-spaces between A and B. Then calculate the time it took the tape to move from A to B.
b) Using a ruler, measure the distance from A to B in mm. Then calculate the average speed of the trolley between A and B, in mm/s.

c) Measure the distance from C to D, then calculate the average speed of the trolley between C and D.
d) Section CD was completed exactly one second after section AB. Calculate the acceleration of the trolley in mm/s².

3 Look at the chart above.
a) Using a ruler, measure the distance travelled by the trolley in the first 0.1 s recorded on the tape.
b) Calculate the trolley's average speed during this first 0.1 s.
c) Measure the distance travelled by the trolley in the last 0.1s recorded on the tape.
d) Calculate the average speed during this last 0.1 s.
e) Calculate the gain in speed during the 0.4 s.
f) Calculate the acceleration of the trolley in mm/s².

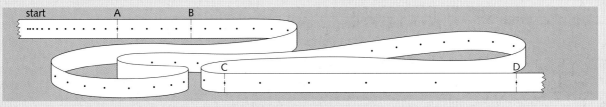

Free fall

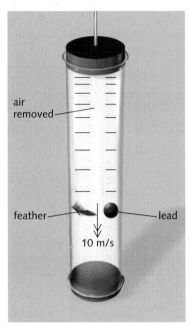

In the experiment above, all the air has been removed from the tube. Without air resistance, a light object falls with the same acceleration as a heavy one.

Experiment to measure g

The acceleration of free fall, g

If you drop a lead weight and a feather, both fall downwards because of gravity. However, the feather is slowed much more by the air.

The diagram on the left shows what would happen if there were no air resistance. Both objects would fall with the same downward acceleration: 9.8 m/s². This is called the **acceleration of free fall**. It is the same for *all* objects falling near the Earth's surface, light and heavy alike.

The acceleration of free fall is represented by the symbol g. Its value varies slightly from one place on the Earth's surface to another, because the Earth's gravitational pull varies. However, the variation is less than 1%. Moving away from the Earth and out into space, g decreases.

Note that the value of g near the Earth's surface is close to 10 m/s². This simple figure is accurate enough for many calculations, and will be the one used in this book.

On the Moon, the acceleration of free fall is only 1.6 m/s². And as there is no atmosphere, a feather would fall with the same acceleration as a lead weight.

Measuring g*

An experiment to find g is shown on the left. The principle is to measure the time taken for a steel ball to drop through a known height, and to calculate the acceleration from this. Air resistance has little effect on a small, heavy ball falling only a short distance, so the ball's acceleration is effectively g.

The ball is dropped by cutting the power to the electromagnet. The electronic timer is automatically switched on when the ball passes through the upper light beam, and switched off when it passes through the lower beam. If the height of the fall is h and the time taken is t, then g can be calculated using this equation (derived from other equations):

$$g = \frac{2h}{t^2}$$

32

Up and down

In the following example, assume that g is 10 m/s², and that there is no air resistance.

The ball on the right is thrown upwards with a velocity of 30 m/s. The diagram shows the velocity of the ball every second as it rises to its highest point and then falls back to where it started.

As an *upward* velocity of 30 m/s is the same as a *downward* velocity of −30 m/s, the motion of the ball can be described like this:

At 0 s....   the downward velocity is −30 m/s
After 1 s....   the downward velocity is −20 m/s
After 2 s....   the downward velocity is −10 m/s
After 3 s....   the downward velocity is   0 m/s
After 4 s....   the downward velocity is +10 m/s
After 5 s....   the downward velocity is +20 m/s
After 6 s....   the downward velocity is +30 m/s

10 m/s is being added to the downward velocity every second

Whether the ball is travelling up or down, it is gaining downward velocity at the rate of 10 m/s per second. So it always has a downward acceleration of 10 m/s², which is g. Even when the ball is moving upwards, or is stationary at its highest point, it still has downward acceleration.

Below, you can see a velocity–time graph for the motion.

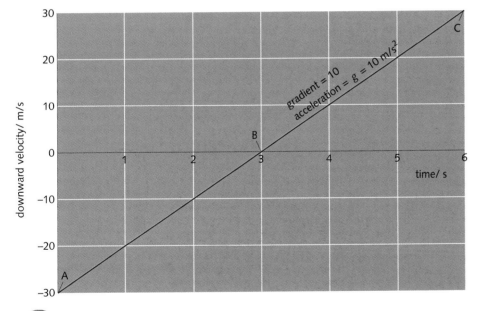

gradient = 10
acceleration = g = 10 m/s²

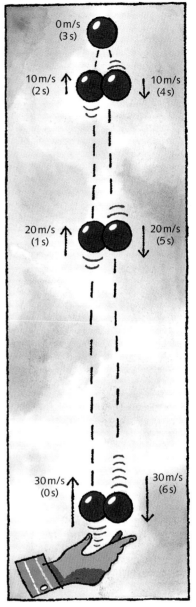

▲ A ball in flight. As g is 10 m/s², the ball's velocity changes by 10 m/s every second.

◄ The velocity–time graph for the ball's motion is shown on the left.

Assume that g = 10 m/s² and that there is no air resistance.

1 A stone is dropped from rest. What is its speed
   a) after 1 s **b)** after 2 s **c)** after 5 s?
2 A stone is thrown downwards at 20 m/s. What is its speed
   a) after 1 s **b)** after 2 s **c)** after 5 s?
3 A stone is thrown upwards at 20 m/s. What is its speed
   a) after 1 s **b)** after 2 s **c)** after 5 s?

4 This question is about the three points, A, B, and C, on the graph above left.
   a) In which direction is the ball moving at point C?
   b) At which point is the ball stationary?
   c) At which point is the ball at its maximum height?
   d) What is the ball's acceleration at point C?
   e) What is the ball's acceleration at point A?
   f) What is the ball's acceleration at point B?
   g) At which point does the ball have the same speed as when it was thrown?

Related topics: acceleration **2.01**; motion graphs **2.02** and **2.05**; gravitational force **2.09**

2.05 # More motion graphs

Motion graph essentials

Here are four examples of velocity–time graphs for a car travelling along a straight line:

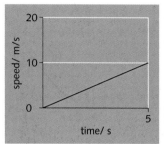 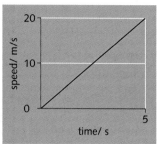 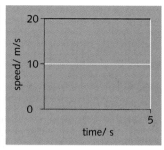 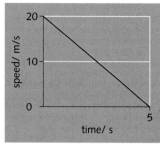

Steady acceleration of 2 m/s^2
The speed of the car increases by 2 m/s every second.

The initial speed is zero, so the car is starting from rest.

Steady acceleration of 4 m/s^2
The speed of the car increases by 4 m/s every second.

The initial speed is zero, so the car is starting from rest.

Zero acceleration
The car has a steady speed of 20 m/s.

Steady retardation (deceleration) of 4 m/s^2
The speed of the car decreases by 4 m/s^2. In other words: the acceleration is –4 m/s^2.

The final speed is zero, so the car comes to rest.

Uniform and non-uniform acceleration

A car is travelling along a straight road. If it has **uniform** acceleration, this means that its acceleration is steady (constant). In other words, it is gaining velocity at a steady rate. In practice, a car's acceleration is rarely steady. For example, as a car approaches its maximum velocity, the acceleration becomes less and less until it is zero, as shown in the example below. Also the car decelerates slightly during gear changes.

If acceleration is not steady then it is **non-uniform**. On a velocity–time graph, as below, the maximum acceleration is where the graph line has its highest gradient (steepness).

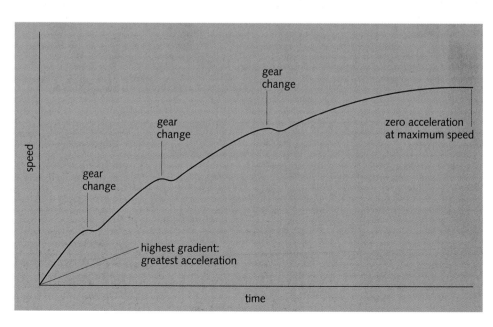

Here are more examples of uniform and non-uniform acceleration:

A stone is dropped from a great height. With no air resistance, the velocity–time graph for the stone would be as shown below left. The acceleration would be uniform. It would be 10 m/s², the acceleration of free fall, *g*.

In practice, there is air resistance on the stone. This affects its motion, producing non-uniform acceleration, as shown below right. At the instant the stone is dropped, it has no velocity. This means that its initial acceleration is *g* because there is not yet any air resistance on it. However, as the velocity increases, air resistance also increases. Eventually, the air resistance is so great that the velocity reaches a maximum and the acceleration falls to zero.

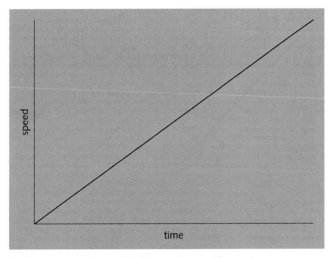

Uniform acceleration of a falling stone with no air resistance acting.

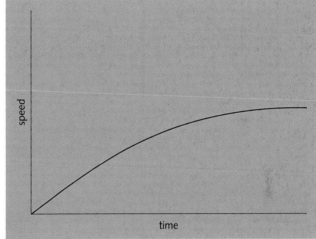

Non-uniform acceleration of a falling stone with air resistance acting.

On a speed-time graph, the area under the line is numerically equal to the distance travelled. This applies whether the motion is uniform or non-uniform – in other words, whether the graph line is straight or curved. With a straight-line graph, the area can be calculated. With a curved-line graph, this may not be possible, although an estimate can be made by counting squares. When doing this, remember that the area must be worked out using the scale numbers on the axis. It isn't the 'real' area on the paper.

Q

1 A boat moves off from its mooring in a straight line. A speed–time graph for its motion is shown on the right. The graph has been divided into sections, AB, BC, CD, and DE. Over which section (or sections) of the graph does the boat
   a) have its greatest speed?
   b) have its greatest acceleration?
   c) have retardation?
   d) have uniform acceleration or retardation?
   e) have non-uniform acceleration or retardation?
   f) travel the greatest distance?
2 Sketch a speed–time graph for a beach-ball falling from a great height. How will this graph differ from that for a falling stone, shown above right?

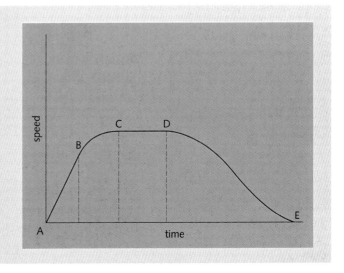

2.06 Forces in balance

A **force** is a push or a pull, exerted by one object on another. It has direction as well as magnitude (size), so it is a vector.

The SI unit of force is the **newton** (**N**). Small forces can be measured using a spring balance like the one below. The greater the force, the more the spring is stretched and the higher the reading on the scale:

Typical forces in newtons	
force to switch on a bathroom light.......	10 N
force to pull open a drinks can......	20 N
force to lift a heavy suitcase.....	200 N
force from a large jet engine....	250 000 N

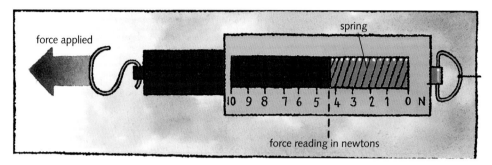

Common forces

Here are some examples of forces:

Upthrust The upward force from a liquid (or gas) that makes some things float.

Weight The gravitational force on an object.

Tension The force in a stretched material.

Friction The force that opposes the motion of one material sliding past another.

Thrust The forward force from an aircraft engine.

Air resistance One type of friction.

Motion without force

On Earth, unpowered vehicles soon come to rest because of friction. But with no friction, gravity, or other external force on it, a moving object will keep moving for ever – at a steady speed in a straight line. It doesn't need a force to keep it moving.

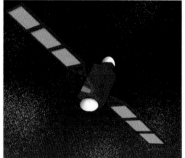

Deep in space with no forces to slow it, a moving object will keep moving for ever.

This idea is summed up in a law first put forward by Sir Isaac Newton in 1687:

If no external force is acting on it, an object will
– if stationary, remain stationary
– if moving, keep moving at a steady speed in a straight line.

This is known as **Newton's first law of motion**.

Balanced forces

An object may have several forces on it. But if the forces are in balance, they cancel each other out. Then, the object behaves as if there is no force on it at all. Here are some examples:

| Stationary gymnast | Skater with steady velocity | Skydiver with steady velocity |

With balanced forces on it, an object is *either* at rest, *or* moving at a steady velocity (steady speed in a straight line). That follows from Newton's first law.

Terminal velocity

When a skydiver falls from a hovering helicopter, as her speed increases, the air resistance on her also increases. Eventually, it is enough to balance her weight, and she gains no more speed. She is at her **terminal velocity**. Typically, this is about 60 m/s, though the actual value depends on air conditions, as well as the size, shape, and weight of the skydiver.

When the skydiver opens her parachute, the extra area of material increases the air resistance. She loses speed rapidly until the forces are again in balance, at a greatly reduced terminal velocity.

If air resistance balances her weight, why doesn't a skydiver stay still? If she wasn't moving, there wouldn't be any air resistance. And with only her weight acting, she would gain velocity.

Surely, if she is travelling downwards, her weight must be greater than the air resistance? Only if is she is gaining velocity. At a steady velocity, the forces must be in balance. That follows from Newton's first law.

If a skydiver is falling at a steady velocity, the forces on her are balanced: her weight downwards is exactly matched by the air resistance upwards.

1 What is the SI unit of force?
2 What does Newton's first law of motion tell you about the forces on an object that is a) stationary b) moving at a steady velocity?
3 The parachutist on the right is descending at a steady velocity.
   a) What name is given to this velocity?
   b) Copy the diagram. Mark in and label another force acting.
   c) How does this force compare with the weight?
   d) If the parachutist used a larger parachute, how would this affect the steady velocity reached? Explain why.

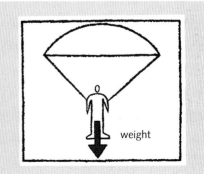

Force, mass, and acceleration

2.07

Inertia and mass

▶ Once a massive ship like this is moving, it is extremely difficult to stop.

If an object is at rest, it takes a force to make it move. If it is moving, it takes a force to make it go faster, slower, or in a different direction. So all objects resist a change in velocity – even if the velocity is zero. This resistance to change in velocity is called **inertia**. The more mass something has, the more inertia it has.

Any change in velocity is an acceleration. So the more mass something has, the more difficult it is to make it accelerate.

Velocity is speed in a particular direction.

Resultant force
In the diagram on the left, the two forces are unbalanced. Together, they are equivalent to a single force. This is called the **resultant force**.

These two forces...

are equivalent to a single force of (5–3) N....

This is the **resultant force**

If forces are balanced, the resultant force is zero and there is no acceleration. Any other resultant force causes an acceleration – in the same direction as the resultant force.

Linking force, mass, and acceleration
There is a link between the resultant force acting, the mass, and the acceleration produced. For example:

If this resultant force...	acts on this mass...	then this is the acceleration...
1 N	1 kg	1 m/s^2
2 N	2 kg	1 m/s^2
4 N	2 kg	2 m/s^2
6 N	2 kg	3 m/s^2

In all cases, the following equation applies:

resultant force = mass × acceleration

In symbols: $F = ma$

This relationship between force, mass, and acceleration is sometimes called **Newton's second law of motion**.

Symbols and units

F = force, in newtons (N)

m = mass, in kilograms (kg)

a = acceleration, in metres/second^2 (m/s^2)

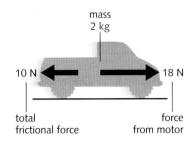
mass
2 kg

10 N   18 N

total
frictional force   force
from motor

Example What is the acceleration of the model car on the right?

First, work out the resultant force on the car. A force of 18 N to the *right* combined with a force of 10 N to the *left* is equivalent to a force of $(18 - 10)$ N to the *right*. So the resultant force is 8 N.

Next, work out the acceleration when $F = 8$ N and $m = 2$ kg:

$$F = ma$$

So: $\quad 8 = 2a \qquad$ (omitting units for simplicity)

Rearranged, this gives $a = 4$. So the car's acceleration is 4 m/s^2.

Finding the link

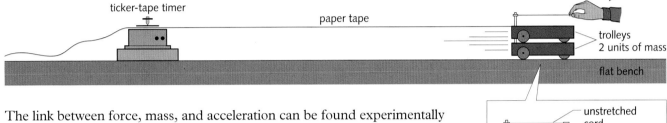

ticker-tape timer

paper tape

trolleys
2 units of mass

flat bench

The link between force, mass, and acceleration can be found experimentally using the equipment above. Different forces are applied to the trolley by pulling it along with one, two, or three elastic cords, stretched to the same length each time. During each run, the ticker-tape timer marks a series of dots on the paper tape. The acceleration can be calculated from the spacing of the dots. To vary the mass, one, two, or three trolleys are used in a stack.

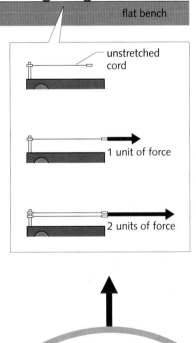

unstretched cord

1 unit of force

2 units of force

Defining the newton

A 1 N resultant force acting on 1 kg produces an acceleration of 1 m/s^2. This simple result is no accident. It arises from the way the newton is defined:

1 newton is the force required to give a mass of 1 kilogram an acceleration of 1 m/s^2.

Further effects of forces

Forces do not only affect motion. If two or more forces act on something, they change its shape or volume (or both). The effect is slight with hard objects, but can be very noticeable with flexible ones, as shown on the right.

Forces causing a shape change

 Q

1 a) What equation links resultant force, mass, and acceleration?
b) Use this equation to calculate the resultant force on each of the stones shown below.

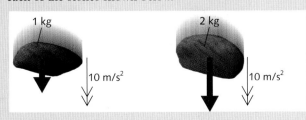

1 kg

10 m/s^2

2 kg

10 m/s^2

2 a) What is the resultant force on the car below?
b) What is the car's acceleration?
c) If the total frictional force rises to 1500 N, what happens to the car?

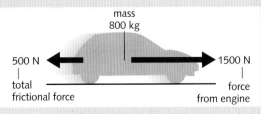

mass
800 kg

500 N   1500 N

total
frictional force   force
from engine

Related topics: mass **1.02**; acceleration **2.01**; using ticker-tape **2.03**; balanced forces **2.06**; stretching and compressing **3.04**

39

Friction and braking★

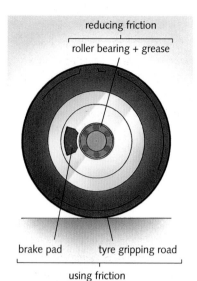

▲ This wheel is mounted on roller bearings to reduce friction.

▼ Air resistance is a form of dynamic friction. When a car is travelling fast, it is the largest of all the frictional forces opposing motion.

Air resistance wastes energy, so less air resistance means better fuel consumption. Car bodies are specially shaped to smooth the air flow past them and reduce air resistance. A low frontal area also helps.

Friction is the force that tries to stop materials sliding across each other. There is friction between your hands when you rub them together, and friction between your shoes and the ground when you walk along. Friction prevents machinery from moving freely and heats up its moving parts. To reduce friction, wheels are mounted on ball or roller bearings, with oil or grease to make the moving surfaces slippery.

Friction is not always a nuisance. It gives shoes and tyres grip on the ground, and it is used in most braking systems. On a bicycle, for example, rubber blocks are pressed against the wheels to slow them down.

Static and dynamic friction

When the block below is pulled gently, friction stops it moving. As the force is increased, the friction rises until the block is about to slip. This is the starting or **static** friction. With a greater downward force on the block, the static friction is higher. Once the block starts to slide, the friction drops: moving or **dynamic** friction is less than static friction.

 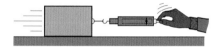

static friction is greater than. . .   . . . dynamic friction

Dynamic friction heats materials up. When something is moved against the force of friction, its energy of motion (called kinetic energy) is changed into thermal energy (heat). Brakes and other machinery must be designed so that they get rid of this thermal energy. Otherwise their moving parts may become so hot that they seize up.

Stopping distance

In an emergency, the driver of a car may have to react quickly and apply the brakes. The car's **stopping distance** is the sum of the following:

- the **thinking distance**. This is how far the car travels *before* the brakes are applied, while the driver is still reacting.
- the **braking distance**. This is how far the car travels *after* the brakes have been applied.

It takes an average driver about half a second to react and press the brake pedal. This is the driver's **reaction time**. During this time, the car does not slow down, and the more speed it has, the further it travels.

The chart below shows how the stopping distance depends on speed:

Speed and energy

A car (or any other moving object) has energy because of its motion. This is called kinetic energy. If a car *doubles* its speed, it has *four times* as much kinetic energy to get rid of, and needs about *four times* the braking distance. It is the greatly increased kinetic energy that makes high-speed crashes so destructive.

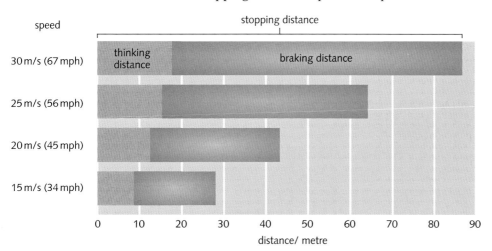

speed

stopping distance

thinking distance | braking distance

- 30 m/s (67 mph)
- 25 m/s (56 mph)
- 20 m/s (45 mph)
- 15 m/s (34 mph)

0 10 20 30 40 50 60 70 80 90

distance/ metre

◄ The stopping distances on the left are average figures for a dry road. The stopping distance is increased if the road is wet or icy, the driver's reaction time is slow, or the car has more mass because it is heavily loaded. And if the brakes are applied too strongly, the car will skid.

◄ The front section of a car is a 'crumple zone', designed to collapse steadily in the event of a crash. It absorbs energy during the impact and reduces the deceleration of the passengers. Seat belts stop the passengers hitting inside parts of the car, while air bags cushion the effects of the impact.

1 In a car, friction is essential in some parts, but needs to be reduced in others. Give *two* examples of where friction is **a)** essential **b)** needs to be reduced.

2 Why are car bodies designed so that air resistance is reduced as much as possible?

3 A car is travelling at 30 m/s. Suddenly, the driver sees a danger ahead and decides to do an emergency stop. The driver's reaction time is 0.6 s.
a) Calculate how far the car travels before the driver starts pressing the brake pedal.

b) Using data from the chart on this page, estimate the car's stopping distance.
c) Re-estimate the stopping distance assuming that the driver is tired and has a reaction time of 1.2 s.
d) Give *two* other factors that could affect the stopping distance.

4 A fast car has kinetic energy (energy of motion). What happens to this energy when the brakes are applied?

5 Explain why, if the speed of a car is doubled, its braking distance is much more than doubled.

Force, weight, and gravity

Gravitational force

If you hang an object from a spring balance, you measure a downward pull from the Earth. This pull is called a **gravitational force**.

No one is sure what causes gravitational force, but here are some of its main features:

- *All* masses attract each other.
- The greater the masses, the stronger the force.
- The closer the masses, the stronger the force.

The pull between small masses is extremely weak. It is less than 10^{-7} N between you and this book! But the Earth is so massive that its gravitational pull is strong enough to hold most things firmly on the ground.

Weight

Weight is another name for the Earth's gravitational force on an object. Like other forces, it is measured in newtons (N).

Near the Earth's surface, an object of mass 1 kg has a weight of 9.8 N, though 10 N is accurate enough for many calculations and will be used in this book. Greater masses have greater weights. Here are some examples:

Near the Earth's surface, a 1 kg mass has a gravitational force on it of about 10 newtons. This is its weight.

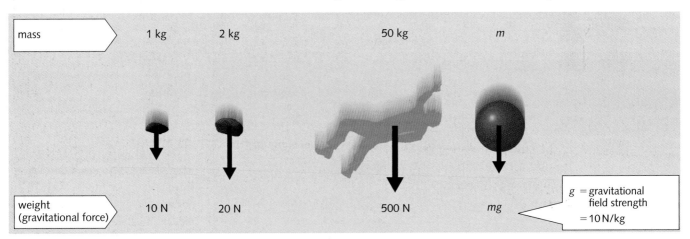

mass	1 kg	2 kg	50 kg	m
weight (gravitational force)	10 N	20 N	500 N	mg

g = gravitational field strength = 10 N/kg

Gravitational field strength, *g*

A **gravitational field** is a region in which a mass experiences a force due to gravitational attraction. The Earth has a gravitational field around it. Near the surface, this exerts a force of 10 newtons on each kilogram of mass: the Earth's **gravitational field strength** is 10 newtons per kilogram (N/kg).

Gravitational field strength is represented by the symbol **g**. So:

$$\text{weight} = \text{mass} \times g \qquad (g = 10 \text{ N/kg})$$

In symbols: $W = mg$

In everyday language, we often use the word 'weight' when it should be 'mass'. Even balances, which detect weight, are normally marked in mass units. But the person in the diagram above doesn't 'weigh' 50 kilograms. He has a *mass* of 50 kilograms and a *weight* of 500 newtons.

Symbols and units

W = weight, in newtons (N)

m = mass, in kilograms (kg)

g = gravitational field strength, 10 N/kg near the Earth's surface

Example What is the acceleration of the rocket on the right?

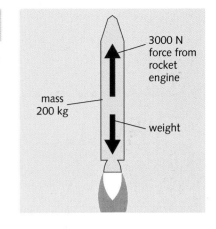

To find the acceleration, you need to know the resultant force on the rocket. And to find that, you need to know the rocket's weight:

$$\text{weight} = mg = 200\,\text{kg} \times 10\,\text{N/kg} = 2000\,\text{N}$$

So: resultant force (upwards) = 3000 N − 2000 N = 1000 N

But :   resultant force = mass × acceleration
So: 1000 N = 200 kg × acceleration

Rearranged, this gives: acceleration = 5 m/s²

Changing weight, fixed mass

On the Moon, your weight (in newtons) would be less than on Earth, because the Moon's gravitational field is weaker.

Even on Earth, your weight can vary slightly from place to place, because the Earth's gravitational field strength varies. Moving away from the Earth, your weight decreases. If you could go deep into space, and be free of any gravitational pull, your weight would be zero.

Whether on the Earth, on the Moon, or deep in space, your body always has the same resistance to a change in motion. So your mass (in kg) doesn't change – at least, not under normal circumstances. But...

According to Einstein's theory of relativity, mass *can* change. For example, it increases when an object gains speed. However, the change is far too small to detect at speeds much below the speed of light. For all practical purposes, you can assume that mass is constant.

Two meanings for *g*

In the diagram opposite, the acceleration of each object can be worked out using the equation force = mass × acceleration. For example, the 2 kg mass has a 20 N force on it, so its acceleration is 10 m/s².

You get the same result for all the other objects. In each case, the acceleration works out at 10 m/s², or *g* (where *g* is the Earth's gravitational field strength, 10 N/kg).

So *g* has *two* meanings:

● *g* is the gravitational field strength (10 newtons per kilogram).
● *g* is the acceleration of free fall (10 metres per second²).

		mass	weight
deep in space		100 kg	zero
on Moon's surface		100 kg	160 N
on Earth's surface		100 kg	1000 N

Assume that *g* = 10 N/kg and there is no air resistance.

5 kg   10 kg

1 The rocks above are falling near the Earth's surface.
a) What is the weight of each rock?
b) What is the acceleration of each rock?
c) What is the gravitational field strength?

2 A spacecraft travels from Earth to Mars, where the gravitational field strength near the surface is 3.7 N/kg. The spacecraft is carrying a probe which has a mass of 100 kg when measured on Earth.
a) What is the probe's weight on Earth?
b) What is the probe's mass in space?
c) What is the probe's mass on Mars?
d) What is the probe's weight on Mars?
e) When the probe is falling, near the surface of Mars, what is its acceleration?

Related topics: acceleration of free fall **2.04**; resultant force and acceleration **2.07**

2.10 Action and reaction★

Action–reaction pairs

A single force cannot exist by itself. Forces are always pushes or pulls between *two* objects. So they always occur in pairs.

The experiment below shows the effect of a pair of forces. To begin with, the two trolleys are stationary. One of them contains a spring-loaded piston which shoots out when a release pin is hit.

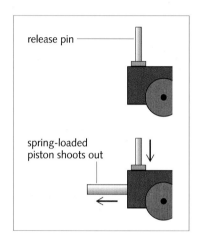

release pin

spring-loaded
piston shoots out

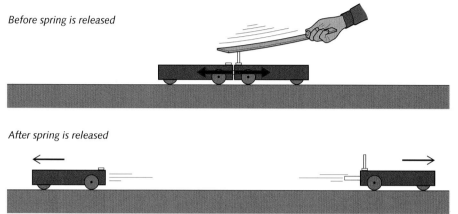

Before spring is released

After spring is released

When the piston is released, the trolleys shoot off in opposite directions. Although the piston comes from one trolley only, two equal but opposite forces are produced, one acting on each trolley. The paired forces are known as the **action** and the **reaction**, but it doesn't matter which you call which. One cannot exist without the other.

Here are some more examples of action–reaction pairs:

Earth pulls downwards
on skydiver

skydiver pulls
upwards on Earth

runner pushes
backwards on
ground

ground pushes
forwards on
runner

forward
force on
bullet: bullet
shoots out

backward
force on gun:
gun recoils

If forces always occur in pairs, why don't they cancel each other out?
The forces in each pair act on *different* objects, not the same object.

If a skydiver is pulled downwards, why isn't the Earth pulled upwards?
It is! But the Earth is so massive that the upward force on it has far too small an effect for any movement to be detected.

Newton's third law of motion

Isaac Newton was the first to point out that every force has an equal but opposite partner acting on a different object. This idea is summed up by **Newton's third law of motion**:

If object A exerts a force on object B, then object B will exert an equal but opposite force on object A.

Here is another way of stating the same law:

To every action there is an equal but opposite reaction.

Rockets and jets

Rockets use the action–reaction principle. A rocket engine gets thrust in one direction by pushing out a huge mass of gas very quickly in the opposite direction. The gas is produced by burning fuel and oxygen. These are either stored as cold liquids, or the fuel may be stored in chemical compounds which have been compressed into solid pellets.

How can a rocket accelerate through space if there is nothing for it to push against? It *does* have something to push against – the huge mass of gas from its burning fuel and oxygen. Fuel and oxygen make up over 90% of the mass of a fully loaded rocket.

Jet engines also get thrust by pushing out a huge mass of gas. But the gas is mostly air that has been drawn in at the front:

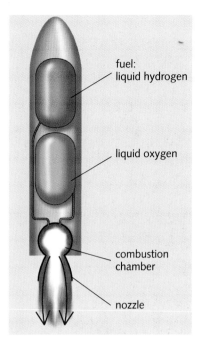

▲ A rocket engine. In the combustion chamber, a huge mass of hot gas expands and rushes out of the nozzle. The gas is produced by burning fuel and oxygen.

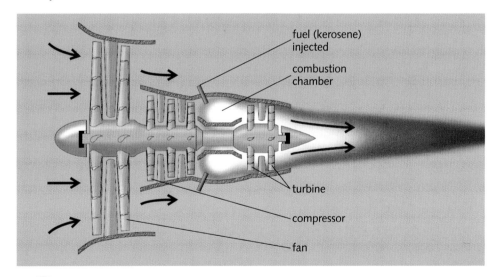

◀ A jet engine. The big fan at the front pushes out a huge mass of air. However, some of the air doesn't come straight out. It is compressed and used to burn fuel in a combustion chamber. As the hot exhaust gas expands, it rushes out of the engine, pushing round a turbine as it goes. The spinning turbine drives the fan and the compressor.

1 The person on the right weighs 500 N. The diagram shows the force of his feet pressing on the ground.
  a) Copy the diagram. Label the size of the force (in newtons).
  b) Draw in the force that the ground exerts on the person's feet. Label the size of this force.
2 When a gun is fired, it exerts a forward force on the bullet. Why does the gun recoil backwards?
3 In the diagram on the opposite page, the forces on the runner and on the ground are equal. Why does the runner move forwards, yet the ground apparently does not move backwards?

More about vectors

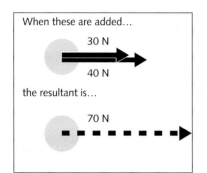

When these are added…

30 N

40 N

the resultant is…

70 N

When these are added…

30 N   40 N

the resultant is…

10 N

Vectors and scalars

Quantities such as force, which have a direction as well as a magnitude (size), are called **vectors**.

Two vectors acting at a point can be replaced by a single vector with the same effect. This is their **resultant**. On the left, you can see how to find it in two simple cases. Finding the resultant of two or more vectors is called *adding* the vectors.

Quantities such as mass and volume, which have magnitude but no direction, are called **scalars**. Adding scalars is easy. A mass of 30 kg added to a mass of 40 kg always gives a mass of 70 kg.

Adding vectors: the parallelogram rule

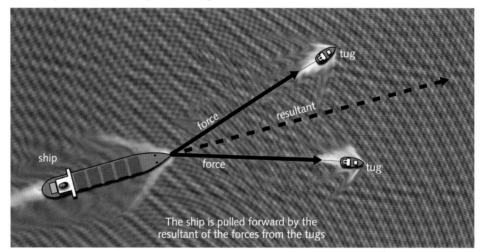

The ship is pulled forward by the resultant of the forces from the tugs

Why the rule works

To see why the parallelogram rule works, consider this simple example using displacement vectors:

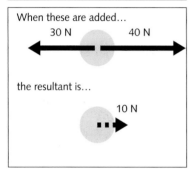

Above, someone starts at O, walks 40 m east, then 30 m north. From Pythagoras' theorem, the person must end up 50 m from O.

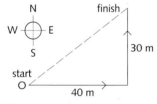

Above, the journey has been shown as the sum of two displacement vectors. When the parallelogram is drawn, its diagonal gives the correct displacement.

The **parallelogram rule** is a method of finding the resultant in situations like the one above, where the vectors are not in line. It works like this:

To find the resultant of two vectors (for example, forces of 30 N and 40 N acting at a point O, as in the diagram below):

1 On paper, draw two lines from O to represent the vectors. The directions must be accurate, and the length of each line must be in proportion to the magnitude of each vector.
2 Draw in two more lines to complete a parallelogram.
3 Draw in the diagonal from O and measure its length. The diagonal represents the resultant in both magnitude and direction. (Below, for example, the resultant is a force of 60 N at 26° to the horizontal.)

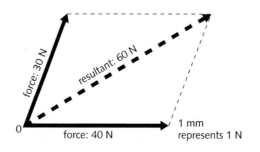

Components of a vector★

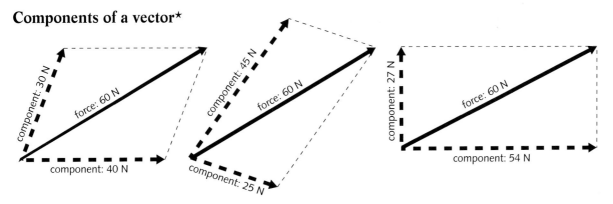

component: 30 N — force: 60 N — component: 40 N

component: 45 N — force: 60 N — component: 25 N

component: 27 N — force: 60 N — component: 54 N

The parallelogram rule also works in reverse: a single vector can be replaced by two vectors having the same effect. Scientifically speaking, a single vector can be **resolved** into two **components**. When using the parallelogram rule in this way, the single vector forms the diagonal.

Above, you can see some of the ways in which a 60 N force can be resolved into two components. There are endless other possibilities.

Components at right angles In working out the effects of a force, it sometimes helps to resolve the force into components at right angles. For example, when a helicopter tilts its main rotor, the force has vertical and horizontal components which lift the helicopter and move it forward:

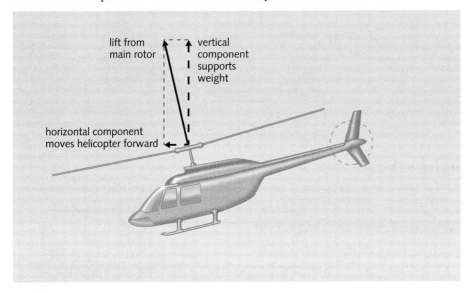

lift from main rotor

vertical component supports weight

horizontal component moves helicopter forward

Calculating components

The horizontal and vertical components of a force F can be calculated using trigonometry:

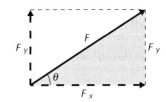

In the tinted triangle above:

$$\cos \theta = \frac{F_x}{F} \quad \text{and} \quad \sin \theta = \frac{F_y}{F}$$

So: $F_x = F \cos \theta$ and $F_y = F \sin \theta$

The horizontal and vertical components of F are therefore as shown below:

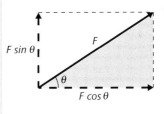

$F \sin \theta$ — F — $F \cos \theta$

 Q

1 How is a scalar different from a vector? Give an example of each.

2 Forces of 12 N and 5 N both act at the same point, but their directions can be varied.
   a) What is their greatest possible resultant?
   b) What is their least possible resultant?
   c) If the two forces are at right angles, find by scale drawing or otherwise the size and direction of their resultant.

3 On the right, someone is pushing a lawnmower.
   a) By scale drawing or otherwise, find the vertical and horizontal components of the 100 N force.

b) If the lawnmower weighs 300 N, what is the total downward force on the ground?

c) If the lawnmower is pulled rather than pushed, how does this affect the total downward force?

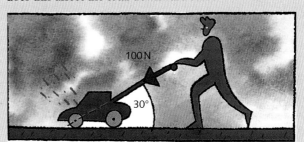

100 N

30°

Moving in circles

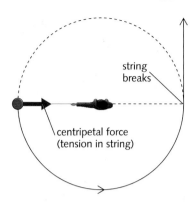

▶ When a motorcycle goes round a corner like this, the sideways friction between the tyres and the road provides the necessary centripetal force.

Centripetal force

On the left, someone is whirling a ball around in a horizontal circle at a steady speed. An inward force is needed to make the ball follow a circular path. The tension in the string provides this force. Without it, the ball would travel in a straight line, as predicted by Newton's first law of motion. This is exactly what happens if the string breaks.

This inward force needed to make an object move in a circle is called the **centripetal force**. *More* centripetal force is needed if:

● the *mass* of the object is *increased*
● the *speed* of the object is *increased*
● the *radius* of the circle is *reduced*.

Changing velocity

Velocity is speed in a particular direction. So a change in velocity can mean *either* a change in speed *or* a change in direction, as shown in the diagrams below. Diagram B shows what happens during circular motion.

If something has a changing velocity, then it has acceleration – in the same direction as the force. So, with circular motion, the acceleration is towards the centre of the circle. It may be difficult to imagine something accelerating towards a point without getting closer to it, but the object is always moving inwards from the position it would have had if travelling in a straight line.

Centripetal force...

Centripetal force isn't produced by circular motion. It is the force that must be *supplied* to make something move in a circle rather than in a straight line.

...and centrifugal force

When you whirl a ball around on the end of some string, you feel an outward pull on your hand. But there is no such thing as a 'centrifugal force' on the ball itself. If the string breaks, the ball moves off at a tangent. It isn't flung outwards.

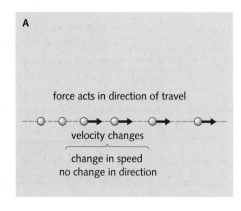

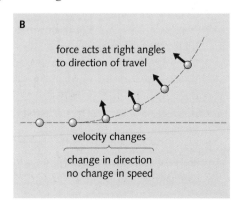

Orbits★

Satellites around the Earth A satellite travels round the Earth in a curved path called an **orbit**, as shown below. Gravitational pull (in other words, the satellite's weight) provides the centripetal force needed. When a satellite is put into orbit, its speed is carefully chosen so that its path does not take it further out into space or back to Earth. Heavy satellites need the same speed as light ones. If the mass is doubled, twice as much centripetal force is required, but that is supplied by the doubled gravitational pull of the Earth.

Planets around the Sun The Earth and other planets move in approximately circular paths around the Sun. The centripetal force needed is supplied by the Sun's gravitational pull.

Electrons around the nucleus In atoms, negatively charged particles called **electrons** are in orbit around a positively charged **nucleus**. The attraction between opposite charges (sometimes called an **electrostatic force**) provides the centripetal force needed.

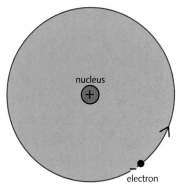

▲ Model of a hydrogen atom: a single electron orbits the nucleus. (According to the quantum theory, electron orbits are much more complicated than that shown in this simple model.)

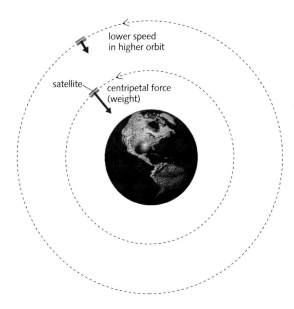

▲ A satellite close to the Earth orbits at a speed of about 29 000 km per hour. The further out the orbit, the lower the gravitational pull, and the less speed is required.

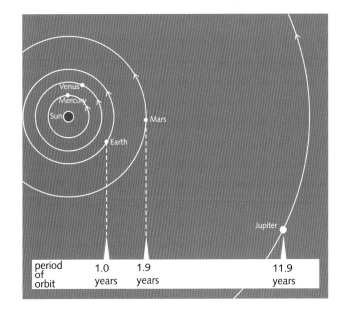

▲ The further a planet is from the Sun, the less speed it has, and the longer it takes to complete one orbit. The time for one orbit is called the **period**.

---

1 A piece of clay is stuck to the edge of a potter's wheel. Draw a diagram to show the path of the clay if it comes unstuck while the wheel is rotating.

2 A car travels round a bend in the road. What supplies the centripetal force needed?

3 In question 2, how does the centripetal force change if the car
a) has less mass
b) travels at a slower speed
c) travels round a tighter curve?

4 What supplies the centripetal force needed for
a) a planet to orbit the Sun
b) an electron to orbit the nucleus in an atom?

5 A satellite is in a circular orbit around the Earth.
a) Draw a diagram to show any forces on the satellite. Show the direction of the satellite's acceleration.
b) If the satellite were in a higher orbit, how would this affect its speed?
c) If the satellite were in a higher orbit, how would this affect the centripetal force required?

Related topics: velocity **2.01**; Newton's 1st law **2.06**; force and acceleration **2.07**; gravity and weight **2.09**; electric charge **8.01**; atoms **11.01**

1 a) Write down, **in words**, the equation connecting speed, distance and time. [1]

b) A car travels at a steady speed of 20 m/s. Calculate the distance travelled in 5 s. [2]

WJEC

2

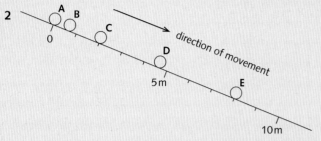

The diagram shows the positions of a ball as it rolled down a track. The ball took 0.5 s to roll from one position to the next. For example, it rolled from **A** to **B** in 0.5 s and from **B** to **C** in 0.5 s and so on.

a) Write down:

  (i) the distance travelled by the ball from **A** to **E**; [1]

  (ii) the time taken by the ball to reach **E**. [1]

b) Calculate the average speed of the ball in rolling from **A** to **E**. Write down the formula that you use and show your working. [3]

c) Explain:

  (i) how you can tell from the diagram that the ball is speeding up; [1]

  (ii) why the ball speeds up. [1]

WJEC

3

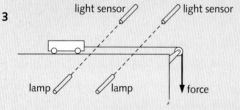

A student measures the acceleration of a trolley using the apparatus above. The light sensors are connected to a computer which is programmed to calculate the acceleration. The results obtained are shown on the acceleration–force graph.

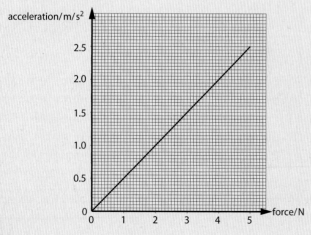

a) (i) Describe how changing the force affects the acceleration. [2]

  (ii) Write down, **in words**, the equation connecting force, mass and acceleration. [1]

  (iii) Use the data from the graph to calculate the mass of the trolley. [2]

b) Sketch the graph and draw the line that would have been obtained for a trolley of larger mass. [1]

WJEC

4 A car has a mass of 900 kg. It accelerates from rest at a rate of 1.2 m/s².

a) Calculate the time taken to reach a velocity of 30 m/s. [3]

b) Calculate the force required to accelerate the car at a rate of 1.2 m/s². [3]

c) Even with the engine working at full power, the car's acceleration decreases as the car goes faster. Why is this? [3]

MEG

5 The diagram below shows some of the forces acting on a car of mass 800 kg.

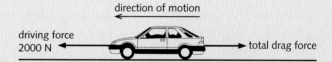

a) State the size of the total drag force when the car is travelling at constant speed. [1]

b) The driving force is increased to 3200 N.

  (i) Find the resultant force on the car at this instant. [1]

  (ii) Write down, **in words**, the equation connecting mass, force and acceleration. [1]

  (iii) Calculate the initial acceleration of the car. [2]

c) Explain why the car will eventually reach a new higher constant speed. [2]

WJEC

6

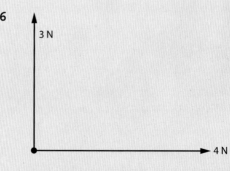

a) Using a scale drawing (for example, on graph paper), find the resultant of the forces above. [3]

b) Draw diagrams to show how, by changing the direction of one of the forces, it is possible to produce a resultant of (i) 7 N (ii) 1 N. [4]

7 This question is about SPEED and ACCELERATION
A cycle track is 500 metres long. A cyclist completes 10 laps (that is, he rides completely round the track 10 times).
a) How many kilometres has the cyclist travelled? [1]
b) On average it took the cyclist 50 seconds to complete one lap (that is, to ride round just once).
(i) What was the average speed of the cyclist? [2]
(ii) How long in minutes and seconds did it take the cyclist to complete the 10 laps? [2]
c) Near the end of the run the cyclist put on a spurt. During this spurt it took the cyclist 2 seconds to increase speed from 8 m/s to 12 m/s. What was the cyclist's acceleration during this spurt? [2]

SEG

8 This question is about FORCE and ACCELERATION.
The driver of a car moving at 20 m/s along a straight level road applies the brakes. The car decelerates at a steady rate of 5 m/s^2.
a) How long does it take the car to stop? [2]
b) What kind of force slows the car down? [1]
c) Where is this force applied? [1]
d) The mass of the car is 600 kg. What is the size of the force slowing the car down? [2]

SEG

9 A girl wearing a parachute jumps from a helicopter. She does not open the parachute straight away. The table shows her speed during the 9 seconds after she jumps.

time in seconds	0	1	2	3	4	5	6	7	8	9
speed in m/s	0	10		30	40	25	17	12	10	10

a) Copy and complete the table by writing down the speed at 2 seconds. [3]
b) Plot a graph of speed against time. [1]
c) How many seconds after she jumped did the girl open her parachute? How do the results show this? [2]
d) (i) What force pulls the girl down? [1]
(ii) What force acts upwards? [1]
(iii)Which of these forces is larger:
at 3 seconds?
at 6 seconds?
at 9 seconds? [3]
e) How will the graph continue after 9 seconds if she is still falling? [1]
f) The girl makes a second jump with a larger area parachute. She falls through the air for the same time before opening her new parachute. How will this affect the graph:
(i) during the first four seconds? [1]
(ii) after this? [1]

SWEB/SEG

10 a) Sketch a velocity–time graph for a car moving with uniform acceleration from 5 m/s to 25 m/s in 15 seconds. [3]
b) Use the sketch graph to find values for (i) the acceleration, (ii) the total distance travelled during acceleration. Show clearly at each stage how you used the graph. [4]

JMB

11 The table below shows how the stopping distance of a car depends on its speed.

stopping distance/ m	0	4	12	22	36	52	72
speed/ m/s	0	5	10	15	20	25	30

a) Write down two factors, apart from speed, that affect the stopping distance of the car. [2]
b) Use the information in the above table to draw a graph of stopping distance against speed. [3]
c) The speed limit in a residential area is 12.5 m/s. Use your graph to estimate the stopping distance of a car travelling at this speed. [2]
d) Describe how the stopping distance changes as the speed of a car increases. [1]

12 In the diagram below, someone is swinging a ball round on the end of a piece of string.

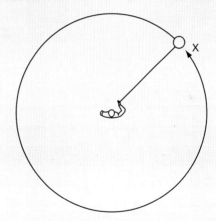

a) What name is given to the force needed to make the ball move in a circle? [1]
b) Copy and complete the diagram to show where the ball will travel if the string breaks when the ball is at point X. [2]
c) Planets move around the Sun in approximately circular orbits. What provides the force necessary for the orbit? [1]

Photocopy the list of topics below and tick the boxes of the ones that are included in your examination syllabus. (Your teacher should be able to tell you which they are.) Use your list when you revise. The spread number in brackets tells you where to find more information.

❑ **1** Measuring speed. (2.01)

❑ **2** The difference between speed and velocity. (2.01)

❑ **3** How acceleration is defined. (2.01)

❑ **4** The meaning of retardation. (2.01)

❑ **5** Representing motion using distance–time and speed–time graphs. (2.02)

❑ **6** Calculating
– speed from a distance–time graph.
– acceleration from a speed–time graph.
– distance travelled from a speed–time graph. (2.02)

❑ **7** Recording the motion of a trolley using ticker-tape, and calculating an acceleration from the data collected. (2.03)

❑ **8** The acceleration of free fall, *g*. (2.04)

❑ **9** Measuring *g*. (2.04)

❑ **10** Recognizing the difference between uniform and non-uniform acceleration from a velocity–time graph. (2.05)

❑ **11** Measuring force. (2.06)

❑ **12** How an object moves if the forces on it are balanced: Newton's first law of motion. (2.06)

❑ **13** Terminal velocity. (2.06)

❑ **14** The meaning of resultant force. (2.07)

❑ **15** The link between force, mass, and acceleration: Newton's second law of motion. (2.07)

❑ **16** Defining the SI unit of force: the newton. (2.07)

❑ **17** How forces can change shape and volume, as well as motion. (2.07)

❑ **18** How friction can be useful. (2.08)

❑ **19** How friction can be a nuisance. (2.08)

❑ **20** How friction can be reduced in machinery. (2.08)

❑ **21** The difference between static and dynamic friction. (2.08)

❑ **22** The factors affecting the stopping distance of a moving vehicle. (2.08)

❑ **23** The meanings of reaction time, thinking distance, and braking distance. (2.08)

❑ **24** The properties of gravitational force. (2.09)

❑ **25** The difference between weight and mass. (2.09)

❑ **26** Defining gravitational field strength. (2.09)

❑ **27** Why the weight of an object can vary but, under normal circumstances, the mass does not change. (2.09)

❑ **28** Two meanings of *g*. (2.09)

❑ **29** How all forces exist in pairs: Newton's third law of motion. (2.10)

❑ **30** How rockets and jets produce a force. (2.10)

❑ **31** The difference between vectors and scalars. (2.11)

❑ **32** Adding vectors using the parallelogram rule. (2.11)

❑ **32** Resolving a vector into two components. (2.11)

❑ **34** The meaning of centripetal force. (2.12)

❑ **35** The factors on which centripetal force depends. (2.12)

❑ **36** Why an object moving in a circle has an inward acceleration. (2.12)

❑ **37** The orbital motions of satellites, planets, and electrons. (2.12)

Forces and Pressure

- TURNING EFFECT OF A FORCE
- CENTRE OF MASS
- BALANCE AND EQUILIBRIUM
- STRETCHING AND COMPRESSING MATERIALS
- PRESSURE
- HYDRAULIC SYSTEMS
- ATMOSPHERIC PRESSURE
- GAS PRESSURE AND VOLUME

Sharks like this are very effective hunters. Their sharp teeth give them a dangerous bite, although because of their long jaws, their biting force is not much more than that of a human. However, when it comes to diving, sharks beat humans easily. Some types can reach depths of over 2000 metres, where the water pressure is far too great for any human diver.

Forces and turning effects

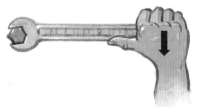

A large force at the end of a long spanner gives a large turning effect.

Moment of a force

It is difficult to tighten a nut with your fingers. But with a spanner, you can produce a larger turning effect. The turning effect is even greater if you increase the force or use a longer spanner. The turning effect of a force is called a **moment**. It is calculated like this:

> moment of a force = force × perpendicular distance
> about a point   from the point

Below, there are some examples of forces and their moments. Moments are described as **clockwise** or **anticlockwise**, depending on their direction. The moment of a force is also called a **torque**.

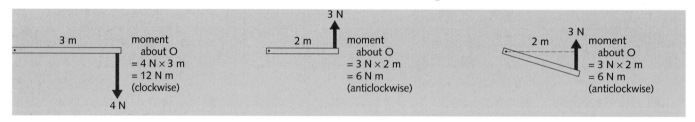

3 m — moment
about O
= 4 N × 3 m
= 12 N m
(clockwise)
4 N

2 m — moment
about O
= 3 N × 2 m
= 6 N m
(anticlockwise)
3 N

2 m — moment
about O
= 3 N × 2 m
= 6 N m
(anticlockwise)
3 N

Unit of force

Force is measured in newtons (N)

Taking moments

Calculating the moments about a point is called *taking moments* about the point.

The principle of moments

In diagram A below, the bar is in a state of balance, or **equilibrium**. Note that the anticlockwise moment about O is equal to the clockwise moment. One turning effect balances the other. In diagram B, there are more forces acting but, once again, the bar is in equilibrium. This time, the *total* clockwise moment about O is equal to the anticlockwise moment.

These examples illustrate the **principle of moments**:

> If an object is in equilibrium:
> the sum of the clockwise moments about any point is equal to the sum of the anticlockwise moments about that point.

A

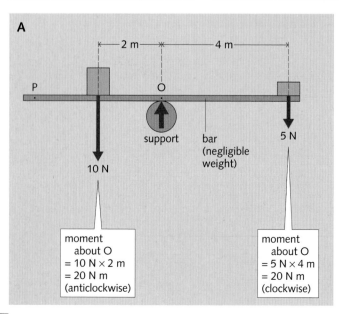

P

2 m — O — 4 m

support

bar
(negligible
weight)

5 N

10 N

moment
about O
= 10 N × 2 m
= 20 N m
(anticlockwise)

moment
about O
= 5 N × 4 m
= 20 N m
(clockwise)

B

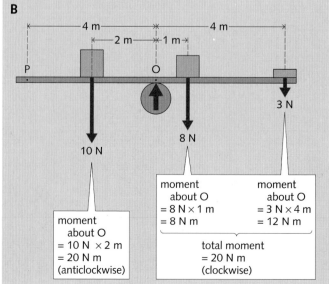

P

4 m — O — 4 m
2 m — 1 m

10 N

8 N

3 N

moment
about O
= 10 N × 2 m
= 20 N m
(anticlockwise)

moment
about O
= 8 N × 1 m
= 8 N m

moment
about O
= 3 N × 4 m
= 12 N m

total moment
= 20 N m
(clockwise)

Conditions for equilibrium

If an object is in equilibrium, the forces on it must balance as well as their turning effects. So:

- The sum of the forces in one direction must equal the sum of the forces in the opposite direction.
- The principle of moments must apply.

For example, in diagram A on the opposite page, the upward force from the support must be 15 N, to balance the 10 N + 5 N total downward force. Also, if you take moments about *any* point – P, for example – the total clockwise moment must equal the total anticlockwise moment.

When taking moments about P, you need to include the moment of the upward force from the support. This doesn't arise when taking moments about O because the force has no moment about that point.

Solving a problem

> *Example* Below right, someone of weight 500 N is standing on a plank supported by two trestles. Calculate the upward forces, *X* and *Y*, exerted by the trestles on the plank. (Assume the plank has negligible weight.)

The system is in equilibrium, so the principle of moments applies. You can take moments about any point. But taking moments about A or B gets rid of one of the unknowns, *X* or *Y*.

Taking moments about A:
clockwise moment = 500 N × 2 m = 1000 N m
anticlockwise moment = *Y* × 5 m
As the moments balance, 5*Y* m = 1000 N m
So: *Y* = 200 N

From here, there are two methods of finding X. *Either* take moments about B and do a calculation like the one above. *Or* use the fact that *X* + *Y* must equal the 500 N downward force. By either method: *X* = 300 N

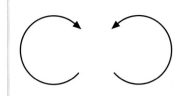

Clockwise .... or anticlockwise?

In the diagram below, the 500 N force has a *clockwise* moment about A, but an *anticlockwise* moment about B.

To decide whether a moment is clockwise or anticlockwise about a point, imagine that the diagram is pinned to the table through the point, then decide which way the force arrow is trying to turn the paper.

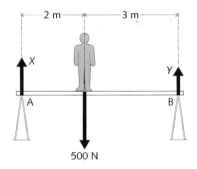

 Q

1 The moment (turning effect) of a force depends on two factors. What are they?

2 What is the principle of moments? What other rule also applies if an object is in equilibrium?

3 Below, someone is trying to balance a plank with stones. The plank has negligible weight.
  a) Calculate the moment of the 4 N force about O.
  b) Calculate the moment of the 6 N force about O.

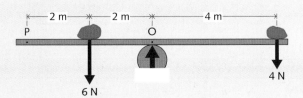

c) Will the plank balance? If not, which way will it tip?
d) What extra force is needed at point P to balance the plank?
e) In which direction must the force at P act?

4 In *diagram* **B** on the opposite page:
  a) What is the upward force from the support?
  b) If moments are taken *about point P,* which forces have clockwise moments? What is the total clockwise moment about P?
  c) Which force or forces have anticlockwise moments about P? What is the total anticlockwise moment about P?
  d) Comparing moments about P, does the principle of moments apply?

Centre of mass

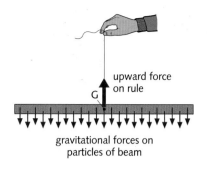

gravitational forces on
particles of beam

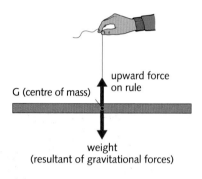

G (centre of mass)

upward force
on rule

weight
(resultant of gravitational forces)

Like other objects, the beam on the left is made up of lots of tiny particles, each with a small gravitational force on it. The beam balances when suspended at one particular point, G, because the gravitational forces have turning effects about G which cancel out.

Together, the small gravitational forces act like a single force at G. In other words, they have a resultant at G. This resultant is the beam's **weight**. G is the **centre of mass** (or **centre of gravity**).

Finding a centre of mass

In diagram 1 below, the card can swing freely from the pin. When the card is released, the forces on it turn the card until its centre of mass is vertically under the pin, as in diagram 2. Whichever point the card is suspended from, it will always hang with its centre of mass vertically under the pin. This fact can be used to find the centre of mass.

In diagram 3, the centre of mass lies somewhere along the plumb line, whose position is marked by the line AB. If the card is suspended at a different point, a second line CD can be drawn. The centre of mass must also lie along this line, so it is at the point where AB crosses CD.

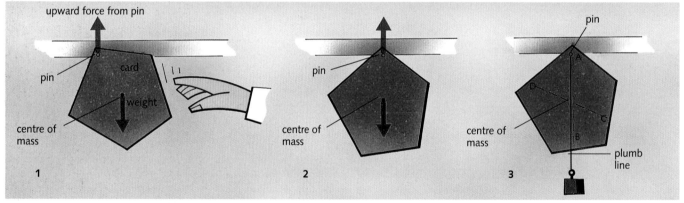

Heavy bar problem

In simple problems, you are often told that a balanced bar has negligible weight. In more complicated problems, you have to include the weight.

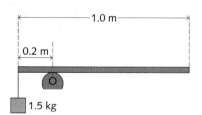

1.0 m

0.2 m

1.5 kg

> *Example* If a uniform bar balances, as on the left, with a 1.5 kg mass attached to one end, what is its weight? (*g* = 10 N/kg)

To solve the problem, redraw the diagram to show all the forces and distances, as in the lower diagram. As *g* = 10 N/kg, the 1.5 kg mass has a weight of 15 N. 'Uniform' means that the bar's weight is evenly distributed, so the centre of mass of the bar (by itself) is at the mid-point, 0.5 m from one end. The bar's weight *W* acts at this point.

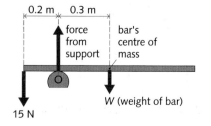

0.2 m   0.3 m

force
from
support

bar's
centre of
mass

W (weight of bar)

15 N

Now take moments about the support, O. The upward force has no moment about this point, but there is an anticlockwise moment of 15 N × 0.2 m and a clockwise moment of *W* × 0.3 m. As the bar is in equilibrium:

$15\,\text{N} \times 0.2\,\text{m} = W \times 0.3\,\text{m}$

So: the bar's weight *W* is 10 N.

Stability

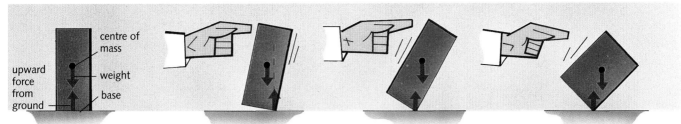

This box is in equilibrium. The forces on it are balanced, and so are their turning effects.

With a small tilt, the forces will turn the box back to its original position.

With a large tilt, the forces will tip the box over.

A box with a wider base and a lower centre of mass can be tilted further before it falls over.

If the box above is pushed a little and then released, it falls back to its original position. Its position was **stable**. If the box is pushed much further, it topples. It starts to topple as soon as its centre of mass passes over the edge of its base. From then on, the forces on the box have a turning effect which tips it even further. A box with a wider base and/or a lower centre of mass is more stable. It can be tilted to a greater angle before it starts to topple.

States of equilibrium

Here are three types of equilibrium:

Stable equilibrium If you tip the cone a little, the centre of mass stays over the base. So the cone falls back to its original position.

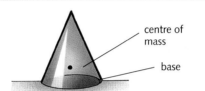

Unstable equilibrium The cone is balanced, but only briefly. Its pointed 'base' is so small that the centre of mass immediately passes beyond it.

Neutral equilibrium Left alone, the ball stays where it is. When moved, it stays in its new position. Wherever it lies, its centre of mass is always exactly over the point which is its 'base'.

He will stay balanced – as long as he keeps his centre of mass over the beam.

1 The stool on the right is about to topple over.
   a) Copy the diagram, showing the position of the centre of mass.
   b) Give *two* features which would make the stool more stable.
2 A uniform metre rule has a 4 N weight hanging from one end. The rule balances when suspended from a point 0.1 m from that end.
   a) Draw a diagram to show the rule and the forces on it.
   b) Calculate the weight of the rule.
3 Draw diagrams to show a drawing pin in positions of stable, unstable, and neutral equilibrium.

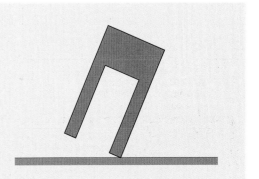

Related topics: resultant force **2.07** and **2.11**; mass, weight, and g **2.09**; turning effects, moments, and equilibrium **3.01**

57

More about moments

Force and moment essentials

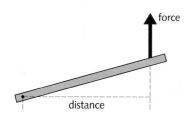

force

distance

A **moment** is the turning effect of a force:

moment of a force = force × perpendicular distance
about a point   from the point

According to the **principle of moments**:

If a system is in equilibrium (balanced), the sum of the clockwise moments about any point is equal to the sum of the anticlockwise moments about that point.

If an object is in equilibrium, the forces on it must balance and also their turning effects. So:

- The sum of the forces in one direction must equal the sum of the forces in the opposite direction.
- The principle of moments must apply.

---

trailer

tractor

centre of mass of trailer •

weight of trailer = 200 kN

hitch

| ← 4 m → | ← 6 m → |

← 4 m → ← 6 m →

X

A

200 kN

Y

B

Force is measured in newtons (N).

1 kN (kilonewton) = 1000 N

Truck problem

Example The articulated truck shown on the left has two parts: the tractor (with the cab, engine, and driver), and the trailer (which carries the load). Use information in the diagram to answer the following.
a) What force does the trailer exert on the hitch?
b) What force do the rear tyres exert on the road?

To answer the question, start by redrawing the trailer, showing all the forces on it. This has been done below left. The downward force of the trailer on the hitch is equal to the upward force of the hitch on the trailer. This has been called X. The downward force of the rear tyres on the road is equal to the upward force of the road on the rear tyres. This has been called Y.

a) As the trailer is in equilibrium, the principle of moments applies. To find X, take moments about point B. Y has zero moment about this point, so it will not appear in the equation.

Taking moments about B:
clockwise moment = $X \times 10$ m
anticlockwise moment = 200 kN × 6 m = 1200 kN m
As the moments balance: $X \times 10$ m = 1200 kN m
So: $X = 120$ kN
So: the downward force on the hitch is 120 kN

b) As the trailer is in equilibrium:
total upward force = total downward force
So $X + Y$ = 200 kN
So   120 kN + Y = 200 kN
So Y = 80 kN
So: the downward force from the rear tyres is 80 kN

Crane problem

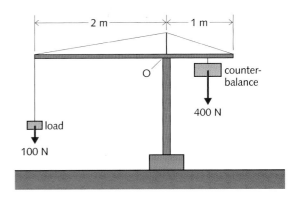

Example The diagram on the right shows a model crane. The crane has a counterbalance weighing 400 N, which can be moved further or closer to O to cope with different loads. (With no load or counterbalance, the top section would balance at point O.)
a) With the 100 N load shown, how far from O should the counterbalance be placed?
b) What is the maximum load the crane can safely lift?

a) To prevent the crane falling over, its top section must balance at point O. So the moment of the 400 N force (the counterbalance) must equal the moment of the 100 N force (the load). That follows from the principle of moments.

Let x be the distance of the 400 N force from O.

Taking moments about O:
clockwise moment = anticlockwise moment
$$400\,\text{N} \times x = 100\,\text{N} \times 2\,\text{m}$$
$$x = 0.5\,\text{m}$$

So: the counterbalance should be placed 0.5 m from O.

b) Let F be the maximum load (in N). With this load on the crane, the counterbalance must produce its maximum moment about O. So it must be the greatest possible distance from O – in other words, 1 m from it. As the crane is in equilibrium, the principle of moments applies:

Taking moments about O:
clockwise moment = anticlockwise moment
$$400\,\text{N} \times 1\,\text{m} = F \times 2\,\text{m}$$
$$F = 200\,\text{N}$$

So: the maximum load is 400 N.

Centre of mass essentials

Although weight is distributed through an object, it acts as a single, downward force from a point called the **centre of mass** (or **centre of gravity**).

For an object to be stable when resting on the ground, its centre of mass must be over its base. If an object is pushed, and its centre of mass passes beyond the edge of its base, it will topple over.

1 In the diagram on the right, a plank weighing 120 N is supported by two trestles at points A and B. A man weighing 480 N is standing on the plank.
a) Redraw the diagram, showing all the forces acting on the plank.
b) Calculate the total clockwise moment of the two weights about A.
c) Use the principle of moments to calculate the upward force from the trestle at B.
d) What is the total downward force on the trestles?
e) What is the upward force from the trestle at A?
f) The man now walks past A towards the left-hand end of the plank. What is the upward force from the trestle at B at the instant the plank starts to tip?
g) How far is the man from A as the plank tips?

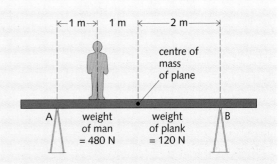

3.04 Stretching and compressing

Elastic and plastic

If you bend a ruler slightly and release it, it springs back to its original shape. Materials that behave like this are **elastic**. However, they stop being elastic if bent or stretched too far. They either break or become permanently deformed (out of shape).

If you stretch or bend Plasticine, it keeps its new shape. Materials that behave like this are **plastic**. (The materials we call 'plastics' were given that name because they are plastic and mouldable when hot.)

Stretching a spring

In the experiment below, a steel spring is stretched by hanging masses from one end. The force applied to the spring is called the **load**. As g is 10 N/kg, the load is 1 N for every 100 g of mass hung from the spring.

As the load is increased, the spring stretches more and more. Its **extension** is the difference between its stretched and unstretched lengths.

> **Force and weight essentials**
>
> Force is measured in newtons (N).
>
> Weight is a force.
>
> On Earth, the weight of an object is 10 N for each kilogram of mass.

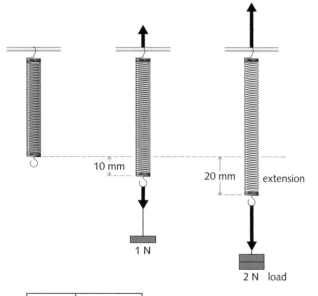

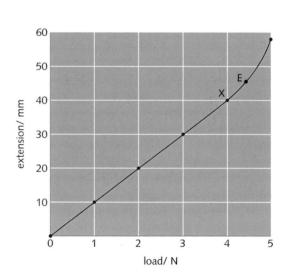

load	extension
N	mm
0	0
1	10
2	20
3	30
4	40
5	58

The readings on the left can be plotted as a graph, as above. Up to point X, the graph line has these features:
- The line is straight, and passes through the origin.
- If the load is doubled, the extension is doubled, and so on.
- Extension ÷ load always has the same value (10 mm/N).
- Every 1 N increase in load produces the same extra extension (10 mm).

Mathematically, these can be summed up as follows:

Up to point X, the extension is proportional to the load. X is the **limit of proportionality**.

Point E marks another change in the spring's behaviour. Up to E, the spring behaves elastically and returns to its original length when the load is removed. E is its **elastic limit**. Beyond E, the spring is left permanently stretched.

Hooke's law

In the 1660s, Robert Hooke investigated how springs and wires stretched when loads were applied. He found that, for many materials, the extension and load were in proportion, provided the elastic limit was not exceeded:

A material obeys Hooke's law if, beneath its elastic limit, the extension is proportional to the load.

Steel wires do not stretch as much as steel springs, but they obey Hooke's law. Glass and wood also obey the law, but rubber does not.

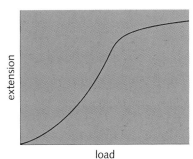

Extension–load graph for rubber

Spring constant

For the spring on the opposite page, up to point X on the graph, dividing the load (force) by the extension always gives the same value, 0.1 N/mm. This is called the **spring constant** (symbol k):

load = spring constant × extension   In symbols: $F = k \times x$

Knowing k, you could use this equation to calculate the extension produced by any load up to the limit of proportionality. For example, for a load of 2.5 N:

2.5 = 0.1 × extension   (omitting units for simplicity)

Rearranged, this gives: extension = 2.5/0.1 = 25 mm

Compressing and bending

Materials can be compressed as well as stretched. If the compression is elastic, the material will return to its original shape when the forces are removed. When a material is bent, the applied forces produce compression on one side and stretching on the other. If the elastic limit is exceeded, the bending is permanent. This can happen when a metal sheet is dented.

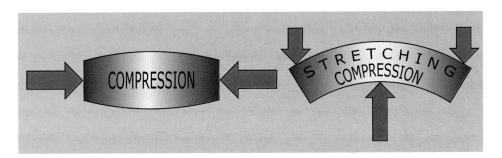

▲ The Oriental Pearl Tower in Shanghai is over half a kilometre high. In high winds, its top can move by a quarter of a metre. But, being elastic, its steel and concrete structure always returns to its original shape.

1 What is meant by an *elastic* material?
2 What is meant by the *elastic limit* of a material?
3 Look at the small graph at the top of the page. Does rubber obey Hooke's law? Explain how you can tell from the graph whether this law is obeyed or not.
4 The table on the right shows the readings taken in a spring-stretching experiment:
    a) What is the unstretched length of the spring?
    b) Copy and complete the table.
    c) Plot a graph of extension against load.
    d) Mark the *elastic limit* on your graph.

e) Over which section of the graph line is the extension proportional to the load?
f) What load would produce a 35 mm extension?
g) What load would make the spring stretch to a length of 65 mm?

load/ N	0	1	2	3	4	5	6
length/ mm	40	49	58	67	76	88	110
extension/ mm							

3.05 Pressure

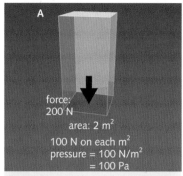

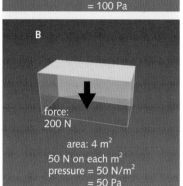

force: 200 N
area: 2 m²
100 N on each m²
pressure = 100 N/m²
= 100 Pa

force: 200 N
area: 4 m²
50 N on each m²
pressure = 50 N/m²
= 50 Pa

Blocks A and B on the left are resting on soft ground. Both weigh the same and exert the same force on the ground. But the force from block B is spread over a larger area, so the force *on each square metre* is reduced. The **pressure** under block B is less than that under block A.

For a force acting at right angles to a surface, the pressure is calculated like this:

$$\text{pressure} = \frac{\text{force}}{\text{area}} \qquad \text{In symbols: } p = \frac{F}{A}$$

If force is measured in newtons (N) and area in square metres (m²), pressure is measured in newton/square metre (N/m²). 1 N/m² is called 1 **pascal** (Pa):

If a 100 N force is spread over an area of   1 m², the pressure is   100 Pa.
If a 100 N force is spread over an area of   2 m², the pressure is   50 Pa.
If a 100 N force is spread over an area of 0.2 m², the pressure is   500 Pa.
If a 200 N force is spread over an area of 0.2 m², the pressure is   1000 Pa.

For most pressure measurements, the pascal is a very small unit. In practical situations, it is often more convenient to use the **kilopascal** (**kPa**).
1 kPa = 1000 Pa

Increasing the pressure by reducing the area

The studs on a football boot have only a small area of contact with the ground. The pressure under the studs is high enough for them to sink into the ground, which gives extra grip.

The area under the edge of a knife's blade is extremely small. Beneath it, the pressure is high enough for the blade to push easily through the material.

Under the tiny area of the point of a drawing pin, the pressure is far too high for the wood to withstand.

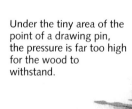

Reducing the pressure by increasing the area

Skis have a large area to reduce the pressure on the snow so that they do not sink in too far.

Wall foundations have a large horizontal area. This reduces the pressure underneath so that the walls do not sink further into the ground.

A load-spreading washer ensures that the nut is not pulled into the wood when tightened up.

Typical pressures

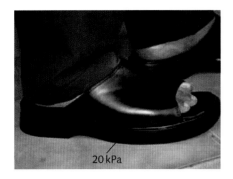

20 kPa

500 kPa

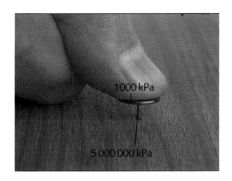

1000 kPa

5 000 000 kPa

Pressure problems

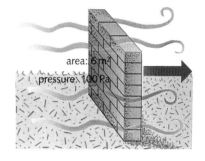

area: 6 m²
pressure: 100 Pa

Example 1 The wind pressure on the wall on the right is 100 Pa. If the wall has an area of 6 m², what is the force on it?

To solve this problem, you need to rearrange the pressure equation:

$$\text{force} = \text{pressure} \times \text{area}$$
$$= 100\,\text{Pa} \times 6\,\text{m}^2 = 600\,\text{N}$$

So the force on the wall is 600 N.

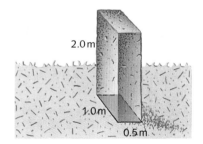

2.0 m

1.0 m

0.5 m

Example 2 A concrete block has a mass of 2600 kg. If the block measures 0.5 m by 1.0 m by 2.0 m, what is the maximum pressure it can exert when resting on the ground? ($g = 10$ N/kg)

As g is 10 N/kg, the 2600 kg block has a weight of 26 000 N, so the force on the ground is also 26 000 N.

To exert *maximum* pressure, the block must be resting on the side with the *smallest* area. This is the side measuring 1.0 m × 0.5 m, as shown on the right. Its area = 1.0 m × 0.5 m = 0.5 m². So:

$$\text{pressure} = \frac{\text{force}}{\text{area}} = \frac{26\,000\,\text{N}}{0.5\,\text{m}^2} = 52\,000\,\text{Pa}$$

So the maximum pressure is 52 000 Pa, or 52 kPa.

 Q

Assume that $g = 10$ N/kg, and that all forces are acting at right angles to any area mentioned.

1 A force of 200 N acts on an area of 4 m².
  a) What pressure is produced?
  b) What would the pressure be if the same force acted on half the area?

2 What force is produced if:
  a) a pressure of 1000 Pa acts on an area of 0.2 m²?
  b) a pressure of 2 kPa acts on an area of 0.2 m²?

3 Explain why a tractor's big tyres stop it sinking too far into soft soil.

4 A rectangular block of mass 30 kg measures 0.1 m by 0.4 m by 1.5 m.
  a) Calculate the weight of the block.
  b) Draw a diagram to show how the block must rest to exert the *maximum* pressure on the ground. Calculate this pressure.
  c) Draw another diagram showing the position of *minimum* pressure. Calculate this pressure.

3.06 Pressure in liquids

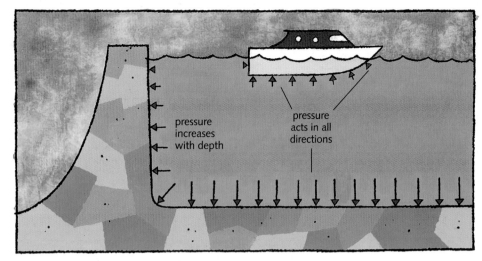

Pressure acts in all directions.

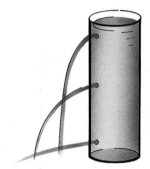

Pressure increases with depth.

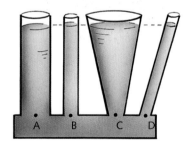

The pressure at points A, B, C, and D is the same.

A liquid is held in its container by its weight. This causes pressure on the container, and pressure on any object in the liquid.

The following properties apply to any stationary liquid in an open container. The experiments on the left demonstrate three of them.

Pressure acts in all directions The liquid pushes on every surface in contact with it, no matter which way the surface is facing. For example, the deep-sea vessel below has to withstand the crushing effect of sea water pushing in on it from all sides, not just downwards.

Pressure increases with depth The deeper into a liquid you go, the greater the weight of liquid above and the higher the pressure. Dams are made thicker at the bottom to withstand the higher pressure there.

Pressure depends on the density of the liquid The more dense the liquid, the higher the pressure at any particular depth.

Pressure doesn't depend on the shape of the container Whatever the shape or width, the pressure at any particular depth is the same.

▶ Deep-sea diving vessels are built to withstand the crushing effect of sea water whose pressure pushes inwards from all directions.

Useful connections

For calculations like those below, you need to know the connections between these:

volume (in m³)   density (in kg/m³)   mass (in kg)   weight (in N)   g (10 N/kg)

For example, you might know the volume and density of a liquid, but need to find its weight. For this, the equations required are:

$$density = \frac{mass}{volume} \qquad weight = mass \times g$$

From these equations, it follows that:

$$mass = density \times volume \qquad weight = density \times volume \times g$$

Pressure and weight essentials

For a force acting at right angles to a surface:

$$pressure = \frac{force}{area}$$

If force is in newtons (N) and area in square metres (m²), then pressure is in pascals (Pa).

Calculating the pressure in a liquid

The container on the right has a base area A. It is filled to a depth h with a liquid of density ρ (Greek letter 'rho'). To calculate the pressure on the base due to the liquid, you first need to know the weight of the liquid on it:

$$volume\ of\ liquid = base\ area \times depth = Ah$$
$$mass\ of\ liquid = density \times volume = \rho Ah$$
$$weight\ of\ liquid = mass \times g \qquad = \rho g Ah \qquad (g = 10\,N/kg)$$

So: force on base $= \rho g Ah$

This force is acting on an area A.

So: $pressure = \dfrac{force}{area} = \dfrac{\rho g Ah}{A} = \rho g h$

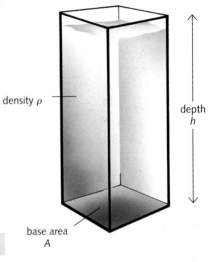

density ρ

depth h

base area A

At a depth h in a liquid of density ρ:

$$pressure = \rho g h$$

Example If the density of water is 1000 kg/m³, what is the pressure due to the water at the bottom of a swimming pool 2 m deep?

$$pressure = \rho g h = 1000\,kg/m^3 \times 10\,N/kg \times 2\,m = 20\,000\,Pa$$

$g = 10$ N/kg; density of water $= 1000$ kg/m³; density of paraffin $= 800$ kg/m³

1 In the diagram on the right:
   a) How does the pressure at A compare with the pressure at B?
   b) How does the pressure at B compare with the pressure at D?
   c) How does the pressure at A compare with the pressure at C?
   d) If the water in the system were replaced with paraffin, how would this affect the pressure at B?
2 A rectangular storage tank 4 m long by 3 m wide is filled with paraffin to a depth of 2 m. Calculate:
   a) the volume of paraffin   **b)** the mass of paraffin
   c) the weight of paraffin   **d)** the pressure at the bottom of the tank due to the paraffin
3 In the diagram on the right, calculate the pressure at B due to the water.

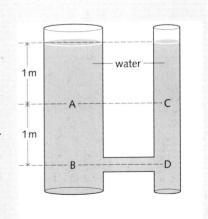

water

1 m

A — — — — — C

1 m

B — — — — — D

Hydraulic systems★

In some machines, the forces are transmitted by liquids under pressure rather than by levers or cogs. Machines like this are called **hydraulic** machines. They make use of these properties of liquids:

- Liquids are virtually incompressible – they cannot be squashed.
- If a trapped liquid is put under pressure, the pressure is transmitted to all parts of the liquid.

Hydraulic brakes

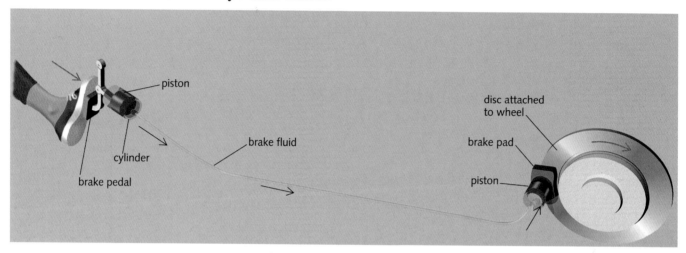

Car brakes work hydraulically. The diagram above shows the principle. When the brake pedal is pressed, a piston forces brake fluid from one cylinder along a connecting pipe to another cylinder. There, the fluid pushes on another piston. This presses a brake pad against a metal disc attached to the rotating wheel of the car. The friction slows the wheel.

In practical braking systems, there are pipes to all four wheels, pads on either side of each disc, and usually 'power assistance' as well.

Hydraulic jack

A load is easier to lift if you use a jack. The diagram at the top of the next page shows a simple hydraulic jack. A downward force on the input piston puts pressure on the oil. The pressure is transmitted by the oil. It produces a larger upward force on the output piston.

Knowing the input force and piston areas, the output force can be calculated:

In the input cylinder:
An input force of 12 N acts on an area of 0.01 m².

So: $\text{pressure on oil} = \dfrac{\text{force}}{\text{area}} = \dfrac{12\,\text{N}}{0.01\,\text{m}^2} = 1200\,\text{Pa}$

In the connecting pipe:
The pressure, 1200 Pa, is transmitted by the oil.

In the output cylinder:
The pressure, 1200 Pa, acts on a piston of area 0.1 m².
So: output force = pressure × area = $1200\,\text{Pa} \times 0.1\,\text{m}^2 = 120\,\text{N}$

Pressure essentials

For a force acting at right angles to a surface:

$$\text{pressure} = \frac{\text{force}}{\text{area}}$$

If force is in newtons (N) and area in square metres (m²), pressure is in pascals (Pa).

simple hydraulic jack

input force

acts on

small area

causing

high pressure

12 N

input...
piston
cylinder
area:
0.01 m²

oil

high pressure transmitted

object being lifted

120 N

output...
piston
cylinder
area:
0.1 m²

high output force

caused because

large area

has high pressure acting on it

A force multiplier With the jack above, you put in a force of 12 N and get out a force of 120 N. The jack is a **force multiplier**. In this case, it multiplies the input force by a factor 10. But there is a price to be paid for the gain in force: the output piston is raised only $\frac{1}{10}$ of the distance that the input piston is pushed down.

The calculation of output force assumes that the jack is frictionless. In a real jack, there is friction to overcome. This reduces the output force.

Hydraulic jack

For a *frictionless* jack:

$$\frac{\text{output force}}{\text{input force}} = \frac{\text{output piston area}}{\text{input piston area}}$$

In the case of the jack above:

$$\frac{120 \text{ N}}{12 \text{ N}} = \frac{0.1 \text{ m}^2}{0.01 \text{ m}^2}$$

Hydraulic press

A hydraulic press is used for compressing (squashing) things. It is like a hydraulic jack, but with a metal plate fixed rigidly above the output piston, so that the gap closes as the piston moves upwards.

◀ The shovel and arm on this digger are operated hydraulically, and the caterpillar tracks are moved by 'hydraulic motors'. The high-pressure oil comes from a pump driven by a diesel engine.

1 The diagram on the right shows a simple hydraulic jack. Assuming that the jack is frictionless:
   a) What is the pressure at A?
   b) What is the pressure at B?
   c) What is the output force?
   d) Explain why the jack can be called a *force multiplier*.
2 In the jack on the right, what would be the effect of
   a) increasing the area of the output piston?
   b) decreasing the area of the input piston?

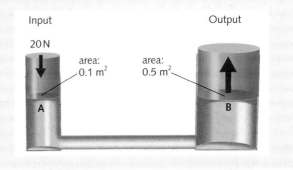

Input

20 N

area: 0.1 m²

A

Output

area: 0.5 m²

B

Related topics: pressure 3.05; pressure in liquids 3.06

3.08 Pressure from the air

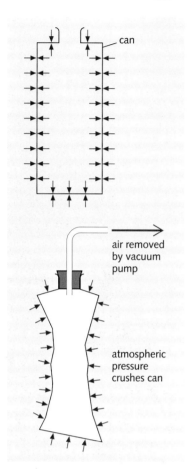

▲ **Demonstrating atmospheric pressure** When the air is removed from the can, there is nothing to resist the outside pressure, and the can is crushed.

The atmosphere is a deep ocean of air which surrounds the Earth. In some ways, it is like a liquid:

● Its pressure acts in all directions.
● Its pressure becomes less as you rise up through it (because there is less and less weight above).

Unlike a liquid, air can be compressed (squashed). This makes the atmosphere denser at lower levels. The atmosphere stretches hundreds of kilometres into space, yet the bulk of the air lies within about 10 kilometres of the Earth's surface.

Atmospheric pressure

At sea level, atmospheric pressure is about 100 kPa (100 000 newtons per square metre) – equivalent to the weight of ten cars pressing on every square metre. But you aren't crushed by this huge pressure because it is matched by the pressure in your lungs and blood system.

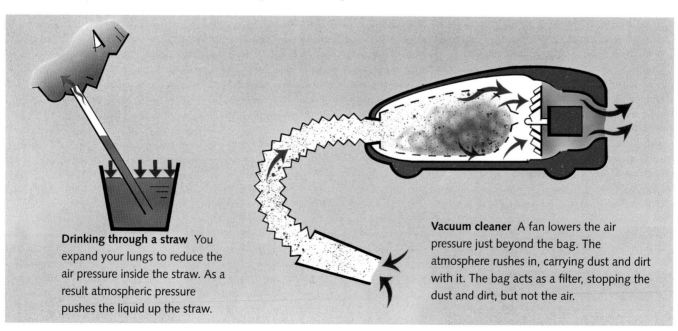

Drinking through a straw You expand your lungs to reduce the air pressure inside the straw. As a result atmospheric pressure pushes the liquid up the straw.

Vacuum cleaner A fan lowers the air pressure just beyond the bag. The atmosphere rushes in, carrying dust and dirt with it. The bag acts as a filter, stopping the dust and dirt, but not the air.

The mercury barometer

Instruments that measure atmospheric pressure are called **barometers**.
The barometer on the right contains the liquid metal mercury. Atmospheric
pressure has pushed mercury up the tube because the space at the top of the
tube has no air in it. It is a **vacuum**. At sea level, atmospheric pressure will
support a column of mercury 760 mm high, on average. For convenience,
scientists sometimes describe this as a pressure of 760 'millimetres of mercury'.
However, it is easily converted into pascals and other units, as you can see below.

The actual value of atmospheric pressure varies slightly depending on the
weather. Rain clouds form in large areas of lower pressure, so a fall in the
barometer reading may mean that rain is on the way. Atmospheric pressure also
decreases with height above sea level. This idea is used in the **altimeter**, an
instrument fitted in aircraft to measure altitude.

Barometer

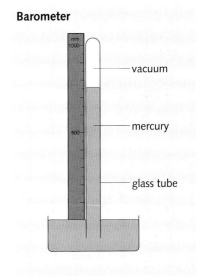

Standard atmospheric pressure

The pressure that will support a column of mercury 760.0 mm high is known as
standard atmospheric pressure, or 1 **atmosphere** (1 atm). Its value in
pascals can be found by calculating the pressure due to such a column.

At a depth h in a liquid of density ρ, the pressure $= \rho gh$, where g is
9.807 N/kg (or 10 N/kg if less accuracy is needed). As the density of mercury is
13 590 kg/m³, and the height of the column is 0.760 0 m:

$$1 \text{ atm} = \rho gh = 13\,590 \text{ kg/m}^3 \times 9.807 \text{ N/kg} \times 0.760\,0 \text{ m} = 101\,300 \text{ Pa}$$

In calculators, for simplicity, you can assume that 1 atm = 100 000 Pa. In
weather forecasting, the **millibar** (**mb**) is often used as a pressure unit.
1 mb = 100 Pa, so standard atmospheric pressure is approximately
1000 millibars.

Manometer

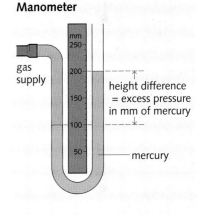

The manometer

A **manometer** measures pressure *difference*. The one in the diagram on the right
is filled with mercury. The height difference shows the *extra* pressure that the gas
supply has in addition to atmospheric pressure. This extra pressure is called the
excess pressure. To find the actual pressure of the gas supply, you add
atmospheric pressure to this excess pressure.

1 Write down *two* ways in which the pressure in the
atmosphere is like the pressure in a liquid.
2 Explain why, when you 'suck' on a straw, the liquid
travels up it.
3 If a mercury barometer were carried up a mountain,
how would you expect the height of the mercury
column to change?
4 Look at the diagram of the manometer on this page. If
atmospheric pressure is 760 mm of mercury:
a) What is the excess pressure of the gas supply (in mm
of mercury)?
b) What is the actual pressure of the gas supply (in mm
of mercury)?

c) What is the actual pressure of the gas supply (in Pa)?
5 If, on a particular day, atmospheric pressure is
730 mm of mercury, what is this **a)** in pascals
b) in atmospheres **c)** in millibars?
6 The density of mercury is 13 590 kg/m³, the density of
water is 1000 kg/m³, and g is 9.81 N/kg.
a) What is the pressure (in Pa) at the bottom of a
column of water 1 metre long?
b) If a barometer is made using water instead of
mercury, and a very long tube, how high is the water
column when atmospheric pressure is 1 atm (760 mm of
mercury)?

Gas pressure and volume

When dealing with a fixed mass of gas, there are always three factors to consider: *pressure, volume, and temperature*. A change in one of these factors always produces a change in at least one of the others. Often all three change at once. This happens, for example, when a balloon rises through the atmosphere, or gases expand in the cylinders of a car engine.

This spread deals with a simpler case: how the pressure of a gas depends on its volume if the temperature is kept constant. The link between the pressure and the volume can be found from the following experiment.

Linking pressure and volume (at constant temperature)

The equipment for the experiment is shown in the diagram below left, where the gas being studied is a *fixed mass* of *dry air*. The air is trapped in a glass tube. Its volume is reduced in stages by pumping air into the reservoir so that oil is pushed further up the tube. Each time the volume is reduced, the pressure of the trapped air is measured on the gauge.

▲ When this balloon rises, the pressure, volume, and temperature can all change.

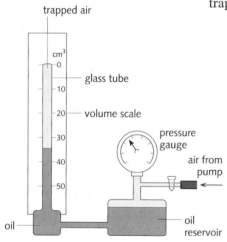

pressure	volume
kPa	cm³
200	50
250	40
400	25
500	20
1000	10

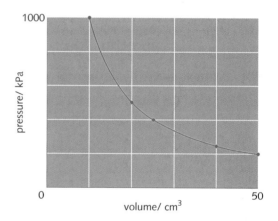

Squashing the air warms it up slightly. So before taking each reading, you have to wait a few moments for the air to return to its original temperature. The gauge actually measures the pressure in the reservoir, but this is the same as in the tube because the oil transmits the pressure.

Above, you can see some typical readings and the graph they produce. Results like this show that the relationship between the pressure and volume is an **inverse proportion**. It has these features:

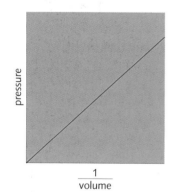

1 If the volume *halves*, the pressure *doubles*, and so on.

2 *Pressure × volume* always has the same value (10 000 in this case).

3 If *pressure* is plotted against $\dfrac{1}{volume}$, the graph is a straight line through the origin, as shown on the left.

The findings can be expressed as a law:

> For a fixed mass of gas at constant temperature, the pressure is inversely proportional to the volume.

This is known as **Boyle's law**.

Here is another way of writing Boyle's law. If the pressure of a gas changes from p_1 to p_2 when the volume is changed from V_1 to V_2:

$$p_1 \times V_1 = p_2 \times V_2 \quad \text{(at constant temperature)}$$

Example An air bubble has a volume of $2\,cm^3$ when released at a depth of $20\,m$ in water. What will its volume be when it reaches the surface? Assume that the temperature does not change and that atmospheric pressure is equivalent to the pressure from a column of water $10\,m$ deep.

In this case: p_1 = atmospheric pressure + pressure due to $20\,m$ of water
$$= 1\,atm + 2\,atm = 3\,atm$$

Also: $p_2 = 1\,atm$, $V_1 = 2\,cm^3$, and V_2 is to be found.

As the temperature does not change, Boyle's law applies. So:

$$p_1 \times V_1 = p_2 \times V_2 \quad \text{(at constant temperature)}$$
So: $\quad 3 \times 2 = 1 \times V_2 \quad$ (omitting units for simplicity)

This gives $V_2 = 6$, so on the surface, the volume of the bubble is $6\,cm^3$.

Explaining Boyle's law★

The **kinetic theory**, summarized on the right, explains Boyle's law like this. In a gas, the molecules are constantly striking and bouncing off the walls of the container. The force of these impacts causes the pressure. If the volume of the gas is halved, as shown below, there are twice as many molecules *in each cubic metre*. So, every second, there are twice as many impacts with each square metre of the container walls. So the pressure is doubled.

A gas that exactly obeys Boyle's law is called an **ideal gas**. Real gases come close to this provided they have a low density, a temperature well above their liquefying point, and are not full of water vapour. Unless these conditions are met, attractions between molecules affect their behaviour. An ideal gas has no attractions between its molecules.

Pressure essentials

$$\text{pressure} = \frac{\text{force}}{\text{area}}$$

If force is measured in newtons (N) and area in square metres (m^2), pressure is measured in pascals (Pa): $1\,Pa = 1\,N/m^2$.

Standard atmospheric pressure, called 1 atmosphere (atm), is approximately $100\,000\,Pa$.

The kinetic theory

According to this theory, a gas is made up of tiny, moving particles (usually molecules). These are spaced out with almost no attractions between them, and move about freely at high speed. The higher the temperature, then on average, the faster they move.

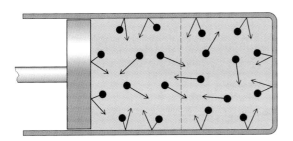

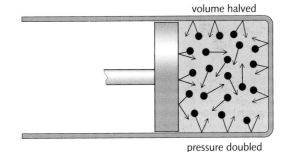

volume halved

pressure doubled

1 If you squash a balloon, the pressure inside it rises. How does the kinetic theory explain this?

2 A balloon contains $6\,m^3$ of helium at a pressure of $100\,kPa$. As the balloon rises through the atmosphere, the pressure falls and the balloon expands. Assuming that the temperature does not change, what is the volume of the balloon when the pressure has fallen to
a) $50\,kPa$ **b)** $40\,kPa$?

3 The readings below are for a fixed mass of gas at constant temperature:

pressure/atm	5.0	4.0	2.0	1.0	0.5	0.4
volume/cm³	4	5	10	20	40	50

a) How can you tell that the gas obeys Boyle's law?
b) Use a calculator to work out values for 1/volume. Plot a graph of pressure against 1/volume and describe its shape.

Related topics: pressure **3.05**; air pressure **3.08**; kinetic theory **5.01**; temperature **5.02**; water vapour **5.09**

3.10 Pressure problems

Pressure essentials

$$\text{pressure} = \frac{\text{force}}{\text{area}}$$

If force is measured in newtons (N) and area in square metres (m²), then pressure is measured in pascals (Pa).

$1\,\text{Pa} = 1\,\text{N/m}^2$

For convenience, air pressure is sometimes measured in atmospheres (atm). 1 atm, standard atmospheric pressure, is about $10^5\,\text{Pa}$.

At a depth h (m) in a liquid of density ρ (kg/m³):

pressure due to liquid $= \rho g h$ (Pa)

where g = Earth's gravitational field strength = 10 N/kg
The pressure acts in all directions.

If the pressure of a gas changes from p_1 to p_2 when its volume changes from V_1 to V_2, *at constant temperature*:

$$p_1 V_1 = p_2 V_2$$

This is **Boyle's Law**.

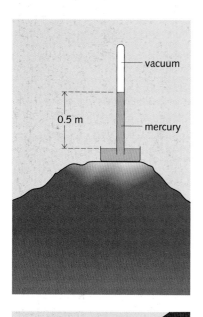

Here are some examples of problems which can be solved using the information in the box above.

Mercury barometer problem

Example A researcher sets up a mercury barometer (shown on the left) at the top of a mountain. She finds that the length of the mercury column is 0.50 m. What is the atmospheric pressure in Pa? (Assume, for simplicity, that the density of mercury is 13 600 kg/m³ and g is 10 N/kg.)

As the atmosphere is supporting the column of mercury, its pressure must equal the pressure due to that column – in other words the pressure at a depth of 0.50 m in mercury. That can be calculated as follows:

pressure $= \rho g h = 13\,600\,\text{kg/m}^3 \times 10\,\text{N/kg} \times 0.50\,\text{m}$

$= 68\,000\,\text{Pa}$

So: atmospheric pressure up the mountain is 69 000 Pa (68 kPa)

Water barometer problem

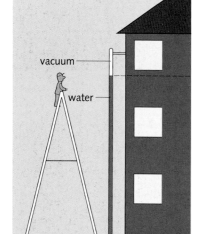

A water barometer would be too tall to be of practical use.

Example A student wants to make a barometer containing water instead of mercury and needs to estimate how tall it should be. Calculate the length of a column of water which can be supported by the atmosphere at sea level. (Assume that atmospheric pressure at sea level is 100 000 Pa, the density of water is 1000 kg/m³, and g is 10 N/kg.)

Let h be the height of the column. Omitting units for simplicity:

pressure due to column $= \rho g h = 1000 \times 10 \times h = 10\,000h$

But this pressure must equal atmospheric pressure, 100 000 Pa.

So: $10\,000h = 100\,000$

This gives: $h = 10$

So, atmospheric pressure will support a column of water 10 m high.

Diving bell problem

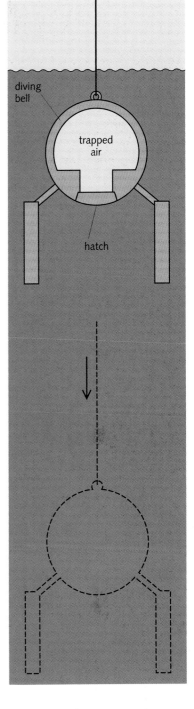

Example The diving bell on the right contains $6\,m^3$ of air. It has a hatch of area $0.5\,m^2$. While the bell is on the surface, at sea level, the hatch is closed and sealed. Then the bell is lowered through the water on a cable.
a) If the diving bell is lowered to a depth of $30\,m$, what is the force on the hatch due to the water pressure? (Assume that the density of water is $1000\,kg/m^3$, and g is $10\,N/kg$.)
b) If water leaks into the bell when it is $30\,m$ deep, what will the volume of air be reduced to? (Assume that atmospheric pressure at sea level will support a $10\,m$ column of water, and that the temperature is constant.)

a) First, the pressure due to the water at a depth of $30\,m$ must be calculated:

pressure $= \rho g h = 1000\,kg/m^3 \times 10\,N/kg \times 30\,m$

$\qquad\qquad = 300\,000\,Pa$

This pressure acts on a hatch of area of $0.5\,m^2$. As pressure = force/area:

force = pressure $\times$ area $= 300\,000\,Pa \times 0.5\,m^2 = 150\,000\,N$.

So the force on the hatch is $150\,000\,N$ ($150\,kN$).

b) When water leaks into the bell, the volume of the air inside is reduced, and its pressure rises until it matches the external pressure. The new volume can be found using Boyle's law. In applying the law, it is simplest to express pressures in 'metres of water'. Atmospheric pressure is $10\,m$ of water. At $30\,m$ depth, the external pressure is $30\,m$ of water plus atmospheric pressure, $10\,m$ of water. So it is $40\,m$ of water in total.

On the surface: pressure of air in bell $= p_1 = 10\,m$ of water

On surface: $\qquad$ volume of air in bell $= V_1 = 6\,m^3$

At $30\,m$ depth: pressure of air in bell $= p_2 = 40\,m$ of water

At $30\,m$ depth: $\quad$ volume of air in bell $= V_2$ (to be found)

As $\quad p_1 V_1 = p_2 V_2$

$\qquad 10 \times 6 = 40 \times V_2$ (omitting units for simplicity)

This gives $V_2 = 1.5$
So, the volume of the air in the bell is reduced to $1.5\,m^3$.

1 A rectangular storage tank, of base area $5\,m^2$, is filled to a depth of $2\,m$ with water. If the density of water is $1000\,kg/m^3$, and g is $10\,N/kg$:
   a) What is the pressure at the bottom of the tank due to the water?
   b) What is the downward force on the base due to the water?
   c) How would your answers to a) and b) be affected if the area of the base were halved?

2 When an upturned beaker is placed on the surface of water, as on the right, it contains $300\,cm^3$ of trapped air at atmospheric pressure. If the beaker is taken $20\,m$ beneath the water surface, what will be the volume of the air inside? (Assume that the temperature is constant, and that atmospheric pressure will support a column of water $10\,m$ high.)

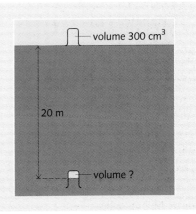

Related topics: pressure **3.05**; pressure in liquids **3.06**; barometers, atmospheric pressure **3.08**, Boyle's law **3.09**

73

1 The diagram shows a pair of nutcrackers. Forces *F* are applied to the handles of the nutcrackers.

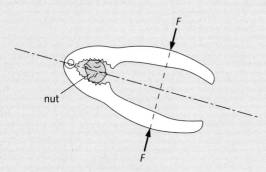

a) The forces on the nut are bigger than *F*. Explain this. [1]

b) The nut does not crack. State **two** changes that could be made to crack the nut. [2]

MEG

2 The diagram below shows a uniform metre rule, weight *W*, pivoted at the 75 cm mark and balanced by a force of 2 N acting at the 95 cm mark.

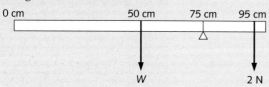

a) Calculate the moment of the 2 N force about the pivot. [2]

b) Use the principle of moments to calculate the value of *W* in N. [2]

UCLES

3 The diving bell below contains trapped air at the same pressure as the water outside. At the surface, air pressure is 100 kPa. As the bell descends, the pressure on it increases by 100 kPa for every 10 m of depth.

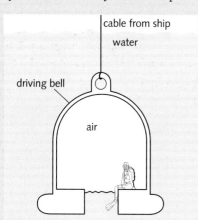

a) What is the pressure on the diver at depths of 0 m, 10 m, 20 m, and 30 m? [2]

b) At the surface, the bell holds 6 m³ of air. If the bell is lowered to a depth of 20 m, and no more air is pumped into it, what will be the volume of the trapped air? (Assume no change in temperature.) [3]

4 The figure shows an empty wheelbarrow which weighs 80 N.

The operator pulls upwards on the handles with a force of 20 N to keep the handles horizontal.

The point marked M is the centre of mass of the wheelbarrow.

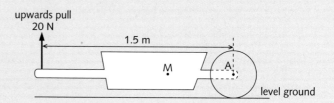

a) Copy the figure and draw arrows to show the other two vertical forces that act on the wheelbarrow. [2]

b) Determine
  (i) the moment of the 20 N force about the centre of the wheel A,
  (ii) the distance between points A and M. [3]

UCLES

5 The following results were obtained when a spring was stretched.

load / N	1.0	3.0	4.5	6.0	7.5
length of spring / cm	12.0	15.5	19.0	22.0	25.0

a) Use the results to plot a graph of length of spring against load. [1]

b) Use the graph to find the
  (i) unloaded length of the spring, [1]
  (ii) extension produced by a 7.0 N load, [1]
  (iii) load required to increase the length of the spring by 5.0 cm. [1]

WJEC

6 a) A glass window pane covers an area of 0.6 m². The force exerted by air pressure on the outside of the glass window pane is 60 000 N. Calculate the pressure of the air. Write down the formula that you use and show your working. [3]

b) Explain why the window does not break under this force. [1]

WJEC

7 A fitness enthusiast is trying to strengthen his calf muscles.

He uses the exercise machine below. His heels apply a force to the padded bar. This lifts the heavy weights.

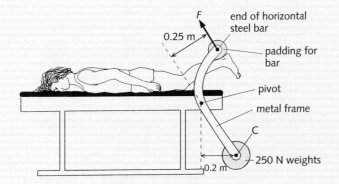

a) The centre of mass of the weights is at C.
Copy the diagram and draw an arrow to show **where** and in which **direction** the force of gravity acts on the weights. Label this arrow *W*. [2]
b) The narrow steel bar is padded. Why does this feel more comfortable when lifting the weights? [2]
c) The heels press against the pad with a force *F* and cause a turning effect about the pivot. Calculate the value of *F* when the weights are in the position shown in the diagram. Show your working. [3]
d) Why does it become harder to lift the weights when they move to the right? [2]

MEG

8 Three concrete blocks can be stacked in two different ways as shown below.

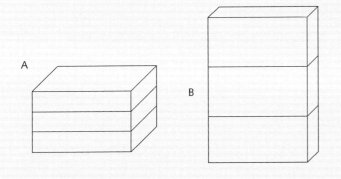

a)

less than	the same as	more than

Copy and complete the paragraph below using a phrase from the boxes above. Each phrase may be used once, more than once or not at all.
The force of stack **A** on the ground is _____ the force of stack **B**.

The pressure on the ground from stack **B** is _____ the pressure from stack **A**, because the area in contact with the ground for **B** is _____ for **A**. [3]
b) Write down, **in words**, the equation connecting pressure, force and area. [1]
c) If the weight of stack **A** is 500 N and the area in contact with the ground is 200 cm², calculate the pressure on the ground in N/cm². [2]

WJEC

9 The figure shows a tyre used on a large earth-moving vehicle.

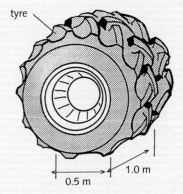

a) When the vehicle is loaded, the area of **each** tyre in contact with the ground is a rectangle of sides 1.0 m and 0.5 m.
  (i) Calculate the area in m² of contact of **one** tyre with the ground.
  (ii) The vehicle has four of these tyres. Calculate the total area in m² of contact with the ground. [4]
b) When the vehicle is loaded, it weighs 100 000 N. Calculate the pressure in N/m² exerted on the ground by the tyres. [3]

NEAB

10 A rectangular storage tank has a base measuring 3 m by 2 m. The tank is filled with water to a depth of 2 m. The density of the water is 1000 kg/m³, and *g* is 10 N/kg. Draw a diagram of the tank with the water in it, and mark all the dimensions on your drawing. Then calculate the following:
a) The volume of water in the tank. [2]
b) The mass of water in the tank. [2]
c) The weight of water in the tank (in N). [2]
d) The pressure at the bottom of the tank. [2]

Photocopy the list of topics below and tick the boxes of the ones that are included in your examination syllabus. (Your teacher should be able to tell you which they are.) Use your list when you revise. The spread number in brackets tells you where to find more information.

❑ **1** The factors affecting the moment of a force. (3.01)

❑ **2** Calculating the moment of a force. (3.01)

❑ **3** The principle of moments. (3.01)

❑ **4** The conditions applying when an object is in a state of equilibrium. (3.01)

❑ **5** The meaning of centre of mass. (3.02)

❑ **6** Finding the centre of mass of an object by experiment. (3.02)

❑ **7** Allowing for weight and the position of the centre of mass in calculations using the principle of moments. (3.02)

❑ **8** The different types of equilibrium. (3.02)

❑ **9** The conditions necessary for stability. (3.02)

❑ **10** Solving problems involving moments. (3.03)

❑ **11** The difference between elastic and plastic behaviour. (3.04)

❑ **12** The meaning of elastic limit. (3.04)

❑ **13** How the extension changes with load when a spring is stretched. (3.04)

❑ **14** Hooke's law, and how it applies to metal springs and wires. (3.04)

❑ **15** The meaning of a spring constant. (3.04)

❑ **16** The behaviour of rubber when stretched. (3.04)

❑ **17** How compression and bending are produced. (3.04)

❑ **18** The definition of pressure. (3.05)

❑ **19** The SI unit of pressure: the pascal. (3.05)

❑ **20** Using the equation linking pressure, force, and area. (3.05)

❑ **21** The factors affecting the pressure in a liquid. (3.06)

❑ **22** Calculating the pressure at a particular depth in a liquid. (3.06)

❑ **23** How hydraulic systems work. (3.07)

❑ **24** Atmospheric pressure and the factors affecting it. (3.08)

❑ **25** The evidence that atmospheric pressure has a high value. (3.08)

❑ **26** Using atmospheric pressure in a drinking straw and in a vacuum cleaner. (3.08)

❑ **27** The mercury barometer. (3.08)

❑ **28** Standard atmospheric pressure. (3.08)

❑ **29** Alternative units of pressure: the mm of mercury, the atmosphere, and the millibar. (3.08)

❑ **30** Using a manometer. (3.08)

❑ **31** Interpreting pressure–volume and pressure–1/volume graphs for a fixed mass of gas at constant temperature. (3.09)

❑ **32** Boyle's law. (3.09)

❑ **33** How the kinetic theory explains Boyle's law. (3.09)

❑ **34** Using the relationship $p_1 V_1 = p_2 V_2$. (3.09 and 3.10)

❑ **35** Solving problems about the pressure in liquids and gases. (3.10)

4 Work and Energy

● WORK ● ENERGY ● CONSERVATION OF ENERGY ● POTENTIAL AND KINETIC ENERGY
● EFFICIENCY ● POWER ● POWER STATIONS ● ENERGY RESOURCES
● RENEWABLE AND NON-RENEWABLE ENERGY ● ENERGY FROM THE SUN

The Niagara Falls, on the USA-Canada border. The photograph shows the highest section of the falls, where the water tumbles over 30 metres to the river below. Nearly three million litres of water flow over the falls every second. Most of the energy is wasted, but some is harnessed by a hydroelectric power station which generates electricity for the surrounding area.

4.01 Work and energy

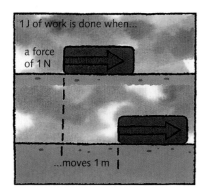

1 J of work is done when...

a force of 1 N

...moves 1 m

Work

In everyday language, work might be writing an essay or digging the garden. But to scientists and engineers, work has a precise meaning: work is done whenever a force makes something move. The greater the force and the greater the distance moved, the more work is done.

The SI unit of work is the **joule (J)**:

> 1 joule of work is done when a force of 1 newton (N) moves an object 1 metre in the direction of the force.

Work is calculated using this equation:

> work done = force × distance moved in the direction of the force

In symbols: $W = F \times d$

For example, if a 4 N force moves an object 3 m, the work done is 12 J.

Energy

Things have energy if they can be used to do work. A compressed spring has energy; so does a tankful of petrol. Like work, energy is measured in joules (J). Although people talk about energy being stored or given out, energy isn't a 'thing'. If, say, a compressed spring stores 100 joules of energy, this is just a measurement of how much work can be done by the spring.

Energy can take different forms. These are described on the opposite page. To understand them, you need to know the following:

- Moving objects have energy. For example, a moving ball can do work by knocking something over.
- Materials are made up of atoms (or groups of atoms). These are constantly in motion. For example, in a solid such as iron, the atoms are vibrating. If the solid is heated and its temperature rises, the atoms move faster. So a material has more energy when hot than when cold.

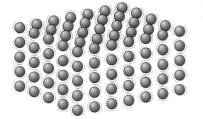

Atoms vibrating in a solid. The atoms have energy because of their motion.

▶ A fully flexed bow stores about 300 joules of energy.

Forms of energy

To describe different forms of energy, these names are sometimes used:

Kinetic energy This is energy due to motion. All moving objects have kinetic energy.

Potential energy This is energy which an object has because of its changed position, shape, or state. There are several different types of potential energy. Here are some of the terms used to describe them:

Gravitational potential energy A stone held up in the air can do work when dropped because gravity will pull it downwards. The stone has gravitational potential energy.

Elastic potential energy (strain energy) A stretched rubber band can do work when released, so can a compressed spring. Both have elastic potential energy.

Chemical potential energy When a fuel burns, its energy is released by chemical reactions. The energy stored in the fuel is called chemical potential energy, or chemical energy for short. Batteries also store it. So do foods. Without it, your muscles could not move.

Electrical potential energy In circuits, the current is a flow of tiny charged particles called electrons. These come from atoms. Electrons can transfer energy from, for example, a battery to a light bulb. They have electrical potential energy, or electrical energy for short.

Nuclear potential energy An atom has a nucleus at its centre. This is made up of particles, held there by strong forces. In some atoms, the particles become rearranged, or the nucleus splits, and energy is released. This is called nuclear potential energy, or nuclear energy for short.

The following terms are sometimes used when describing energy which is being transferred from one place to another, or from one object to another:

Thermal energy When hot objects cool down, their atoms and molecules slow down and lose energy. This is known as thermal energy, or heat. Engines use thermal energy to do work. For example, in a car engine, burning fuel produces hot gases which expand, push on pistons, and make them move. The motion is used to turn the wheels of the car.

Radiated energy The Sun radiates light. Loudspeakers radiate sound. Light and sound both travel in the form of waves. These carry energy.

Typical energy values

kinetic energy of a football when kicked	50 J
gravitational potential energy of a skier at the top of a ski jump	15 000 J
chemical energy in a chocolate biscuit	300 000 J
kinetic energy of a car travelling at 70 mph (30 m/s)	500 000 J
thermal energy needed to boil a kettleful of water	700 000 J
electrical energy supplied by a fully charged car battery	2 000 000 J
chemical energy in all the food you eat in one day	11 000 000 J
chemical energy in one litre of petrol	35 000 000 J

1 kilojoule (kJ)	= 1000 J (10^3 J)
1 megajoule (MJ)	= 1 000 000 J (10^6 J)

---

1 How much work is done if a force of 12 N moves an object a distance of 5 m?
2 If you use a 40 N force to lift a bag, and do 20 J of work, how far do you lift it?
3 Express the following amounts of energy in joules:
   a) 10 kJ   **b)** 35 MJ   **c)** 0.5 MJ   **d)** 0.2 kJ
4 Using information in the chart of energy values on this page, estimate how many fully charged car batteries are needed to store the same amount of energy as one litre of petrol.
5 **a)** Write down three forms of energy which the apple on the right has.
   b) Using the energy chart on this page as a guide, decide in which form you think the apple has most energy.

Related topics: scientific notation **1.01**; SI units **1.02**; force **2.06**; particles **5.01**; electrons in circuits **8.04**

Energy transformation

Conservation of energy

To do work, you have to spend energy. But, like money, energy doesn't vanish when you spend it. It goes somewhere else! People talk about 'using energy', but energy is never used up. It just changes into different forms, as in the example below.

A stone is thrown upwards ...   ... and falls to the ground

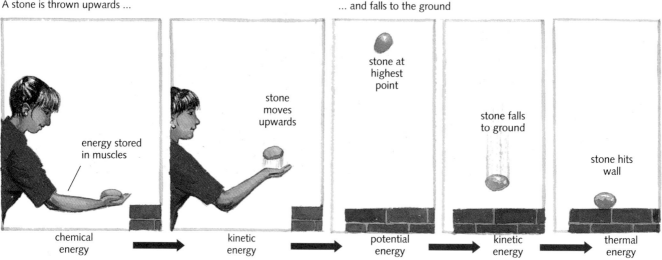

energy stored in muscles

stone at highest point

stone moves upwards

stone falls to ground

stone hits wall

| chemical energy | kinetic energy | potential energy | kinetic energy | thermal energy |

Transform or transfer?

When energy changes form, some scientists describe this as an energy 'transfer'. However, in this book, 'transfer' will only be used if energy moves from one place to another – for example, radiant energy travelling from the Sun to the Earth. A change in form will be a 'transformation'.

Work and energy essentials

Work is done whenever a force makes something move.

work done
= force × distance moved

Things have energy if they can be used to do work.

Work and energy are both measured in joules (J).

When energy changes from one form to another, scientists say that energy is **transformed**. The diagram above shows a sequence of energy transformations. The last one is from kinetic energy into thermal energy (heat). When the stone hits the ground, it makes the atoms and molecules in the stone and the ground move faster, so the materials warm up a little.

During each transformation, the *total amount* of energy stays the same. This is an example of the **law of conservation of energy**:

> Energy cannot be made or destroyed, but it can change from one form to another.

Wasting energy

The above diagram shows the energy transformations as a simple chain. In reality, energy is lost from the system at different stages. For example, muscles convert less than $1/5$th of the stored energy in food into kinetic energy. The rest is wasted as thermal energy – which is why exercise makes you sweat. And when objects move through the air, some of their kinetic energy is changed into thermal energy because of friction (air resistance). Even sound is eventually 'absorbed', which leaves the absorbing materials a little warmer than before.

The diagram at the top of the next page shows how all of the original energy of the thrower eventually ends up as thermal energy – although most of it is far too spread out to detect. Despite the apparent loss of energy from the system, the law of conservation of energy still applies. The *total amount* of energy is unchanged.

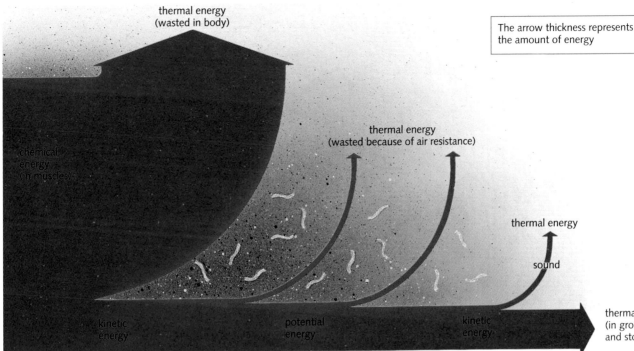

The arrow thickness represents the amount of energy

thermal energy
(wasted in body)

chemical energy (in muscles)

thermal energy
(wasted because of air resistance)

thermal energy

sound

kinetic energy

potential energy

kinetic energy

thermal energy
(in ground and stone)

stone thrown upwards   stone at highest point   stone hits ground

Work done and energy transformed

Whenever work is done, energy is transformed. In the diagram on the right, for example, a falling brick loses 20 J of potential energy. Assuming no air resistance, this is changed into 20 J of kinetic energy. So 20 J of work is done in accelerating the brick. If the brick hits the ground and comes to rest, 20 J of kinetic energy is changed into thermal energy. Again 20 J of work is done as the brick flattens the ground beneath it.

In all cases:

work done = energy transformed

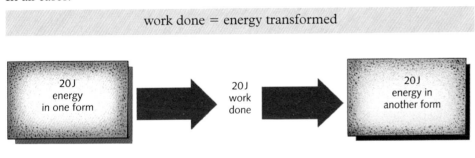

20 J
energy
in one form

20 J
work
done

20 J
energy in
another form

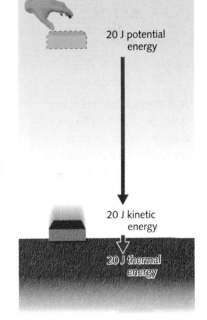

20 J potential energy

20 J kinetic energy

20 J thermal energy

Q

1 50 J of work must be done to lift a vase from the ground up on to a shelf.
  a) When the vase is on the shelf, what is its gravitational potential energy?
  b) If the vase falls from the shelf, how much kinetic energy does it have just before it hits the ground? (Assume that air resistance is negligible.)
  c) What happens to this energy after the vase has hit the ground?
2 What is the law of conservation of energy?
3 On the right, you can see someone's idea for an electric fan that costs nothing to run. The electric motor which turns the fan also turns a generator. This produces electricity for the motor, so no battery or mains supply is needed! Explain why this idea will not work.

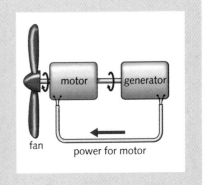

motor   generator

fan   power for motor

4.03 Calculating PE and KE

The ball below has potential energy because of the Earth's gravitational pull on it and its position above the ground. This is called **gravitational potential energy** (**PE**). If the ball falls, it gains **kinetic energy** (**KE**). Both PE and KE can be calculated.

Calculating PE

The gravitational potential energy of the ball on the left is equal to the work which would be done if the ball were to fall to the ground. Assuming no air resistance, it is also equal to the work done in lifting the ball a distance h up from the ground in the first place:

$$\text{downward force on ball} = \text{weight of ball} = mg$$

So: $$\text{upward force needed to lift ball} = mg$$

So: $$\text{work done in lifting ball} = \text{force} \times \text{distance moved}$$
$$= mgh$$

For an object of mass m at a vertical height h above the ground:

gravitational potential energy $= mgh$

For example, if a 2 kg mass is 3 m above the ground, and g is 10 N/kg:
gravitational PE $= 2\,\text{kg} \times 3\,\text{m} \times 10\,\text{N/kg} = 60\,\text{J}$

Calculating KE

The kinetic energy of the ball above is equal to the work which the ball could do by losing all of its speed. Assuming no air resistance, it is also equal to the work done on the ball in increasing its speed from zero to v in the first place:

$$\begin{aligned}
\text{work done} &= \text{force} && \times \text{distance moved} \\
&= \text{mass} \times \text{acceleration} && \times \text{distance moved} \\
&= \text{mass} \times \frac{\text{gain in speed}}{\text{time taken}} && \times \text{average speed} \times \text{time taken} \\
&= \text{mass} \times \text{gain in speed} && \times \text{average speed} \\
&= m \quad \times \quad v && \times \quad \tfrac{1}{2}v \\
&= \tfrac{1}{2}mv^2
\end{aligned}$$

For an object of mass m and speed v:

kinetic energy $= \frac{1}{2}mv^2$

For example, if a 2 kg mass has a speed of 3 m/s:
kinetic energy $= \frac{1}{2} \times 2\,\text{kg} \times (3\,\text{m/s})^2 = \frac{1}{2} \times 2 \times 3^2\,\text{J} = 9\,\text{J}$

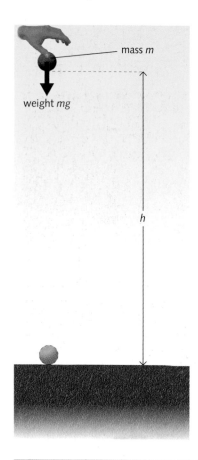

mass m

weight mg

h

Units

Mass is measured in kilograms (kg).

Force is measured in newtons (N).
Weight is a force.

Work is measured in joules (J).
Energy is measured in joules (J).

Useful equations

$$\text{average speed} = \frac{\text{distance moved}}{\text{time taken}}$$

$$\text{acceleration} = \frac{\text{gain in speed}}{\text{time taken}}$$

force = mass x acceleration

weight = mass x g (g = 10 N/kg)

work done = force x distance moved

work done = energy transformed

Scalar energy

Energy is a **scalar** quantity: it has magnitude (size) but no direction. So you do not have to allow for direction when doing energy calculations.

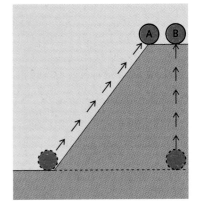

On the right, objects A and B have the same mass and are at the same height above the ground. B was lifted vertically but A was moved up a smooth slope. Although A had to be moved further, less force was needed to move it, and the work done was the same as for B. As a result, both objects have the same PE. The PE (*mgh*) depends on the vertical gain in height *h* and not on the particular path taken to gain that height.

KE and PE problems

Example If the stone on the right is dropped, what is its kinetic energy when it has fallen half-way to the ground? ($g = 10$ N/kg)

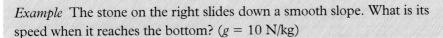

In problems like this, you don't necessarily have to use KE = $\frac{1}{2}mv^2$. When the stone falls, its *gain in KE* is equal to its *loss in PE*, so you can calculate that instead:

height lost by stone = 2 m

So: gravitational PE lost by stone = mgh = 4 kg × 10 N/kg × 2 m = 80 J

So: KE gained by stone = 80 J

As the stone started with no KE, this is the stone's KE half-way down.

Example The stone on the right slides down a smooth slope. What is its speed when it reaches the bottom? ($g = 10$ N/kg)

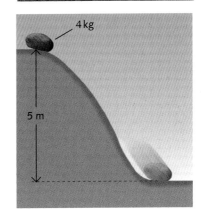

This problem can also be solved by considering energy changes. At the top of the slope, the stone has extra gravitational PE. When it reaches the bottom, all of this PE has been transformed into KE.

gravitational PE at top of slope = mgh = 4 kg × 10 N/kg × 5 m = 200 J

So: kinetic energy at bottom of slope = 200 J

So: $\frac{1}{2}mv^2$ = 200 J

So: $\frac{1}{2}$ × 4 kg × v^2 = 200 J

This gives: v = 10 m/s

So the stone's speed at the bottom of the slope is 10 m/s.

Note: if the stone fell vertically, it would start with the same gravitational PE and end up with the same KE, so its final speed would still be 10 m/s.

 Q

Assume that *g* is 10 N/kg and that air resistance and other frictional forces are negligible.

1 An object has a mass of 6 kg. What is its gravitational potential energy
   a) 4 m above the ground   **b)** 6 m above the ground?

2 An object of mass 6 kg has a speed of 5 m/s.
   a) What is its kinetic energy?
   b) What is its kinetic energy if its speed is doubled?

3 A ball of mass 0.5 kg has 100 J of kinetic energy. What is the speed of the ball?

4 A ball has a mass of 0.5 kg. Dropped from a cliff top, the ball hits the sea below at a speed of 10 m/s.
   a) What is the kinetic energy of the ball as it is about to hit the sea?
   b) What was the ball's gravitational potential energy before it was dropped?
   c) From what height was the ball dropped?
   d) A stone of mass 1 kg also hits the sea at 10 m/s. Repeat stages a, b, and c above to find the height from which the stone was dropped.

4.04 Efficiency and power

Engines and motors do work by making things move. Petrol and diesel engines spend the energy stored in their fuel. Electric motors spend energy supplied by a battery or generator. The human body is also a form of engine. It spends the energy stored in food.

Efficiency★

An engine does useful work with some of the energy supplied to it, but the rest is wasted as thermal energy (heat). The **efficiency** of an engine can be calculated as follows:

$$\text{efficiency} = \frac{\text{useful work done}}{\text{total energy input}} \quad or \quad \text{efficiency} = \frac{\text{useful energy output}}{\text{total energy input}}$$

For example, if a petrol engine does 25 J of useful work for every 100 J of energy supplied to it, then its efficiency is $1/4$, or 25%. In other words, its useful energy output is $1/4$ of its total energy input.

energy supplied		useful work done	efficiency
100 J	petrol engine	25 J	25%
100 J	diesel engine	35 J	35%
100 J	electric motor	80 J	80%
100 J	human body	15 J	15%

The chart above shows the efficiencies of some typical engines and motors. The low efficiency of fuel-burning engines is not due to poor design. When a fuel burns, it is impossible to transform its thermal energy into kinetic (motion) energy without wasting much of it.

Power

A small engine can do just as much work as a big engine, but it takes longer to do it. The big engine can do work at a faster rate. The rate at which work is done is called the **power**.

The SI unit of power is the **watt** (**W**). A power of 1 watt means that work is being done (or energy transformed) at the rate of 1 joule per second. Power can be calculated as follows:

$$\text{power} = \frac{\text{work done}}{\text{time taken}} \quad or \quad \text{power} = \frac{\text{energy transformed}}{\text{time taken}}$$

For example, if an engine does 1000 joules of useful work in 2 seconds, its power output is 500 watts (500 joules per second).

Force, work, and energy essentials

Work is measured in joules (J).

Energy is measured in joules (J).

work done = energy transformed

Force is measured in newtons (N).

work done = force × distance moved

Inputs and outputs

In any system, the *total* energy output must equal the *total* energy input. That follows from the law of conservation of energy. Therefore, the equations on the right could also be written with 'total energy output' replacing 'total energy input'.

Typical power outputs

washing machine motor	250 W
athlete	400 W
small car engine	35 000 W
large car engine	150 000 W
large jet engine	75 000 000 W

---

1 kilowatt (kW) = 1000 W

The **horsepower (hp)** is a power unit which dates back to the days of the early steam engines:

1 hp = 746 W (about $3/4$ kilowatt)

As energy and power are related, there is another way of calculating the efficiency of an engine:

$$\text{efficiency} = \frac{\text{useful power output}}{\text{total power input}}$$

Power problems⋆

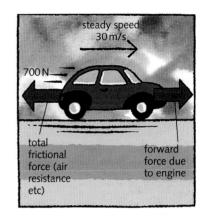

Example 1 The car on the right has a steady speed of 30 m/s. If the total frictional force on the car is 700 N, what useful power output does the engine deliver to the driving wheels?

steady speed 30 m/s

700 N

total frictional force (air resistance etc)

forward force due to engine

As the speed is steady, the engine must provide a forward force of 700 N to balance the total frictional force. In 1 second, the 700 N force moves 30 m, so:
work done = force × distance = 700 N × 30 m = 21 000 J.

As the engine does 21 000 J of useful work in 1 second, its useful power output must be 21 000 W, or 21 kW.

Problems of this type can also be solved with this equation:

$$\text{useful power output} = \text{force} \times \text{speed}$$

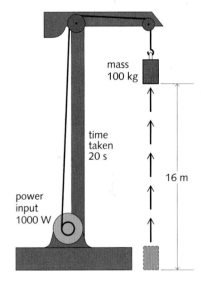

Example 2 The crane on the right lifts a 100 kg block of concrete through a vertical height of 16 m in 20 s. If the power input to the motor is 1000 W, what is the efficiency of the motor?

mass 100 kg

time taken 20 s

16 m

power input 1000 W

On Earth, *g* is 10 N/kg, so a 100 kg block has a weight of 1000 N. Therefore, a force of 1000 N is needed to lift the block. When the block is lifted:

$$\text{work done} = \text{force} \times \text{distance} = 1000\,\text{N} \times 16\,\text{m} = 16\,000\,\text{J}$$

$$\text{useful power output} = \frac{\text{useful work done}}{\text{time taken}} = \frac{16\,000\,\text{J}}{20\,\text{s}} = 800\,\text{W}$$

$$\text{efficiency} = \frac{\text{useful power output}}{\text{total power input}} = \frac{800\,\text{W}}{1000\,\text{W}} = 0.8$$

So the motor has an efficiency of 80%.

g = 10 N/kg

1 An engine does 1500 J of useful work with each 5000 J of energy supplied to it.
 a) What is its efficiency?
 b) What happens to the rest of the energy supplied?
2 If an engine does 1500 J of work in 3 seconds, what is its useful power output?
3 A motor has a useful power output of 3 kW.
 a) What is its useful power output in watts?
 b) How much useful work does it do in 1 s?
 c) How much useful work does it do in 20 s?
 d) If the power input to the motor is 4 kW, what is the efficiency?

4 Someone hauls a load weighing 600 N through a vertical height of 10 m in 20 s.
 a) How much useful work does she do?
 b) How much useful work does she do in 1 s?
 c) What is her useful power output?
5 A crane lifts a 600 kg mass through a vertical height of 12 m in 18 s.
 a) What weight (in N) is the crane lifting?
 b) What is the crane's useful power output?
6 With frictional forces acting, a forward force of 2500 N is needed to keep a lorry travelling at a steady speed of 20 m/s along a level road. What useful power is being delivered to the driving wheels?

Related topics: SI units **1.02**; force, mass, weight, and *g* **2.09**; law of conservation of energy **4.02**; work and energy **4.02–4.03**

85

4.05 Energy for electricity (1)

▶ Part of a thermal power station. The large, round towers with clouds of steam coming from them are cooling towers.

Industrial societies spend huge amounts of energy. Much of it is supplied by electricity which comes from **generators** in **power stations**.

Thermal power stations

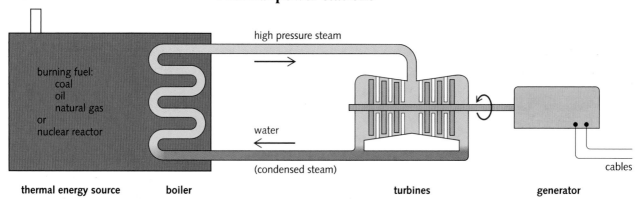

high pressure steam

burning fuel:
  coal
  oil
  natural gas
or
  nuclear reactor

water

(condensed steam)

cables

thermal energy source   boiler   turbines   generator

▲ A turbine

▶ Block diagram of what happens in a thermal power station

In most power stations, the generators are turned by **turbines**, blown round by high pressure steam. To produce the steam, water is heated in a boiler. The thermal energy comes from burning fuel (coal, oil, or natural gas) or from a **nuclear reactor**. Nuclear fuel does not burn. Its energy is released by nuclear reactions which split uranium atoms. The process is called **nuclear fission**.

Once steam has passed through the turbines, it is cooled and condensed (turned back into a liquid) so that it can be fed back to the boiler. Some power stations have huge cooling towers, with draughts of air up through them. Others use the cooling effect of nearby sea or river water.

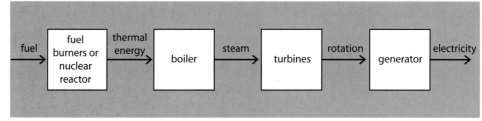

fuel → fuel burners or nuclear reactor → thermal energy → boiler → steam → turbines → rotation → generator → electricity

Energy spreading★

Thermal power stations waste more energy than they deliver. Most is lost as thermal energy in the cooling water and waste gases. For example, the efficiency of a typical coal-burning power station is only about 35% – in other words, only about 35% of the energy in its fuel is transformed into electrical energy. The diagram below shows what happens to the rest:

$$\text{efficiency} = \frac{\text{useful energy output}}{\text{energy input}}$$

$$= \frac{\text{useful power output}}{\text{power input}}$$

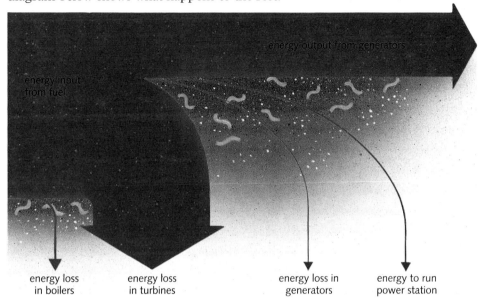

energy output from generators

energy input from fuel

energy loss in boilers

energy loss in turbines

energy loss in generators

energy to run power station

◀ Typical energy-flow chart for a thermal power station. A chart like this is called a **Sankey diagram**. The thickness of each arrow represents the amount of energy.

Engineers try to make power stations as efficient as possible. But once energy is in thermal form, it cannot all be used to drive the generators. Thermal energy is the energy of randomly moving particles (such as atoms and molecules). It has a natural tendency to spread out. As it spreads, it becomes less and less useful. For example, the concentrated energy in a hot flame could be used to make steam for a turbine. But if the same amount of thermal energy were spread through a huge tankful of water, it would only warm the water by a few degrees. This warm water could not be used as an energy source for a turbine.

District heating The unused thermal energy from a power station does not have to be wasted. Using long water pipes, it can heat homes, offices, and factories in the local area. This works best if the power station is run at a slightly lower efficiency so that hotter water is produced.

Combined cycle gas turbine power stations

These are smaller units which can be brought up to speed or shut off very quickly, as the demand for electricity varies. In them, natural gas is used as the fuel for a jet engine. The shaft of the engine turns one generator. The hot gases from the jet are used to make steam to drive another generator.

1 Write down *four* different types of fuel used in thermal power stations.

2 In a thermal power station:
   a) What is the steam used for?
   b) What do the cooling towers do?

3 The table on the right gives data about the power input and losses in two power stations, X and Y.
   a) Where is most energy wasted?
   b) In what form is this wasted energy lost?
   c) What is the electrical power output of each station? (You can assume that the table shows all the power losses in each station.)
   d) What is the efficiency of each power station?

	power station	
	X	**Y**
	coal	nuclear
power input from fuel in MW	5600	5600
power losses in MW:		
– in reactors/boilers	600	200
– in turbines	2900	3800
– in generators	40	40
power to run station in MW	60	60
electrical power output in MW	?	?

4.06 Energy for electricity (2)

Reactions for energy★

When fuels burn, they combine with oxygen in the air. With most fuels, including fossil fuels, the energy is released by this chemical reaction:

$$\underbrace{\text{fuel} + \text{oxygen}}_{\substack{\text{these are}\\\text{used up}}} \xrightarrow{\text{burning}} \underbrace{\text{carbon dioxide} + \text{water}}_{\substack{\text{these waste gases}\\\text{are made}}} + \textit{thermal energy}$$

There may be other waste gases as well. For example, burning coal produces some sulphur dioxide. Natural gas, which is mainly methane, is the 'cleanest' (least polluting) of the fuels burned in power stations.

In a nuclear power station, the nuclear reactions produce no waste gases like those above. However, they do produce radioactive waste.

Pollution problems★

Thermal power stations can cause pollution in a variety of ways:

- Fuel-burning power stations put carbon dioxide gas into the atmosphere. The carbon dioxide traps the Sun's energy and may be adding to **global warming**.
- The sulphur dioxide from coal-burning power stations causes acid rain. This can damage stonework and harm wildlife. One solution is to burn expensive low-sulphur coal. Another is to fit costly flue gas desulphurization (FGD) units to the power stations.
- Transporting fuels can cause pollution. For example, there may be a leak from an oil tanker.
- The radioactive waste from nuclear power stations is highly dangerous. It must be carried away and stored safely in sealed containers for many years – in some cases, thousands of years.
- Nuclear accidents are rare. But when they do occur, radioactive gas and dust can be carried thousands of kilometres by winds.

One effect of acid rain

Power from water and wind

Some generators are turned by the force of moving water or wind. There are three examples on the next page. Power schemes like this have no fuel costs, and give off no polluting gases. However, they can be expensive to build, and need large areas of land. Compared with fossil fuels, moving water and wind are much less concentrated sources of energy:

1 kWh of electrical energy can be supplied using...

...0.5 litres of oil (burning)

...5000 litres of fast-flowing water (20 m/s)

Hydroelectric power scheme River and rain water fill up a lake behind a dam. As water rushes down through the dam, it turns turbines which turn generators.

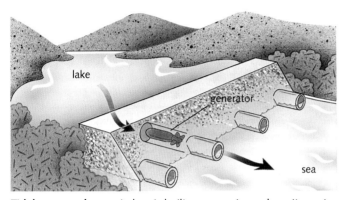

Tidal power scheme A dam is built across a river where it meets the sea. The lake behind the dam fills when the tide comes in and empties when the tide goes out. The flow of water turns the generators.

Pumped storage scheme This is a form of hydroelectric scheme. At night, when power stations have spare capacity, power is used to pump water from a lower reservoir to a higher one. During the day, when extra electricity is needed, the water runs down again to turn generators.

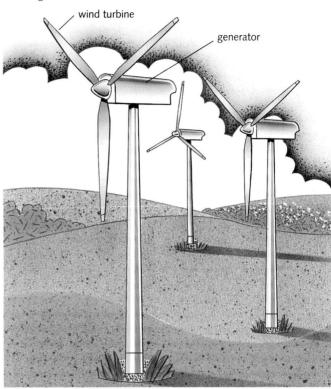

Wind farm This is a collection of **aerogenerators** – generators driven by giant wind turbines ('windmills').

power station	A	B	C	D	E
(1 MW = 1 000 000 W)	coal (non-FGD)	combined cycle gas	nuclear	wind farm	large tidal scheme
power output in MW	1800	600	1200	20	6000
efficiency (fuel energy ⟶ electrical energy)	35%	45%	25%	–	–
The following are on a scale 0–5					
build cost per MW output	2	1	5	3	4
fuel cost per kWh output	5	4	2	0	0
atmospheric pollution per kWh output	5	3	<1	0	0

1 What is the source of energy in a hydroelectric power station?

2 The table above gives data about five different power stations, A–E.

a) C has an efficiency of 25%. What does this mean?

b) Which power station has the highest efficiency? What are the other advantages of this type of power station?

c) Which power station cost most to build?

d) Which power station has the highest fuel cost per kWh output?

e) Which power station produces most atmospheric pollution per kWh output? What can be done to reduce this problem?

f) Why do two of the power stations have a zero rating for fuel costs and atmospheric pollution?

Related topics: efficiency and power **4.04**; energy resources **4.07–4.08**; calculating energy in kWh **8.14**

Energy resources

How energy is used in a typical industrialized country

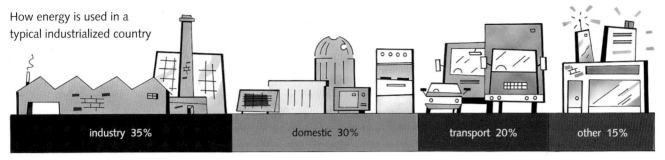

| industry 35% | domestic 30% | transport 20% | other 15% |

Most of the energy that we use comes from fuels that are burned in power stations, factories, homes, and vehicles. Nearly all of this energy originally came from the Sun. To find out how, see the next spread, 4.08.

The Sun is 75% hydrogen. It releases energy by a process called **nuclear fusion**, in which the nuclei (centres) of hydrogen atoms are pushed together to form helium. One day, it may be possible to harness this process on Earth (see Spread 11.07), but until this can be done, we shall have to manage with other resources.

The energy resources we use on Earth can be **renewable** or **non-renewable**. For example, wood is a renewable fuel. Once used, more can be grown to replace it. Oil, on the other hand, is non-renewable. It took millions of years to form in the ground, and cannot be replaced.

Non-renewable energy resources

Fossil fuels Coal, oil, and natural gas are called fossil fuels because they formed from the remains of plants and tiny sea creatures that lived millions of years ago. They are a very concentrated source of energy. Oil is especially useful because petrol, diesel, and jet fuel can be extracted from it. It is also the raw material from which most plastics are made.

Problems Supplies are limited. When the fuels burn, their waste gases pollute the atmosphere. Probably the most serious concern is the amount of carbon dioxide produced. This may be adding to global warming.

Nuclear fuels Most contain uranium. 1 kg of nuclear fuel stores as much energy as 55 tonnes of coal. In nuclear power stations, the energy is released by fission, a process in which the nuclei of uranium atoms are split.

Problems High safety standards are needed. The waste from nuclear fuel is very dangerous and stays radioactive for thousands of years. Nuclear power stations are expensive to build, and expensive to decommission (close down and dismantle at the end of their working life).

Renewable energy resources

Hydroelectric energy A river fills a lake behind a dam. Water flowing down from the lake turns generators.

Problems Expensive to build. Few areas of the world are suitable. Flooding land and building a dam causes environmental damage.

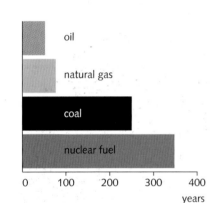

▲ This chart shows how long our *known* reserves of non-renewable fuels might last at present rates of consumption. However, it does not allow for the results of further exploration.

Where to find out more	
For more detailed information on...	*see spread...*
hydroelectric energy	4.06
tidal energy	4.06
wind energy	4.06
solar panel	5.08
energy and mass	11.06
nuclear fission	11.06
nuclear fusion	11.07

Tidal energy Similar to hydroelectric energy, but a lake fills when the tide comes in and empties when it goes out.

Problems As for hydroelectric energy.

Wind energy Generators are driven by wind turbines ('windmills').

Problems Large, remote, windy sites are needed. Winds are variable. The wind turbines are noisy and can spoil the landscape.

Wave energy Generators are driven by the up-and-down motion of waves at sea.

Problems Difficult to build – few devices have been successful.

Geothermal energy 'Geothermal' means heat from the Earth. Water is pumped down to hot rocks deep underground and rises as steam. In areas of volcanic activity, the steam comes naturally from hot springs.

Problems Deep drilling is difficult and expensive.

Solar energy (energy radiated from the Sun) Solar panels absorb this energy and use it to heat water. Solar cells are made from materials that can deliver an electric current when they absorb the energy in light.

Problems Variable amounts of sunshine in some countries. Solar cells are expensive, and must be large to deliver useful amounts of power. A cell area of around 10 m² is needed to power an electric kettle.

Biofuels These are fuels made from plant or animal matter, sometimes called **biomass**. They include wood, alcohol made from sugar cane, and methane gas from rotting waste.

Problems Huge areas of land are needed to grow plants.

Saving energy

Burning fossil fuels causes pollution. But the alternatives have their own environmental problems. That is why many people think that we should be less wasteful with energy. Methods could include using public transport and bicycles instead of cars, manufacturing goods that last longer, and recycling more waste materials. Also, better insulation in buildings would mean less need for heating in cold countries and for air conditioning in hot ones.

Power start-up

The demand for electricity varies through the day. When more power is needed, extra generators must be brought 'on line' quickly.

Small, gas-burning power stations can come up to speed very rapidly. Hydroelectric power stations are also quick to start up. Large fuel-burning power stations take longer. And nuclear power stations take longest of all. With a 'cold' reactor, a nuclear power station takes about two days to reach full power.

In Brazil, many cars use alcohol as a fuel instead of petrol. The alcohol is made from sugar cane, which is grown as a crop.

To answer these questions, you may need information from the illustration on the next spread, 4.08.

1 Some energy resources are *non-renewable*. What does this mean? Give *two* examples.
2 Give *two* ways of generating electricity in which no fuel is burned and the energy is renewable.
3 The energy in petrol originally came from the Sun. Explain how it got into the petrol.
4 Describe *two* problems caused by using fossil fuels.
5 Describe *two* problems caused by the use of nuclear energy.
6 What is *geothermal* energy? How can it be used?
7 What is solar energy? Give *two* ways in which it can be used.
8 Three of the energy resources described in this spread make use of moving water. What are they?
9 Give *four* practical methods of saving energy so that we use less of the Earth's energy resources.

Related topics: power stations **4.05–4.06**; energy from the Sun **4.08**; solar panel **5.08**; nuclear reactors **11.06–11.07**

4.08 How the world gets its energy

Solar panels

These absorb energy radiated from the Sun and use it to heat water.

Solar cells

These use the energy in sunlight to produce small amounts of electricity.

Energy in food

We get energy from the food we eat. The food may be from plants, or from animals which fed on plants.

Biofuels from plants

Wood is an important fuel in many countries. When wood is burned, it releases energy that the tree once took in from the Sun. In some countries, sugar cane is grown and fermented to make alcohol. This can be used as a fuel instead of petrol.

Biofuels from waste

Rotting animal and plant waste can give off methane gas. This is similar to natural gas and can be used as a fuel. Marshes, rubbish tips, and sewage treatment works are all sources of methane. Some waste can also be used directly as fuel by burning it.

Batteries

Some batteries (e.g. car batteries) have to be given energy by charging them with electricity. Others are manufactured from chemicals which already store energy. But energy is needed to produce the chemicals in the first place.

The Sun

The Sun radiates energy because of nuclear fusion reactions deep inside it. Its output is equivalent to that from 3×10^{26} electric hotplates. Just a tiny fraction of this reaches the Earth.

Energy in plants

Plants take in energy from sunlight falling on their leaves. They use it to turn water and carbon dioxide from the air into new growth. The process is called photosynthesis. Animals eat plants to get the energy stored in them.

Fossil fuels

Fossil fuels (coal, oil, and natural gas) were formed from the remains of plants and tiny sea creatures which lived many millions of years ago. Industrial societies rely on fossil fuels for most of their energy. Many power stations burn fossil fuels.

Fuels from oil
Many fuels can be extracted from oil (crude). These include: petrol, diesel fuel, jet fuel, paraffin, central heating oil, bottled gas.

The tides

The gravitational pull of the Moon (and to a lesser extent, the Sun) creates gentle bulges in the Earth's oceans. As the Earth rotates, different places have high and low tides as they pass in and out of the bulges. The motion of the tides carries energy with it.

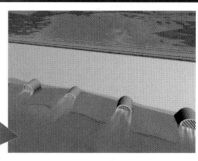

Tidal energy

In a tidal energy scheme, an estuary is dammed to form an artificial lake. Incoming tides fill the lake; outgoing tides empty it. The flow of water in and out of the lake turns generators.

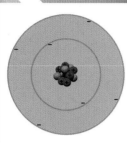

Nucleus of the atom

Radioactive materials have atoms with unstable nuclei (centres) which break up and release energy. The material gives off the energy slowly as thermal energy. Energy can be released more quickly by splitting heavy nuclei (fission). Energy can also be released by joining light nuclei (fusion), as happens in the Sun.

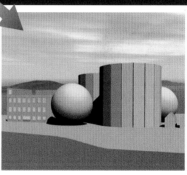

Nuclear energy

In a reactor, nuclear fission reactions release energy from the nuclei of uranium atoms. This heats water to make steam for driving generators.

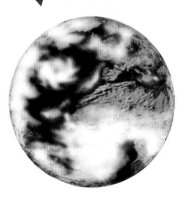

Weather systems

These are driven by energy radiated from the Sun. Heated air rising above the equator causes belts of wind around the Earth. Winds carry water vapour from the oceans and bring rain and snow.

Geothermal energy

Deep underground, the rocks are hotter than they are on the surface. The thermal energy comes from radioactive materials naturally present in the rocks. It can make steam for heating buildings or driving generators.

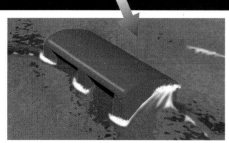

Wave energy

Waves are caused by the wind (and partly by tides). Waves cause a rapid up-and-down movement on the surface of the sea. This movement can be used to drive generators.

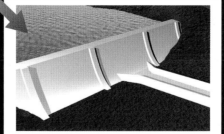

Hydroelectric energy

An artificial lake forms behind a dam. Water rushing down from this lake is used to turn generators. The lake is kept full by river water which once fell as rain or snow.

Wind energy

For centuries, people have been using the power of the wind to move ships, pump water, and grind corn. Today, huge wind turbines are used to turn generators.

1

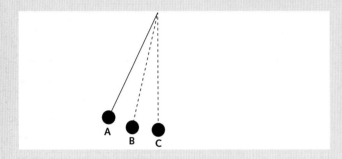

wound up spring	batteries connected to motors
rotating flywheel	stretched rubber bands

a) State which of the above **change shape** when their stored energy is transferred. [2]

b) Describe how the energy from a rotating flywheel can be transferred to moving parts of a child's toy. [2]

WJEC

2 The diagram below shows a pendulum which was released from position **A**.

a) What form(s) of energy did the pendulum have at
(i) **A**, (ii) **B**, (iii) **C**? [3]

b) Eventually the pendulum would stop moving. Explain what has happened to the initial energy of the pendulum. [2]

UCLES

3

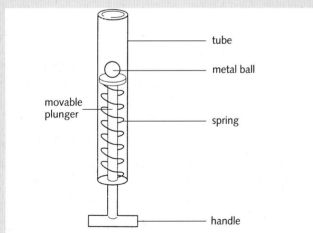

A type of toy catapult consists of a movable plunger which has a spring attached as shown above. The handle was pulled down to fully compress the spring and on release the metal ball of mass 0.1 kg (weight 1 N) was projected 0.75 m vertically.

a) (i) What type of energy is stored in a compressed spring? [1]

(ii) What happens to this stored energy when the handle of the plunger is released? [2]

b) Calculate the maximum potential energy acquired by the metal ball from the catapult. Write down the formula that you use and show your working. Take the acceleration due to gravity to be 10 m/s². [3]

c) Explain why the maximum potential energy gained by the metal ball is less than the original stored energy of the spring. [3]

WJEC

4 a) Name **four** renewable energy sources that are used to generate electricity. [4]

b) The main advantage of using renewable sources to generate electricity is that there are no fuel costs.
Give **two** major **disadvantages** of using renewable energy. [2]

c) The fuel costs for nuclear energy are low. State the main **financial** drawbacks in the use of nuclear energy to generate electricity. [2]

WJEC

5 A drop hammer is used to drive a hollow steel post into the ground. The hammer is placed inside the post by a crane. The crane lifts the hammer and then drops it so that it falls onto the baseplate of the post.

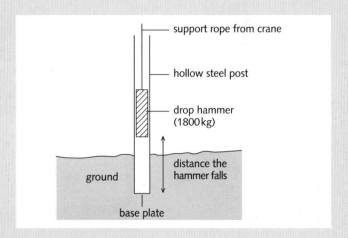

The hammer has a mass of 1800 kg. Its velocity is 5 m/s just before it hits the post.

a) Calculate the kinetic energy of the hammer just before it hits the post. [3]

b) How much potential energy has the hammer lost as it falls? Assume that it falls freely. [1]

c) Calculate the distance the hammer has fallen.
(Assume $g = 10$ N/kg) [3]

MEG

6 A crate of mass 300 kg is raised by an electric motor through a height of 60 m in 45 s. Calculate:

a) The weight of the crate. [2]

b) The work done by the motor. [2]

c) The useful power of the motor. [2]
(Assume $g = 10$ N/kg)

7

electrical appliance	power rating/ kW	power rating/ W
television	0.2	200
electric kettle		2000
food mixer	0.6	

The table above shows the power rating of three electrical appliances.

a) Copy the table and fill in the blank spaces. [2]

b) State which appliance transfers the least amount of energy per second. [1]

c) State which appliance converts electrical energy into heat and kinetic energy. [1]

WJEC

8 a) Explain what you understand by the phrase *non-renewable energy resources*. [2]

b) Explain why most non-renewable energy resources are burned. [1]

c) Name a non-renewable energy resource which is not burned. [1]

WJEC

9 a) The chemical energy stored in a fossil fuel produces heat energy when the fuel is burned. Describe how this heat energy is then used to produce electrical energy at a power station. [2]

b) Identify and compare the financial and environment costs of generating electricity using fossil fuels and wind. [4]

WJEC

10 a)

fuel/energy resource	renewable fuel	must be burned to release energy	found in the Earth's crust
coal	no	yes	yes
wood		yes	
uranium			yes

The table above shows that coal is **not** a renewable fuel. It releases energy when burned and is found in the Earth's crust.

Copy and complete the table for the other fuels/energy resources named. [2]

b) (i) Explain how fossil fuels were produced. [1]

(ii) State **two** reasons why we should use less fossil fuels. [2]

WJEC

11 a) Most of the energy available on Earth comes, or has come, from the Sun. Some energy resources on Earth store the Sun's energy from millions of years ago. Name one of these resources. [1]

b) Copy and complete the sentence below to say what a fuel does. [2]
A fuel is a material which supplies _____ when it _____

c) Explain the difference between renewable and non-renewable fuels. [1]

d) Copy and complete the following table to give examples of some fuels and their uses. The first one has been done for you. [?]

description	example	use
a gaseous fuel	hydrogen	rocket fuel
a liquid fuel		
a solid fuel		
a renewable fuel		
a non-renewable fuel		

NEAB

12 a) Copy and complete the following sentences about household electrical devices. Use words from the list below. Each word may be used once, more than once or not at all.

chemical   electrical   heat   kinetic   light   sound

(i) In an iron, electrical energy is transferred into mainly _____ energy.

(ii) In a vacuum cleaner, electrical energy is transferred into mainly _____ energy and unwanted _____ energy.

(iii) In a torch, electrical energy is transferred into mainly _____ and _____ energy.

(iv) In a hi-fi system, electrical energy is transferred into mainly _____ energy, and some unwanted _____ energy is also produced. [7]

b) The list below contains some types of potential energy.

chemical   elastic   gravitational   nuclear

Copy and complete the table below by naming the potential energy stored in each one. Use words from the list. Each word may be used once, more than once or not at all. [3]

a bow about to fire an arrow	
water at the top of a waterfall	
a birthday cake	

NEAB

95

Photocopy the list of topics below and tick the boxes of the ones that are included in your examination syllabus. (Your teacher should be able to tell you which they are.) Use your list when you revise. The spread number in brackets tells you where to find more information.

❏ **1** Calculating work done. (4.01)

❏ **2** Defining energy. (4.01)

❏ **3** The SI unit of work and energy: the joule. (4.01)

❏ **4** The different forms of energy. (4.01)

❏ **5** Energy transformations. (4.02)

❏ **6** The law of conservation of energy. (4.02)

❏ **7** How energy transformations eventually result in thermal energy. (4.02)

❏ **8** The link between work done and energy transformed. (4.02)

❏ **9** Calculating gravitational potential energy (PE) and kinetic energy (KE). (4.03)

❏ **10** Energy a scalar quantity. (4.03)

❏ **11** Solving problems on PE and KE. (4.03)

❏ **12** Calculating efficiency. (4.04)

❏ **13** Defining power. (4.04)

❏ **14** The SI unit of power: the watt. (4.04)

❏ **15** Carrying out calculations on power. (4.04)

❏ **16** The link between efficiency and power. (4.04)

❏ **17** How thermal power stations (nuclear power stations and fuel-burning power stations) work. (4.05)

❏ **18** The limited efficiency of thermal power stations. (4.05)

❏ **19** How thermal energy wasted by a power station can be used. (4.05)

❏ **20** The pollution problems caused by thermal power stations. (4.06)

❏ **21** Why fuel-burning power stations may be contributing to global warming. (4.06)

❏ **22** The alternatives to thermal power stations, including hydroelectric and tidal power stations and wind farms. (4.06-4.08)

❏ **23** The difference between renewable and non-renewable energy resources. (4.07)

❏ **24** Non-renewable energy resources:
– fossil fuels
– nuclear fuels
The advantages and disadvantages of each type. (4.07)

❏ **25** Renewable energy resources:
– hydroelectric energy
– tidal energy
– wind energy
– wave energy
– geothermal energy
– solar energy
– biofuels
The advantages and disadvantages of each type. (4.07)

❏ **26** The benefits of energy saving. (4.07)

❏ **27** Why the Sun is the main source of energy on Earth. (4.08)

Thermal Effects

A Boeing 747 flies slowly forward prior to landing. The glow comes from hot gases in the core of its jet engines, where the temperature can reach more than 1500°C. At high altitudes, jet aircraft like this leave 'vapour trails' across the sky. However, the trails are not really vapours, but millions of tiny droplets, formed when water vapour from the engines condenses in the cold atmosphere.

5.01 Moving particles

▶ Water can exist in three forms: solid, liquid, and gas. (The gas is called water vapour, and it is present in the air.) Like all materials, water is made up of tiny particles. Which form it takes depends on how firmly its particles stick together.

Solids, liquids, and gases

Every material is a solid, a liquid, or a gas. Scientists have developed a model (description) called the **kinetic theory** to explain how solids, liquids, and gases behave. According to this theory, matter is made up of tiny particles which are constantly in motion. The particles attract each other strongly when close, but the attractions weaken if they move further apart.

Solid A solid, such as iron, has a fixed shape and volume. Its particles are held closely together by strong forces of attraction called **bonds**. They vibrate to and fro but cannot change positions.

Liquid A liquid, such as water, has a fixed volume but can flow to fill any shape. The particles are close together and attract each other. But they vibrate so vigorously that the attractions cannot hold them in fixed positions, and they can move past each other.

Gas A gas, such as hydrogen, has no fixed shape or volume and quickly fills any space available. Its particles are well spaced out, and virtually free of any attractions. They move about at high speed, colliding with each other and the walls of their container.

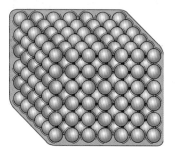

Solid Particles vibrate about fixed positions.

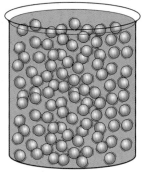

Liquid Particles vibrate, but can change positions.

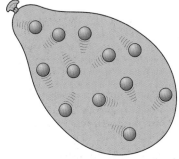

Gas Particles move about freely.

What are the particles?

Everything is made from about 100 simple substances called **elements**. An **atom** is the smallest possible amount of an element. In some materials, the 'moving particles' of the kinetic theory are atoms. However, in most materials, they are groups of atoms called **molecules**. Below, each atom is shown as a coloured sphere. This is a simplified model (description) of an atom. Atoms have no colour or precise shape.

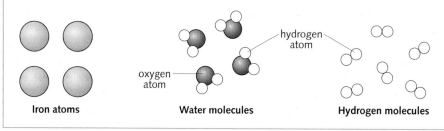

Iron atoms   Water molecules   Hydrogen molecules

Brownian motion: evidence for moving particles

Smoke is made up of millions of tiny bits of ash or oil droplets. If you look at smoke through a microscope, as on the right, you can see the bits of smoke glinting in the light. As they drift through the air, they wobble about in zig-zag paths. This effect is called **Brownian motion**, after the scientist Robert Brown who first noticed the wobbling, wandering motion of pollen grains in water, in 1827.

The kinetic theory explains Brownian motion as follows. The bits of smoke are just big enough to be seen, but have so little mass that they are jostled about as thousands of particles (gas molecules) in the surrounding air bump into them at random.

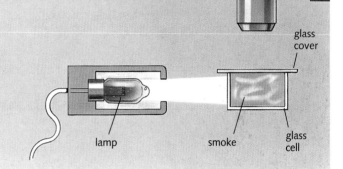

microscope

glass cover

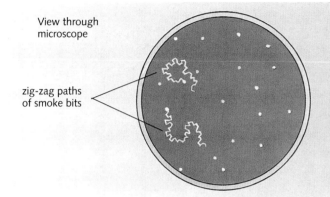

View through microscope

zig-zag paths of smoke bits

lamp   smoke   glass cell

Energy of particles

The particles (atoms or molecules) in solids, liquids, and gases have kinetic energy because they are moving. They also have potential energy because their motion keeps them separated and opposes the bonds trying to pull them together. The particles in gases have the most potential energy because they are furthest apart.

The total kinetic and potential energies of all the atoms or molecules in a material is called its **internal energy**. The hotter a material is, the faster its particles move, and the more internal energy it has.

If a hot material is in contact with a cold one, the hot one cools down and loses internal energy, while the cold one heats up and gains internal energy. The energy transferred is known as **heat**.

The term **thermal energy** is often used for both internal energy and heat.

Kinetic energy

Energy because of motion.

Potential energy

Energy stored because of a change in position or shape.

 Q

1 Say whether each of the following describes a *solid*, a *liquid*, or a *gas*:
   a) Particles move about freely at high speed.
   b) Particles vibrate and cannot change positions.
   c) Fixed shape and volume.
   d) Particles vibrate but can change positions.
   e) No fixed shape or volume.
   f) Fixed volume but no fixed shape.
   g) Virtually no attractions between particles.

2 Smoke is made up of millions of tiny bits of ash or oil droplets.
   a) What do you see when you use a microscope to study illuminated smoke floating in air?
   b) What is the effect called?
   c) How does the kinetic theory explain the effect?

3 If a gas is heated up, how does this affect the motion of its particles?

4 What is meant by the *internal energy* of an object?

5.02 Temperature (1)

Sun's centre	15 000 000 °C
Sun's surface	6000 °C
bulb filament	2500 °C
bunsen flame	1500 °C
boiling water	100 °C
human body	37 °C
warm room	20 °C
melting ice	0 °C
food in freezer	−18 °C
liquid oxygen	−180 °C
absolute zero	−273 °C

The Celsius scale

A **temperature scale** is a range of numbers for measuring the level of hotness. Everyday temperatures are normally measured on the **Celsius** scale (sometimes called the 'centigrade' scale). Its unit of temperature is the **degree Celsius** (**°C**). The numbers on the scale were specially chosen so that pure ice melts at 0 °C and pure water boils at 100 °C (under standard atmospheric pressure of 101 325 pascals). These are its two **fixed points**. Temperatures below 0 °C have negative (−) values.

Thermometers

Temperature is measured using a **thermometer**. One simple type is shown below. The glass bulb contains a liquid – either mercury or coloured alcohol – which expands when the temperature rises and pushes a 'thread' of liquid further along the scale.

alcohol (or mercury)   'thread'   narrow tube

Every thermometer depends on some *property* (characteristic) of a material that varies with temperature. For example, the thermometer above contains a liquid whose volume increases with temperature. The two thermometers below use materials whose electrical properties vary with temperature.

All thermometers agree at the fixed points. However, at other temperatures, they may not agree exactly because their chosen properties may not vary with temperature in quite the same way.

▼ **Clinical thermometers** like the one below measure the temperature of the human body very accurately. Their range is only a few degrees either side of the average body temperature of 37 °C. When removed from the body, they keep their reading until reset.

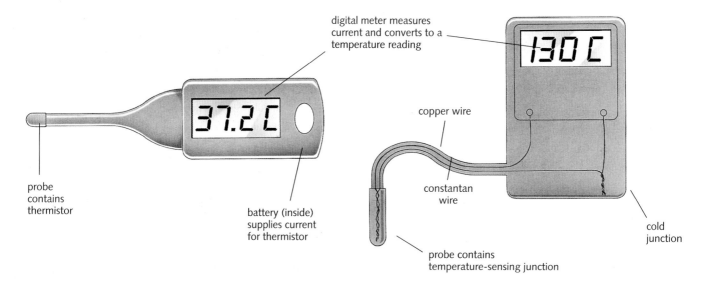

digital meter measures current and converts to a temperature reading

37.2 C

130 C

probe contains thermistor

battery (inside) supplies current for thermistor

copper wire

constantan wire

cold junction

probe contains temperature-sensing junction

Thermistor thermometer The thermistor is a device which becomes a much better electrical conductor when its temperature rises. This means that a higher current flows from the battery, causing a higher reading on the meter.

Thermocouple thermometer Two different metals are joined to form two junctions. A temperature difference between the junctions causes a tiny voltage which makes a current flow. The greater the temperature difference, the greater the current.

What is temperature?

In any object, the particles (atoms or molecules) are moving, so they have kinetic energy. They move at varying speeds, but the higher the temperature, then – on average – the faster they move.

If a hot object is placed in contact with a cold one, as on the right, there is a transfer of thermal energy from one to the other. As the hot object cools down, its particles lose kinetic energy. As the cold object heats up, it particles gain kinetic energy. When both objects reach the same temperature, the transfer of energy stops because the average kinetic energy per particle is the same in both:

Objects at the *same temperature* have the *same average kinetic energy per particle*. The higher the temperature, the greater the average kinetic energy per particle.

Temperature is not the same as heat. For example, a spoonful of boiling water has exactly the same temperature (100 °C) as a saucepanful of boiling water, but you could get far less thermal energy (heat) from it.

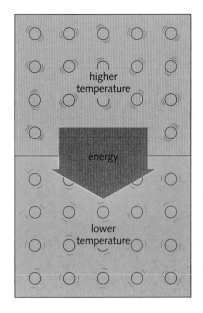

Absolute zero and the Kelvin scale★

As the temperature falls, the particles in a material lose kinetic energy and move more and more slowly. At −273 °C, they can go no slower. This is the lowest temperature there is, and it is called **absolute zero**. The rules of atomic physics do not allow particles to have zero energy, but at absolute zero they would have the minimum energy possible.

In scientific work, temperatures are often measured using the **Kelvin scale**. Its temperature unit, the **kelvin** (**K**), is the same size as the degree Celsius, but the scale uses absolute zero as its zero (0 K). You convert from one scale to the other like this:

> Kelvin temperature/K = Celsius temperature/°C + 273

The Kelvin scale is a **thermodynamic** scale. It is based on the average kinetic energy of particles, rather than on a property of a particular substance.

The **constant volume hydrogen thermometer** contains trapped hydrogen gas whose pressure increases with temperature. It gives the closest match to the thermodynamic scale and is used as a standard against which other thermometers are calibrated (marked).

	absolute zero	melting ice	boiling water
Celsius scale	−273 °C	0 °C	100 °C
Kelvin scale	0 K	273 K	373 K

1 −273 0 100 273 373

Say which of the above is the temperature of
a) boiling water in °C **b)** boiling water in K
c) absolute zero in °C **d)** absolute zero in K
e) melting ice in °C **f)** melting ice in K.

2 Every thermometer depends on some property of a material that varies with temperature. What property is used in each of the following?
a) A mercury-in-glass thermometer.
b) A thermistor thermometer.

3 Blocks A and B above are identical apart from their temperature.
a) How does the motion of the particles in A compare with that in B?
b) In what direction is thermal energy transferred?
c) When does the transfer of thermal energy cease?

Related topics: kinetic energy **4.01** and **4.03**; motion of particles in solids, liquids, and gases **5.01**; expansion of liquids **5.04**; Kelvin scale and particles in a gas **5.05**; thermistors **8.06** and **10.04**

5.03 Temperature (2)

Fixing a temperature scale

To create a temperature scale, two standard temperatures must be chosen against which others can be judged. These **fixed points** need to be defined so that they can be reproduced in laboratories anywhere in the world:

> On the Celsius scale:
>
> 0 degrees Celsius (0 °C) is defined as the melting point of pure ice. This is the **lower fixed point**, known as the **ice point**.
>
> 100 degrees Celsius (100 °C) is defined as the boiling point of pure water, where the water is boiling under standard atmospheric pressure (101 325 Pa). This is the **upper fixed point**, known as the **steam point**.

Putting a scale on an instrument, so that it gives accurate readings, is called **calibrating** the instrument. The diagrams on the left show how the fixed points can be used to calibrate an unmarked thermometer.

To find the 0 °C point, the unmarked thermometer is placed in pure, melting ice, as in the upper diagram on the left. The ice needs to be pure because its melting point is lowered if any impurities are present.

To find the 100 °C point, the thermometer is placed in steam above boiling water, as in the lower diagram on the left. Boiling must take place under standard atmospheric pressure because a change in pressure alters the boiling point of the water. Impurities also affect the boiling point but, in most cases, they do not affect the temperature of the steam just above the water.

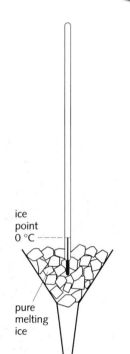

ice
point
0 °C

pure
melting
ice

Finding the lower fixed point

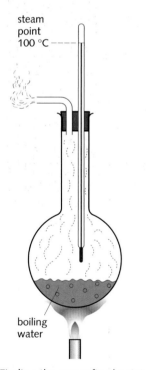

steam
point
100 °C

boiling
water

Finding the upper fixed point

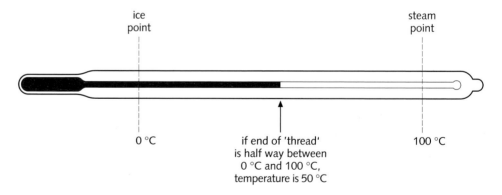

ice
point

steam
point

0 °C

if end of 'thread'
is half way between
0 °C and 100 °C,
temperature is 50 °C

100 °C

Once the 0 °C and 100 °C points have been fixed, the rest of the scale is made by dividing the distance between them into 100 equal divisions, or degrees ('centigrade' means 'one hundred divisions'). The idea can can be extended to produce a scale going above 100 °C and below 0 °C, although for some thermometers, additional fixed points are used (see the top of the next page).

Putting equal divisions on a thermometer *defines* the temperature scale for that particular type of thermometer. For example, if the end of the 'thread' is exactly half way between the ice points and the steam points, as in the diagram above, then by *definition*, the temperature is exactly half way between 0 °C and 100 °C. So the temperature is 50 °C. If a scale has equal divisions, it is described as a **linear** scale.

Liquid-in-glass thermometers

Nearly all liquids expand slightly when heated. This property is used in liquid-in-glass thermometers, which are normally filled with alcohol or mercury.

Sensitivity Some thermometers are more sensitive to temperature change than others. The 'thread' of liquid moves further. The diagrams on the right show how tube width affects the sensitivity. The narrower the tube, the higher the sensitivity of the thermometer.

Mercury expands less than alcohol (for the same volume and same temperature rise). So a mercury thermometer must have a narrower tube than an alcohol thermometer to give the same sensitivity.

Range Mercury freezes at −39 °C; alcohol freezes at a much lower temperature, −115 °C. However, some mercury thermometers have an upper limit of 500 °C, which is much higher than that of any alcohol thermometer.

Responsiveness Some thermometers respond more quickly to a change in temperature than others. A thermometer with a larger bulb, or thicker glass round the bulb, is less responsive because it takes longer for the alcohol or mercury to reach the temperature of the surroundings.

Linearity Although mercury and alcohol thermometers must agree at the fixed points, they do not exactly agree at other temperatures. That is because the expansion of one liquid is not quite linear compared with the other. However, within the 0−100 °C range, the disagreement is very small.

Thermocouple thermometer

For a diagram and brief description, see the previous spread, 5.02. Compared with a liquid-in-glass thermometer, a thermocouple thermometer is robust, quick to respond to temperature change, has a wide range (−200 °C to 1100 °C), and can be linked to other electrical circuits or a computer.

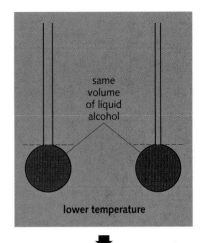

same volume of liquid alcohol

lower temperature

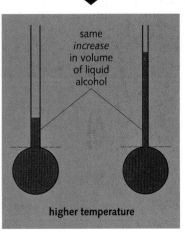

same *increase* in volume of liquid alcohol

higher temperature

The narrower the tube, the further the liquid moves up it when the temperature rises.

1 The thermometer on the right has the ice and steam points marked on it.
   a) On the Celsius scale, what is the temperature of
      (i) the ice point   (ii) the steam point?
   b) What is the temperature reading in °C, if the end of the 'thread' is at
      (i) point A   (ii) point B   (iii) point C?
   c) Explain why reading C would not be possible with a mercury thermometer.
2 *A − smaller bulb   B − thicker glass round bulb   C − thinner tube*
   For a liquid-in-glass thermometer, which of the above would
   a) increase the sensitivity?   b) increase the responsiveness?

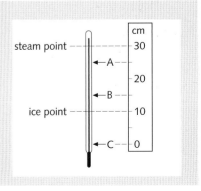

Expanding solids and liquids

If a concrete or steel bar is heated, its volume will increase slightly. The effect is called **thermal expansion**. It is usually too small to notice, but unless space is left for it, it can produce enough force to crack the concrete or buckle the steel. Most solids expand when heated. So do most liquids – and by more than solids. If a liquid is stored in a sealed container, a space must be left at the top to allow for expansion.

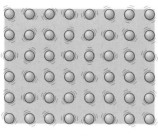 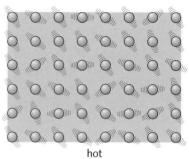

cold   hot

The kinetic theory explains thermal expansion as follows. When, say, a steel bar is heated, its particles speed up. Their vibrations take up more space, so the bar expands slightly in all directions. If the temperature falls, the reverse happens and the material contracts (gets smaller).

Comparing expansions

The chart on the left shows how much 1 metre lengths of different materials expand when their temperature goes up by 100 °C. For greater lengths and higher temperature increases, the expansion is more.

When choosing materials for particular jobs, it can be important to know how much they will expand. Here are two examples:

Steel rods can be used to reinforce concrete because both materials expand equally. If the expansions were different, the steel might crack the concrete on a hot day.

If an ordinary glass dish is put straight into a hot oven, the outside of the glass expands before the inside and the strain cracks the glass. Pyrex expands much less than ordinary glass, so should not crack.

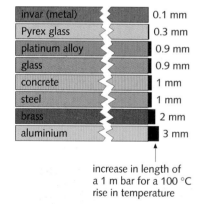

invar (metal)	0.1 mm
Pyrex glass	0.3 mm
platinum alloy	0.9 mm
glass	0.9 mm
concrete	1 mm
steel	1 mm
brass	2 mm
aluminium	3 mm

increase in length of a 1 m bar for a 100 °C rise in temperature

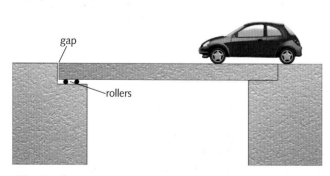

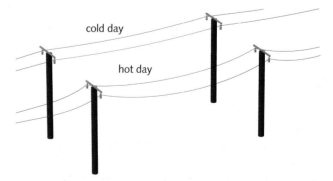

Allowing for expansion...
Gaps are left at the ends of bridges to allow for expansion. One end of the bridge is often supported on rollers so that movement can take place.

... and contraction
When overhead cables are suspended from poles or pylons, they are left slack, partly to allow for the contraction that would happen on a very cold day.

Using expansion

alcohol (or mercury)   'thread'   narrow tube

In the thermometer above, the liquid in the bulb expands when the temperature rises. The tube is made narrow so that a small increase in volume of the liquid produces a large movement along the tube, as explained in the previous spread, 5.03.

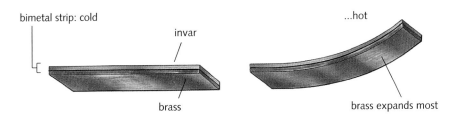

bimetal strip: cold   invar   ...hot

brass   brass expands most

In the **bimetal strip** above, thin strips of two different metals are bonded together. When heated, one metal expands more than the other, which makes the bimetal strip bend. Bimetal strips are used in some **thermostats** – devices for keeping a steady temperature. The thermostat shown on the right is controlling an electric heater.

Water and ice⋆

When hot water cools, it contracts. However, when water freezes it *expands* as it turns into ice. The force of the expansion can burst water pipes and split rocks with rainwater trapped in them.

Water expands on freezing for the following reason. In liquid water, the particles (water molecules) are close together. But in ice, the molecules link up in a very open structure that actually takes up *more* space than in the liquid – as shown in the diagram on the right.

Ice has a lower *density* than liquid water – in other words, each kilogram has a greater volume. Because of its lower density, ice floats on water. When liquid water is cooled, the molecules start forming into an open structure at 4 °C, just before freezing point is reached. As a result, water expands very slightly as it is cooled from 4 °C to 0 °C. It takes up least space, and therefore has its maximum density, at 4 °C.

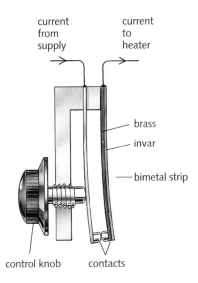

current from supply   current to heater

brass

invar

bimetal strip

control knob   contacts

▲ **Bimetal thermostat** When the temperature rises, the bimetal strip bends, the contacts separate, and the current to the heater is cut off. When the temperature falls, the bimetal strip straightens, and the current is switched on again. In this way, an approximately steady temperature is maintained.

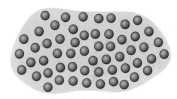

molecules in liquid water

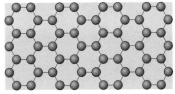

molecules in ice

Q

1 Explain the following:
  a) A metal bar expands when heated.
  b) Overhead cables are hung with plenty of slack in them.
  c) It would *not* be a good idea to reinforce concrete with aluminium rods.
  d) A bimetal strip bends when heated.
  e) Water expands when it freezes.

2 This question is about the thermostat in the diagram at the top of the page.
  a) Why does the power to the heater get cut off if the temperature rises too much?
  b) To maintain a *higher* temperature, which way would you move the control knob? – to the *right* so that it moves towards the contacts, or to the *left*? Explain your answer.

Related topics: density 1.04; kinetic theory and particles 5.01; thermometers 5.02

5.05 Heating gases

Kinetic theory essentials

According to the kinetic theory, a gas is made up of tiny, moving particles (usually molecules). These move about freely at high speed and bounce off the walls of their container. The higher the temperature, then on average, the faster they move.

Unlike a solid or a liquid, a gas does not necessarily expand when heated. That is because its volume depends on the container it is in. When dealing with a fixed mass of gas, there are always three factors to consider: *pressure*, *volume*, and *temperature*. Depending on the circumstances, a change in temperature can produce a change in pressure, or volume, or both.

Linking pressure and temperature (at constant volume)

In the experiment below left, air is trapped in a flask of fixed volume. The temperature of the air is changed in stages by heating the water – or putting a hotter or colder material (melting ice, for example) in the container. At each stage, the pressure is measured on the gauge.

As the temperature of the air rises, so does the pressure. This is because the molecules move faster, so they hit the sides of the flask with greater force.

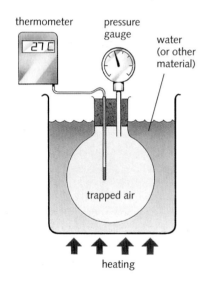

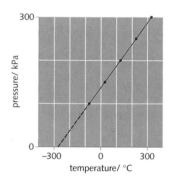

pressure	temperature	temperature
kPa	°C	K
100	−73	200
150	27	300
200	127	400
250	227	500
300	327	600

Above, you can see some typical readings and the graph they produce. The graph is a straight line which, if extended backwards, cuts the temperature axis at −273 °C. Other gases give the same result – provided conditions are such that they behave as ideal gases (explained on the next page).

Absolute zero and the kinetic theory★ According to the kinetic theory, if the temperature of a gas is reduced, the molecules move more slowly. As a result, they strike the container walls with less force, so the pressure drops. If you could go on cooling a gas, its molecules would eventually stop moving and cause no pressure. The graph shows that this would happen at −273 °C. This is the lowest possible temperature and it is called **absolute zero**. (Because of attractions between their molecules, real gases turn liquid before absolute zero is reached.)

Kelvin temperature scale★ This uses absolute zero as its zero (0 K). Its unit, the kelvin (K), is the same size as the degree Celsius (°C). To convert from °C to kelvin (K), you just add 273.

The pressure law★ On the left, you can see what happens if the graph above is redrawn using Kelvin temperatures. The straight line now passes through the origin and the features of a **direct proportion** can be seen:
1 If the Kelvin temperature *doubles*, the pressure *doubles*, and so on.
2 *pressure ÷ Kelvin temperature* always has the same value (0.5 in this case).

These results are summed up by the **pressure law**:

> For a fixed mass of gas at constant pressure, the volume is directly proportional to the Kelvin temperature.

This only applies exactly to an **ideal gas**, a gas with no attractions between its molecules. Real gases come close to this provided they have a low density, a temperature well above their liquefying point, and are dry.

There is another way of expressing the pressure law:
If a fixed mass of gas has a pressure p_1 at temperature T_1 (in kelvin), and a pressure p_2 when its temperature is changed to T_2:

$$\frac{P_1}{T_1} = \frac{P_2}{T_2} \qquad \text{(at constant volume)}$$

For example, if air trapped in a container has a pressure of 3 atm at 27 °C, you could use the equation to work out its pressure if it is heated to 127 °C.

First, change the temperatures into kelvin by adding 273:

27 °C = 300 K and 127 °C = 400 K

So, P_1 = 3 atm, T_1 = 300 K, T_2 = 400 K, and P_2 is to be found.

Putting the values in the above equation gives:

$$\frac{3}{300} = \frac{P_2}{400} \quad \text{Rearranged, this give } P_2 = \frac{400}{300} \times 3 = 4 \text{ atm}$$

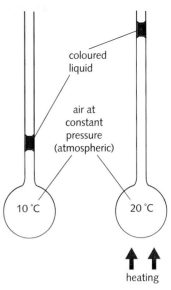

coloured liquid

air at constant pressure (atmospheric)

10 °C   20 °C

heating

In this experiment, the *pressure* of the gas is kept constant. As the temperature increases, so does the volume.

Linking volume and temperature (at constant pressure)
In the experiment above right, trapped gas (air) is heated at constant pressure. This is atmospheric pressure because only the short length of liquid separates the air from the atmosphere outside. As the temperature rises, the volume of the gas increases.

Charles's law★ When dry air is heated and allowed to expand at constant pressure, experiments show that *volume ÷ Kelvin temperature* always has the same value for any particular sample. This result is often called **Charles's law**. Again, it only applies exactly to an ideal gas:

> For a fixed mass of gas at constant pressure, the volume is directly proportional to the Kelvin temperature.

Comparing expansions of a solid, liquid, and gas

For the same volume of material and the same rise in temperature (starting at room temperature):

Water expands 7 times as much as steel

Air (at constant pressure) expands 16 times as much as water.

1 How does the kinetic theory explain the following?
   a) A gas exerts a pressure on its container walls.
   b) The pressure increases with temperature (assuming that the volume does not change).
2 If a gas is heated at constant pressure, what happens to its volume?
3 A gas in a fixed container is at a pressure of 4 atm and a temperature of 27 °C. What will its pressure be if it is heated to a temperature of 177 °C?

4 The readings below are for a fixed mass of gas at constant volume:

pressure/ atm	0.78	0.96	1.13	1.31	1.48
temperature/°C	-50	0	50	100	150

a) By plotting a graph, estimate a value for absolute zero.
b) Does the graph obey the pressure law? Explain your answer.

Related topics: ideal gas, pressure and volume **3.09**; kinetic theory **5.01**; temperature, Kelvin scale, constant volume hydrogen thermometer **5.02**; expansion of solids and liquids **5.04**

5.06 Thermal conduction

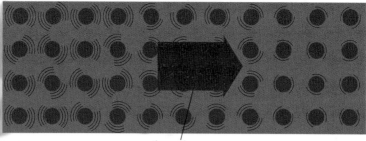

high temperature

lower temperature

thermal energy transferred by conduction

All materials are made up of tiny, moving particles (atoms or molecules). The higher the temperature, the faster the particles move.

If one end of a metal bar is heated as above, the other end eventually becomes too hot to touch. Thermal energy (heat) is transferred from the hot end to the cold end as the faster particles pass on their extra motion to particles all along the bar. The process is called **conduction**.

More thermal energy is transferred every second if:
● the temperature difference across the ends of the bar is *increased*
● the cross-sectional ('end-on') area of the bar is *increased*
● the length of the bar is *reduced*.

Thermal conductors and insulators

Some materials are much better **conductors** of thermal energy than others. Poor conductors are called **insulators**.

Metals are the best thermal conductors. Non-metal solids tend to be poor conductors; so do most liquids. Gases are the worst of all. Many materials are insulators because they contain tiny pockets of trapped air. You use this idea when you put on lots of layers of clothes to keep you warm. There are some more examples at the top of the next page.

You can sometimes tell how well something conducts just by touching it. A metal door handle feels cold because it quickly conducts thermal energy away from your hand, which is warmer. A polystyrene tile feels warm because it insulates your hand and stops it losing thermal energy.

Good conductors

metals e.g.	copper
	aluminium
	iron
silicon	
graphite	

Poor conductors (insulators)

glass

water

plastics

rubber

wood

materials containing trapped air — wool / glass wool (fibreglass) / plastic foam / expanded polystyrene

---

The materials above are arranged in order of conducting ability starting with the best.

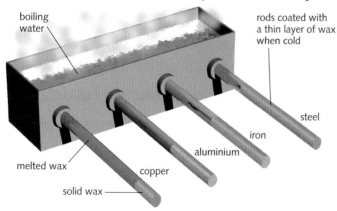

boiling water

rods coated with a thin layer of wax when cold

steel

iron

aluminium

copper

melted wax

solid wax

▲ Comparing four good thermal conductors. Ten minutes or so after the boiling water has been tipped into the tank, the length of melted wax shows which material is the best conductor.

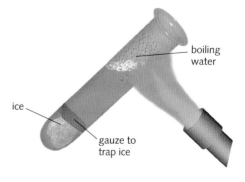

boiling water

ice

gauze to trap ice

▲ This experiment shows that water is a poor thermal conductor. The water at the top of the tube can be boiled without the ice melting.

Using insulating materials

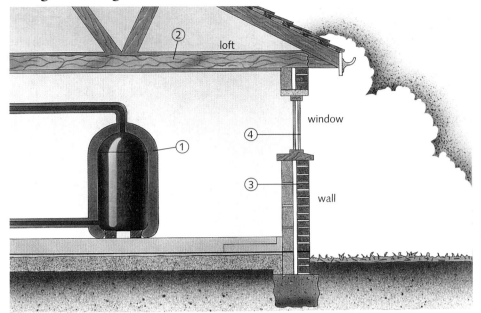

▲ Feathers give good thermal insulation, especially when fluffed up to trap more air.

In countries where buildings need to be heated, good insulation means lower fuel bills. Above are some of the ways in which insulating materials are used to reduce heat losses from a house:

1 Plastic foam lagging round the hot water storage tank.
2 Glass or mineral wool insulation in the loft.
3 Wall cavity filled with plastic foam, beads, or mineral wool.
4 Double-glazed windows: two sheets of glass with air between them.

How materials conduct

When a material is heated, the particles move faster, push on neighbouring particles, and speed those up too. All materials conduct like this but, in metals, energy is also transferred by another, much quicker method.

In atoms, there are tiny particles called **electrons**. Most are firmly attached, but in metals, some are 'loose' and free to drift between the atoms. When a metal is heated, these **free electrons** speed up. As they move randomly within the metal, they collide with atoms and make them vibrate faster. In this way, thermal energy is rapidly transferred to all parts.

An electric current is a flow of electrons – so metals are good electrical conductors as well as good thermal conductors.

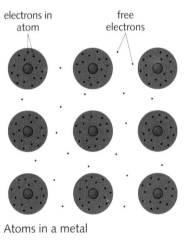

Atoms in a metal

Q

1 Explain each of the following:
   a) A saucepan might have a copper bottom but a plastic handle.
   b) Wool and feathers are good insulators.
   c) An aluminium window frame feels colder than a wooden window frame when you touch it.
   d) It is much safer picking up hot dishes with a dry cloth than a wet one.
2 Give *three* ways in which insulating materials are used to reduce thermal energy losses from a house.

3 A hot water tank loses thermal energy even when lagged. How could the energy loss be reduced?
4 Look at the experiment shown on the opposite page, comparing four thermal conductors.
   a) Which of the metals is the best conductor?
   b) In experiments like this, it is important to make sure that the test is fair. Write down *three* features of this experiment which make it a fair test.
5 Why are metals much better thermal conductors than most other materials?

Related topics: energy **4.01**; particles of matter **5.01**; temperature **5.02**; electrical conductors **8.01**

Convection

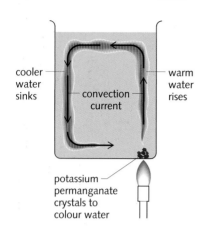

cooler water sinks

convection current

warm water rises

potassium permanganate crystals to colour water

Liquids and gases are poor thermal conductors, but if they are free to circulate, they can carry thermal energy (heat) from one place to another very quickly.

Convection in a liquid

In the experiment on the left, the bottom of the beaker is being gently heated in one place only. As the water above the flame becomes warmer, it expands and becomes less dense. It rises upwards as cooler, denser water sinks and displaces it (pushes it out of the way). The result is a circulating stream, called a **convection current**. Where the water is heated, its particles (water molecules) gain energy and vibrate more rapidly. As the particles circulate, they transfer energy to other parts of the beaker.

Convection does not occur if the water is heated at the top rather than at the bottom. The warmer, less dense water stays at the top.

Convection in air

Convection can occur in gases as well as liquids. For example, warm air rises when it is displaced by cooler, denser air sinking around it.

Heated by the Sun, warm air rises above the equator as it is displaced by cooler, denser air sinking to the north and south. The result is huge convection currents in the Earth's atmosphere. These cause winds across all oceans and continents. Convection also causes the onshore and offshore breezes which sometimes blow at the coast during the summer:

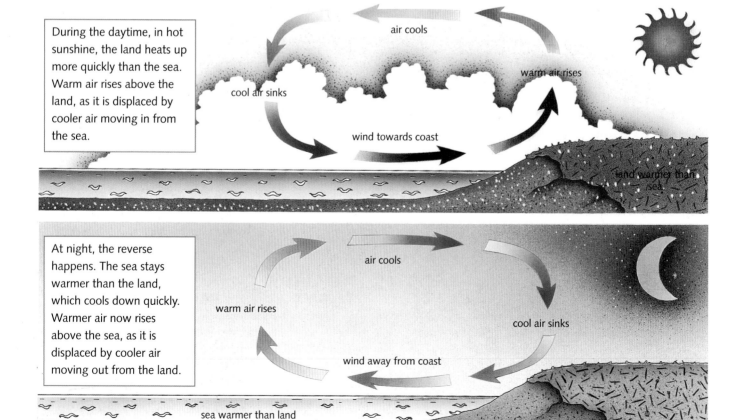

During the daytime, in hot sunshine, the land heats up more quickly than the sea. Warm air rises above the land, as it is displaced by cooler air moving in from the sea.

air cools

warm air rises

cool air sinks

wind towards coast

land warmer than sea

At night, the reverse happens. The sea stays warmer than the land, which cools down quickly. Warmer air now rises above the sea, as it is displaced by cooler air moving out from the land.

air cools

warm air rises

cool air sinks

wind away from coast

sea warmer than land

Using convection in the home

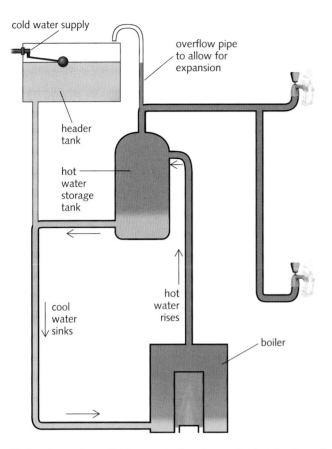

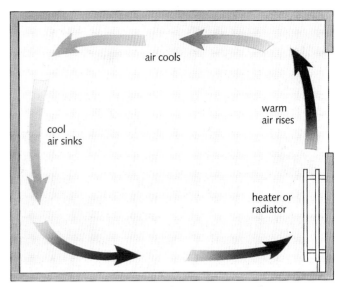

Room heating Warm air rising above a convector heater or radiator carries thermal energy all around the room – though unfortunately, the coolest air is always around your feet.

Hot water system Cold water in the storage tank sinks down to the boiler, where it is heated. The heated water in the boiler rises to the top of the storage tank. In this way, a supply of hot water collects in the storage tank from the top down. The storage tank is insulated to reduce thermal energy losses by conduction and convection. The header tank provides the pressure to push the water out of the taps.

Practical systems are more complicated than that shown. There may be separate circuits for the taps and radiators, and a pump to assist the flow of water.

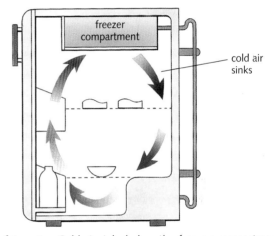

Refrigerator Cold air sinks below the freezer compartment. This sets up a circulating current of air which cools all the food in the refrigerator.

1 Explain the following:
 a) A radiator quickly warms all the air in a room, even though air is a poor thermal conductor.
 b) The smoke from a bonfire rises upwards.
 c) Anyone standing near a bonfire feels a draught.
 d) The freezer compartment in a refrigerator is placed at the top.
 e) A refrigerator does not cool the food inside it properly if the food is too tightly packed.
2 On a hot summer's day, coastal winds often blow in from the sea.
 a) What causes these winds?
 b) Why do the winds change direction at night?

3 Some hot water systems have an immersion heater – an electrical heating element in the storage tank. In the tank below, should the heating element be placed at A or at B? Explain your answer.

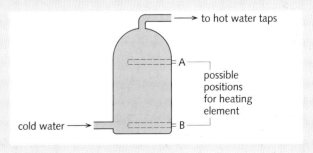

5.08 Thermal radiation

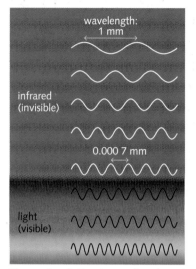

Thermal radiation is mainly infrared waves, but very hot objects also give out light waves.

On Earth, we are warmed by the Sun. Its energy travels to us in the form of **electromagnetic waves**. These include invisible **infrared waves** as well as light, and they can travel through a vacuum (empty space). They heat up things that absorb them, so are often called **thermal radiation**.

All objects give out some thermal radiation. The higher their surface temperature and the greater their surface area, the more energy they radiate per second. Thermal radiation is a mixture of different **wavelengths**, as shown on the left. Warm objects radiate infrared. But if they become hotter, they also emit shorter wavelengths which may include light. That is why a radiant heater or grill starts to glow 'red hot' when it heats up.

Emitters and absorbers

Some surfaces are better at emitting (sending out) thermal radiation than others. For example, a black saucepan cools down more quickly than a similar white one because it emits energy at a faster rate.

Good emitters of thermal radiation are also good absorbers, as shown in the chart below. White or silvery surfaces are poor absorbers because they reflect most of the thermal radiation away. That is why, in hot, sunny countries, houses are often painted white to keep them cool inside.

▶ This chart shows how some surfaces compare as emitters, reflectors, and absorbers of thermal radiation.

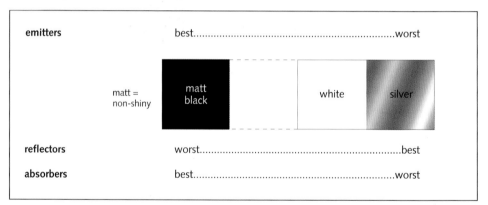

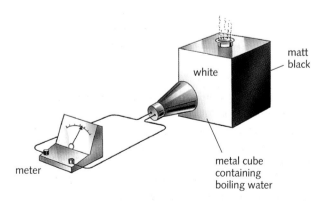

Comparing emitters The metal cube is filled with boiling water which heats the surfaces to the same temperature. The thermal radiation detector is placed in turn at the same distance from each surface and the meter readings compared.

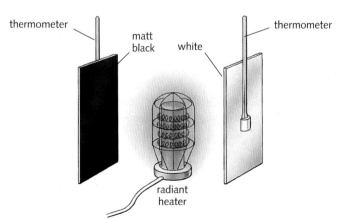

Comparing absorbers The metal plates are placed at the same distance from a radiant heater. To find out which surface absorbs thermal radiation most rapidly, the rises in temperature are compared.

Greenhouse effects★

When the Sun's thermal radiation reaches the Earth, the atmosphere acts as a 'heat trap'. This happens because some gases (notably water vapour, carbon dioxide, and methane) absorb energy strongly at certain wavelengths in the infrared region of the spectrum. The heat-trapping action of the atmosphere is called the **greenhouse effect**. Without it, the Earth's surface would be around 25 °C cooler than it is. The present concern is that extra carbon dioxide from burning fuels may be adding to the effect and causing global warming

Greenhouses act as heat traps, which is how the greenhouse effect got its name. However, they work in a different way. Thermal radiation from the Sun passes easily through the glass or plastic. The ground inside warms up and heats the air. But the hot air is trapped. It cannot escape by rising and flowing away.

The Sun's thermal radiation passes easily into a greenhouse. But unless you leave the door open, the heated air inside cannot escape.

The solar panel

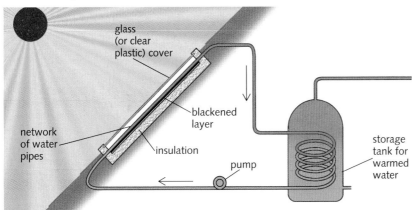

Some houses have a **solar panel** on the roof like the one above. It uses the Sun's thermal radiation to warm up water for the house. The blackened layer absorbs the radiant energy and warms up the water flowing through the pipes.

The vacuum flask

A vacuum flask can keep drinks hot (or cold) for hours. It has these features for reducing the rate at which thermal energy flows out (or in):

1 An insulated stopper to reduce *conduction* and *convection*.
2 A double-walled container with a gap between the walls. Air has been removed from the gap to reduce *conduction* and *convection*.
3 Walls with silvery surfaces to reduce *thermal radiation*.

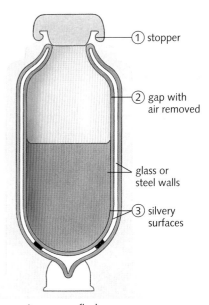

A vacuum flask

1 *white*   *silvery*   *matt black*
   Which of the above surfaces is the best at
   a) absorbing thermal radiation
   b) emitting thermal radiation
   c) reflecting thermal radiation?
2 When a warm object is heated up, the thermal radiation it emits changes. Give *two* ways in which the thermal radiation changes.
3 What feature does a vacuum flask have to reduce the transfer of heat by thermal radiation?

4 In experiments like those on the opposite page, it is important to make sure that each test is fair.
   a) Write down *three* features of the *Comparing emitters* experiment that make it a fair test.
   b) Repeat for the *Comparing absorbers* experiment.
5 Why, on a sunny day, is it normally hotter inside a greenhouse than it is inside a wooden shed?
6 In the solar panel above, why does the panel have
   a) a blackened layer at the back
   b) a network of water pipes?

Related topics: energy **4.01**; global warming **4.06**; solar energy **4.07–4.08**; thermal energy **5.01**; conduction **5.06**; convection **5.07**; electromagnetic waves **7.11–7.12**

5.09 Liquids and vapours

▶ When a liquid **evaporates**, faster particles escape from its surface to form a gas. However, unless the gas is removed, some of the particles will return to the liquid.

Evaporation

Even on a cool day, rain puddles can vanish and wet clothes dry out. The water becomes an invisible gas (called water vapour) which drifts away in the air. When a liquid below its boiling point changes into a gas, this is called **evaporation**. It happens because some particles in the liquid move faster than others. The faster ones near the surface have enough energy to escape and form a gas.

There are several ways of making a liquid evaporate more quickly:

Increase the temperature Wet clothes dry faster on a warm day because more of the particles (water molecules) have enough energy to escape.

Increase the surface area Water in a puddle dries out more quickly than water in a cup because more of its molecules are close to the surface.

Reduce the humidity If air is very *humid*, this means that it already has a high water vapour content. In humid air, wet washing dries slowly because molecules in the vapour return to the liquid almost as fast as those in the liquid escape. In less humid air, wet washing dries more quickly.

Blow air across the surface Wet clothes dry faster on a windy day because the moving air carries escaping water molecules away before many of them can return to the liquid.

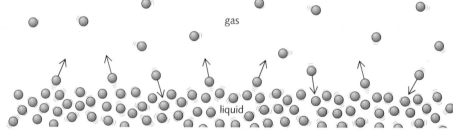

Boiling

Boiling is a very rapid form of evaporation. When water boils, as in the photograph on the left, vapour bubbles form deep in the liquid. They expand, rise, burst, and release large amounts of vapour.

Even cold water has tiny vapour bubbles in it, but these are squashed by the pressure of the atmosphere. At 100 °C, the vapour pressure in the bubbles is strong enough to overcome atmospheric pressure, so the bubbles start to expand and boiling occurs. At the top of Mount Everest, where atmospheric pressure is less, water would boil at only 70 °C.

The cooling effect of evaporation

Evaporation has a cooling effect. For example, if you wet your hands, the water on them starts to evaporate. As it evaporates, it takes thermal energy away from your skin. So your hands feel cold.

The kinetic theory explains the cooling effect like this. If faster particles escape from the liquid, slower ones are left behind, so the temperature of the liquid is less than before.

Refrigerators use the cooling effect of evaporation. In the refrigerator on the right, the process works like this:

1 In the pipes in the freezer compartment, a liquid called a **refrigerant** evaporates and takes thermal energy from the food and air.

2 The vapour is drawn away by the pump, which compresses it and turns it into a liquid. This releases thermal energy, so the liquid heats up.

3 The hot liquid is cooled as it passes through the pipes at the back, and the thermal energy is carried away by the air.

Overall, thermal energy is transferred from the things inside the fridge to the air outside.

Sweating also uses the cooling effect of evaporation. You start to sweat if your body temperature rises more than about 0.5 °C above normal. The sweat, which is mainly water, comes out of tiny pores in your skin. As it evaporates, it takes thermal energy from your body and cools you down.

On a humid ('close') day, sweat cannot evaporate so easily, so it is more difficult to stay cool and comfortable.

Condensation

When a gas changes back into a liquid, this is called **condensation**. For example, cold air can hold less water vapour than warm air, so if humid air is suddenly cooled, some of the water vapour may condense. It may become billions of tiny water droplets in the air – we see these as clouds, mist, or fog. Or it may become condensation on windows or other surfaces. If condensation freezes, the result is frost.

> **Gas and vapour**
>
> A gas is called a vapour if it can be turned back into a liquid by compressing it.

Condensation can be seen   ...on mirrors   ...as clouds in the sky   ...and as clouds of 'steam' from a kettle (the vapour itself is invisible)

1 A puddle and a small bowl are next to each other. There is the same amount of water in each.
  a) Explain why the puddle dries out more rapidly than the water in the bowl.
  b) Give *two* changes that would make the puddle dry out even more rapidly.
2 If you are wearing wet clothes, and the water evaporates, it cools you down. How does the kinetic theory explain the cooling effect?

3 Give *two* practical uses of the cooling effect of evaporation.
4 Explain why, on a humid day
  a) you may feel hot and uncomfortable
  b) you do not feel so uncomfortable if there is a breeze blowing.
5 What is the difference between evaporation and boiling?
6 Why does condensation form on cold windows?

5.10 Specific heat capacity

If a material absorbs thermal energy, then unless it is melting or boiling, its temperature rises. However, some materials have a greater capacity for absorbing thermal energy than others. For example, if you heat a kilogram each of water and aluminium, the water must be supplied with nearly five times as much energy as the aluminium for the same rise in temperature:

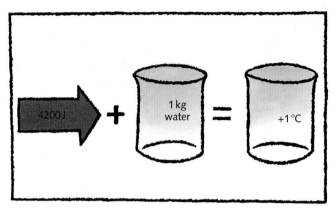

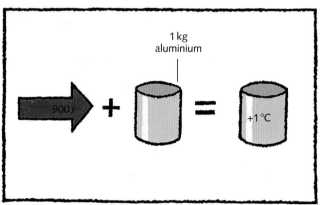

▲ 4200 joules of energy are needed to raise the temperature of 1 kg of water by 1 °C.

▲ 900 joules of energy are needed to raise the temperature of 1 kg of aluminium by 1 °C.

Units

Energy is measured in joules (J).

Temperature is measured in °C or in kelvin (K). Both scales have the same size 'degree', so a 1 °C *change* in temperature is the same as a 1 K change.

	specific heat capacity
	J/(kg °C)
water	4200
alcohol	2500
ice	2100
aluminium	900
concrete	800
glass	700
steel	500
copper	400

Scientifically speaking, water has a **specific heat capacity** of 4200 J/(kg °C). Aluminium has a specific heat capacity of only 900 J/(kg °C). Other specific heat capacities are shown in the table below left.

The energy that must be transferred to an object to increase its temperature can be calculated using this equation:

energy transferred = mass × specific heat capacity × temperature change

In symbols: energy transferred = $mc\Delta T$

where m is the mass in kg, c is the specific heat capacity in J/(kg °C), and ΔT represents the temperature *change* in °C (or in K).

The same equation can also be used to calculate the energy transferred when a hot object cools down.

Example If 2 kg of water cools from 70 °C to 20 °C, how much thermal energy does it lose?

In this case, the temperature change is 50 °C.
So: energy transferred = $mc\Delta T$ = 2 × 4200 × 50 J
                        = 420 000 J

Thermal capacity

The quantity *mass × specific heat capacity* is called the **thermal capacity** (or **heat capacity**). For example, if there is 2 kg of water in a kettle:
thermal capacity of the water = 2 kg × 4200 J/(kg °C) = 8400 J/°C

This means that, for each 1 °C rise in temperature, 8400 joules of energy must be supplied to the water in the kettle. A greater mass of water would have a higher thermal capacity.

<table>
<tr><td colspan="2">

Linking energy and power

$$power = \frac{energy}{time}$$

So: energy = power × time

</td><td>

Energy is measured in joules (J).

Power is measured in watts (W).

Time is measured in seconds (s).

</td></tr>
</table>

Measuring specific heat capacity

Water A typical experiment is shown on the right. Here, the beaker contains 0.5 kg of water. When the 100 watt electric heater is switched on for 230 seconds, the temperature of the water rises by 10 °C. From these figures, a value for the specific heat capacity of water can be calculated:

(Omitting some of the units for simplicity)

energy transferred to water $= mc\Delta T = 0.5 \times c \times 10$

energy supplied by heater $=$ power × time $= 100 \times 230 = 23\,000$ J

so: $0.5 \times c \times 10 = 23\,000$

Rearranged and simplified, this gives $c = 4600$

so the specific heat capacity of water is 4600 J/(kg °C).

This method makes no allowance for any thermal energy lost to the beaker or the surroundings, so the value of c is only approximate.

Aluminium (or other metal) The method is as above, except that a block of aluminium is used instead of water. The block has holes drilled in it for the heater and thermometer. As before, c is calculated from this equation:

$$power \times time = mc\Delta T \quad \text{(assuming no thermal energy losses)}$$

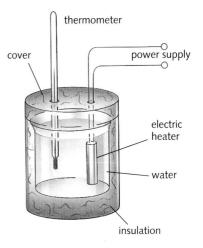

Storing thermal energy

Because of its high specific heat capacity, water is a very useful substance for storing and carrying thermal energy. For example, in central heating systems, water carries thermal energy from the boiler to the radiators around the house. In car cooling systems, water carries unwanted thermal energy from the engine to the radiator.

Night storage heaters use concrete blocks to store thermal energy. Although concrete has a lower specific heat capacity than water, it is more dense, so the same mass takes up less space. Electric heating elements heat up the blocks overnight, using cheap, 'off-peak' electricity supplied through a special meter. The hot blocks release thermal energy through the day as they cool down.

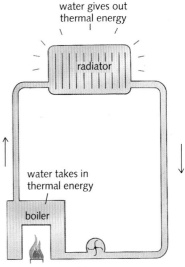

In most central heating systems, water is used to carry the thermal energy.

The specific heat capacities of copper and water are given in the table on the opposite page.

1 Water has a very high specific heat capacity. Give *two* practical uses of this.

2 **a)** How much thermal energy is needed to raise the temperature of 1 kg of copper by 1 °C?
   b) If a 10 kg block of copper cools from 100 °C to 50 °C, how much thermal energy does it give out?

c) If, in part b, the copper were replaced by water, how much thermal energy would this give out?

3 A 210 W heater is placed in 2 kg of water and switched on for 200 seconds.
   a) How much energy is needed to raise the temperature of 2 kg of water by 1 °C?
   b) How much energy does the heater supply?
   c) Assuming that no thermal energy is lost, what is the temperature rise of the water?

Related topics: density **1.04**; thermal energy **4.01** and **5.01**; temperature **5.02**; electrical power **8.11**

5.11 Latent heat

Water can be a solid (ice), a liquid, or a gas called water vapour (or steam). These are its three **phases**, or **states**.

Latent heat of fusion

If ice from a cold freezer is put in a warm room, it absorbs thermal energy. The graph on the left shows what happens to its temperature. While melting, the ice goes on absorbing energy, but its temperature does not change. The energy absorbed is called the **latent heat of fusion**. It is needed to separate the particles so that they can form the liquid. If the liquid changes back to a solid, the energy is released again.

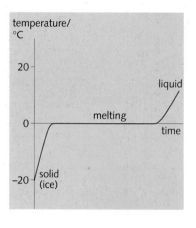

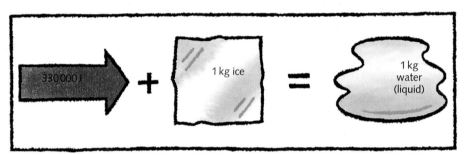

Ice has a **specific latent heat of fusion** of 330 000 J/kg. This means that 330 000 joules of energy must be transferred to change each kilogram of ice into liquid water at the same temperature (0 °C). For any known mass, the energy transferred can be calculated using this equation:

$$\text{energy transferred} = \text{mass} \times \text{specific latent heat}$$

In symbols: energy transferred = mL

For example, if 2 kg of ice is melted (at 0 °C):
energy transferred = mL = 2 kg × 330 000 J/kg = 660 000 J

Measuring the specific latent heat of fusion of ice In the experiment on the left, a 100 watt heater is switched on for 300 seconds. By weighing the water collected in the beaker, it is found that 0.10 kg of ice has melted. From these figures, a value for L can be calculated:

(Omitting some of the units for simplicity)
energy transferred when ice melts = mL = 0.10L
energy supplied by heater = power × time = 100 W × 300 s = 30 000 J
So: 0.10L = 30 000, which gives L = 300 000
So the specific latent heat of fusion of ice is 300 000 J/kg.

This method makes no allowance for any thermal energy received from the funnel or surroundings, so the value of L is only approximate.

Kinetic theory essentials

According to the kinetic theory, materials are made up of tiny, moving particles (usually molecules). In solids, the particles are held together by strong attractions. In liquids, they have more energy and are less strongly held. In gases, they have enough energy to overcome the attractions, stay spaced out, and move around freely.

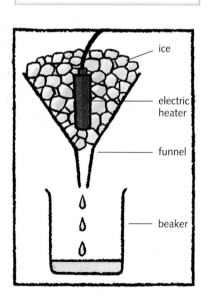

Linking energy and power

$$\text{power} = \frac{\text{energy}}{\text{time}}$$

So: energy = power × time

Energy is measured in joules (J).

Power is measured in watts (W).

Time is measured in seconds (s).

Latent heat of vaporization

If you heat water in a kettle, the temperature rises until the water is boiling at 100 °C, then stops rising. If the kettle is left switched on, the water absorbs more and more thermal energy, but this just turns more and more of the boiling water into steam, still at 100 °C. The energy absorbed is called **latent heat of vaporization**. Most is needed to separate the particles so that they can form a gas, but some is required to push back the atmosphere as the gas forms.

▲ A jet of steam releases latent heat when it condenses (turns liquid). This idea can be used to heat drinks quickly.

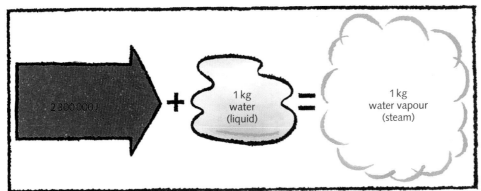

Water has a **specific latent heat of vaporization** of 2 300 000 J/kg. This means that 2 300 000 joules of energy must be transferred to change each kilogram of liquid water into steam at the same temperature (100 °C).

To calculate the energy transferred when any known mass of liquid changes into a gas at the same temperature, you use the equation on the opposite page. However, L is now the specific latent heat of *vaporization*.

Measuring the specific latent heat of vaporization of water

In the experiment on the right, the can contains boiling water. When the 100 watt heater has been switched on for 500 seconds, the change in the mass balance's reading shows that 0.020 kg of water has boiled away. From these figures, a value for L can be calculated:

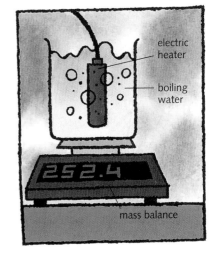

(Omitting some of the units for simplicity)
energy transferred when water is vaporized = mL = 0.020L
energy supplied by heater = power × time = 100 W × 500 s = 50 000 J
So: 0.020L = 50 000, which gives L = 2 500 000
So the specific latent heat of vaporization of water is 2 500 000 J/kg.

This method makes no allowance for any thermal energy lost to the surroundings, so the value of L is only approximate.

 Q

Specific latent heat of fusion of ice = 330 000 J/kg; specific latent heat of vaporization of water = 2 300 000 J/kg

1 Some crystals were melted to form a hot liquid, which was then left to cool. As it cooled, the readings in the table below were taken.
a) What was happening to the liquid between 10 and 20 minutes after it started to cool?
b) What is the melting point of the crystals in °C?

2 Energy is needed to turn water into water vapour (steam). How does the kinetic theory explain this?

3 How much energy is needed to change
a) 10 kg of ice into water at the same temperature
b) 10 kg of water into water vapour at the same temperature?

4 A 460 watt water heater is used to boil water. Assuming no thermal energy losses, what mass of steam will it produce in 10 minutes?

Time/minutes	0	5	10	15	20	25	30
Temperature/°C	90	75	68	68	68	62	58

Related topics: kinetic theory and thermal energy **5.01**; evaporation, boiling, and condensation **5.09**; electrical power **8.11**

119

1 Explain in terms of molecules:
a) the process of evaporation; [3]
b) why the pressure of the air inside a car tyre increases when the car is driven at high speed. [2]

WJEC

2 Which of the following describes particles in a solid at room temperature?
A Close together and stationary.
B Close together and vibrating.
C Close together and moving around at random.
D far apart and moving at random. [1]

LEAG

3 In sunny countries, some houses have a solar heater on the roof. It warms up water for the house. The diagram below shows a typical arrangement.

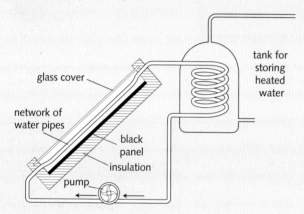

a) Why is the panel in the solar heater black? [1]
b) Why is there an insulating layer behind the panel? [1]
c) How does the water in the tank get heated? [2]
d) On average, each square metre of the solar panel above receives 1000 joules of energy from the Sun every second. Use this figure to calculate the power input (in kW) of the panel if its surface area is 2 m². [2]
e) The solar heater in the diagram has an efficiency of 60% (it wastes 40% of the solar energy it receives). What area of panel would be needed to deliver heat at the same rate, on average, as a 3 kW electric immersion heater? [2]
f) (i) What are the advantages of using a solar heater instead of an immersion heater? [2]
  (ii) What are the disadvantages? [2]

4 The scale of a mercury-in-glass thermometer is linear. One such thermometer has a scale extending from −10 °C to 110 °C. The length of that scale is 240 mm.
a) What is meant by the statement that *the scale is linear*? [2]
b) Calculate the distance moved by the end of the mercury thread when the temperature of the thermometer rises
  (i) from 0.0 °C to 1.0 °C
  (ii) from 1.0 °C to 100.0 °C [3]

UCLES

5 a) The table gives the melting and boiling points for lead and oxygen.

	melting point in °C	boiling point in °C
lead	327	1744
oxygen	−219	−183

(i) At 450 °C will the lead be a solid, a liquid or a gas? [1]

(ii) At −200 °C will the oxygen be a solid, liquid or a gas? [1]

b) The graph shows how the temperature of a pure substance changes as it is heated.

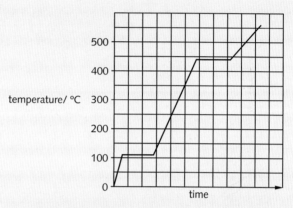

(i) At what temperature does the substance boil? [1]
(ii) Sketch the graph and mark with an **X** any point where the substance exists as both a liquid and gas at the same time. [1]
c) (i) All substances consist of particles. What happens to the average kinetic energy of these particles as the substance changes from a liquid to a gas? [1]
(ii) Explain, in terms of particles, why energy must be given to a liquid if it is to change to a gas. [2]

SEG

6 The diagram below shows a refrigerator. In and around a refrigerator, heat is transferred by conduction, by convection, and by evaporation. Decide which process is mainly responsible for the heat transfer in each of the examples listed at the top of the next page:

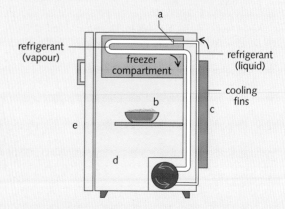

a) Heat is absorbed as liquid refrigerant changes to vapour in the pipework. [1]
b) Cool air descending from the freezer compartment takes away heat from the food. [1]
c) Heat is lost to the outside air through the cooling fins at the back. [1]
d) Some heat from the kitchen enters the refrigerator through its outer panels. [1]
e) Some heat enters the refrigerator every time the door is opened. [1]

7 The diagram below shows a hot water storage tank. The water is heated by an electric immersion heater at the bottom.

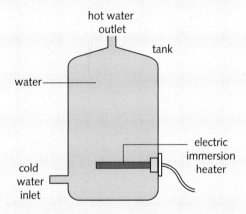

a) How could heat loss from the tank be reduced? What materials would be suitable for the job? [2]
b) Why is the heater placed at the bottom of the tank rather than the top? [2]
c) The heater has a power output of 3 kW.
  (i)   What does the 'k' stand for in 'kW'? [1]
  (ii) How much energy (in joules) does the heater deliver in one second? [1]
  (iii) How much energy (in joules) does the heater deliver in 7 minutes? [2]
d) The tank holds 100 kg of water. The specific heat capacity of water is 4200 J/(kg °C).
  (i)   How much energy (in joules) is needed to raise the temperature of 1 kg of water by 1 °C? [1]
  (ii) How much energy (in joules) is needed to raise the average temperature of all the water in the tank by 1 °C? [2]
  (iii) If the heater is switched on for 7 minutes, what is the average rise in temperature of the water in the tank (assuming that no heat is lost)? [2]

8 The diagram below shows a type of heater used in schools.

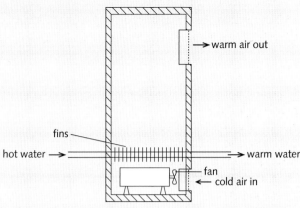

Hot water is pumped from the boiler into pipes inside the heater. Fins are attached to those pipes. Cold air is drawn into the base of the heater by an electric fan.
a) Why are fins attached to the pipes inside the heater? [2]
b) 600 kg of water pass through the heater every hour. The temperature of the water falls by 5 °C as it passes through the heater.
Calculate the amount of heat energy transferred from the water every hour. The specific heat capacity of water is 4200 J/(kg °C). [3]

MEG

9 The graph below shows how the temperature of some liquid in a beaker changed as it was heated until it was boiling.

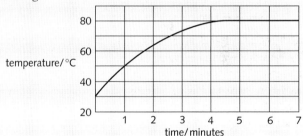

a) What was the boiling point of the liquid? [1]
b) State and explain what difference, if any, there would be in the final temperature if the liquid was heated more strongly. [2]
c) State **one** difference between boiling and evaporation. [1]

UCLES

Photocopy the list of topics below and tick the boxes of the ones that are included in your examination syllabus. (Your teacher should be able to tell you which they are.) Use your list when you revise. The spread number in brackets tells you where to find more information.

❏ 1 The kinetic theory of matter. (5.01)

❏ 2 The different properties of solids, liquids, and gases. (5.01)

❏ 3 The motion and arrangement of particles in solids, liquids, and gases. (5.01)

❏ 4 Atoms and molecules. (5.01)

❏ 5 Brownian motion. (5.01)

❏ 6 The meaning of internal energy. (5.01)

❏ 7 The Celsius temperature scale. (5.02)

❏ 8 The principles used in

    – liquid-in-glass thermometers

    – thermistor thermometers.

    – thermocouple thermometers. (5.02 and 5.03)

❏ 9 The difference between heat (thermal energy) and temperature. (5.02)

❏ 10 The Celsius and Kelvin temperature scales. (5.02)

❏ 11 Defining a temperature scale using the ice point and the steam point. (5.03)

❏ 12 Calibrating an unmarked thermometer. (5.03)

❏ 13 The sensitivity, range, responsiveness, and linearity of liquid-in-glass thermometers. (5.03)

❏ 14 Why solids and liquids normally expand when heated. (5.04)

❏ 15 Effects and uses of thermal expansion. (5.04)

❏ 16 How water changes volume when it freezes, and the effects this can produce. (5.04)

❏ 17 How the expansions of solids, liquids, and gases compare. (5.05)

❏ 18 The law linking the pressure and temperature of a fixed mass of gas at constant volume. (5.05)

❏ 19 How the kinetic theory explains absolute zero. (5.05)

❏ 20 Charles's law. (5.05)

❏ 21 The factors affecting thermal conduction. (5.06)

❏ 22 Examples of good and poor thermal conductors. (5.06)

❏ 23 Uses of thermally insulating materials. (5.06)

❏ 24 Why some materials are better thermal conductors than others. (5.06)

❏ 25 Convection currents and why they occur. (5.07)

❏ 26 Examples and uses of convection in liquids and gases. (5.07)

❏ 27 The nature of thermal radiation. (5.08)

❏ 28 How the energy radiated by an object depends on its surface temperature and surface area. (5.08)

❏ 29 How different surfaces compare as emitters, reflectors, and absorbers of thermal radiation. (5.08)

❏ 30 How a vacuum flask works. (5.08)

❏ 31 The difference between evaporation and boiling. (5.09)

❏ 32 Factors affecting the rate at which a liquid evaporates. (5.09)

❏ 33 The cooling effect of evaporation, and its uses. (5.09)

❏ 34 Why condensation occurs. (5.09)

❏ 35 Thermal capacity and specific heat capacity. (5.10)

❏ 36 Calculating the thermal energy changes that occur when the temperature of an object changes. (5.10)

❏ 37 Using materials to store thermal energy. (5.10)

❏ 38 Specific latent heat of fusion (of ice) and its measurement. (5.11)

❏ 39 Specific latent heat of vaporization (of water) and its measurement. (5.11)

Waves, Sounds, and Vibrations

- TRANSVERSE AND LONGITUDINAL WAVES

- DIFFRACTION AND OTHER WAVE EFFECTS

- SOUND WAVES

- SPEED, FREQUENCY, AND WAVELENGTH

- ECHOES

- FREQUENCY AND PITCH

- AMPLITUDE AND LOUDNESS

- ULTRASOUND

This hyla tree frog from South America uses the large, inflatable sac under its throat to amplify the sound of its voice. Only the males can do this, and their calls can travel ten times further than sounds from other frogs. The sound itself is generated when air from the sac is blown past two stretched membranes in the bottom of the frog's mouth, making them vibrate.

Transverse and longitudinal waves

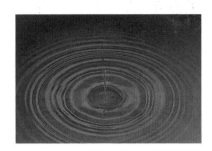

If you drop a stone into a pond, ripples spread across the surface. The tiny waves carry energy, as you can tell from the movements they cause at the water's edge. But there is no flow of water across the pond. The wave effect is just the result of up-and-down motions in the water.

Waves are not only found on water. Sound travels as waves, so does light. Waves can also travel along stretched springs like those in the experiments below. These show that there are two main types of waves.

Transverse waves

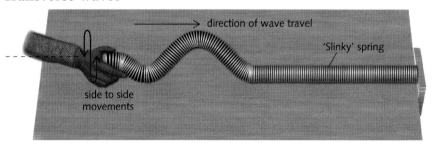

When the end coil of the spring is moved sideways, it pulls the next coil sideways a fraction of a second later... and so on along the spring. In this way, the sideways motion (and its energy) is passed from coil to coil, and a travelling wave effect is produced.

The to-and-fro movements of the coils are called **oscillations**. When the oscillations are up and down or from side to side like those above, the waves are called **transverse waves**. In transverse waves, the oscillations are at right-angles to the direction of travel.

Light waves are transverse waves, although it is electric and magnetic fields which oscillate, rather than any material.

Longitudinal waves

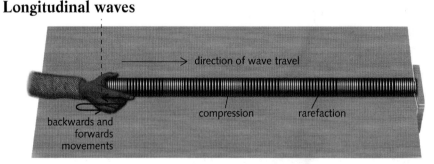

Moving the end coil of the spring backwards and forwards also produces a travelling wave effect. However, the waves are bunched-up sections of coils with stretched-out sections in between. These sections are known as **compressions** and **rarefactions**.

When the oscillations are backwards-and-forwards like those above, the waves are called **longitudinal waves**. In longitudinal waves, the oscillations are in the direction of travel.

Sound waves are longitudinal waves. When you speak, compressions and rarefactions travel out through the air.

Drawing waves

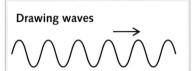

Transverse waves can be drawn as above.

Waves can also be drawn using lines called **wavefronts**. You can think of each wavefront as the 'peak' of a transverse wave or the compression of a longitudinal wave.

Examples of...

transverse waves

electromagnetic waves:

    radio waves

    microwaves

    infrared rays

    light

    ultraviolet rays

    X-rays

    gamma rays

longitudinal waves

sound waves

Describing waves

On the right, transverse waves are being sent along a rope. Here are some of the terms used to describe these and other waves:

Speed The speed of the waves is measured in metres per second (m/s).

Frequency This is the number of waves passing any point per second. The SI unit of frequency is the **hertz** (**Hz**). For example, if the hand on the right makes 4 oscillations per second, then 4 waves pass any point per second, and the frequency is 4 Hz. The time for one oscillation is called the **period**. It is equal to 1/frequency. If the frequency is 4 Hz, the period is 1/4 s (0.25 s)

Wavelength This is the distance between any point on a wave and the equivalent point on the next.

Amplitude This is the maximum distance a point moves from its rest position when a wave passes.

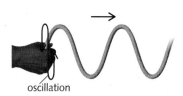
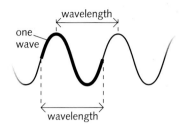
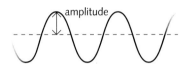

The wave equation

The speed, frequency, and wavelength of any set of waves are linked by this equation:

$$\text{speed} = \text{frequency} \times \text{wavelength}$$

In symbols: $v = f\lambda$ (λ = Greek letter *lambda*)

where speed is in m/s, frequency in Hz, and wavelength in m.

The following example shows why the equation works:

> Frequency (in Hz) is the number of oscillations per second.
>
> Period (in seconds) is the time for one oscillation.
>
> $\text{frequency} = \dfrac{1}{\text{period}}$

The waves on the right are travelling across water.

Each wave is 2 m long, so the wavelength is 2 m.

One second later...

3 waves have passed the flag, so the frequency is 3 Hz.

The waves have moved 3 wavelengths (3 × 2 m) to the right so their speed is 6 m/s.

Therefore:

6 m/s	=	3 Hz	×	2 m
(speed)		(frequency)		(wavelength)

one second later

1 The waves in A below are travelling across water.
a) Are the waves *transverse* or *longitudinal*?
b) What is the wavelength of the waves?
c) What is the amplitude of the waves?
d) If 2 waves pass the flag every second, what is
i) the frequency ii) the period?

e) Use the wave equation to calculate the speed of the waves in A.
f) What is the wavelength of the waves in diagram B below?
g) If the waves in B have the same speed as those in A, what is their frequency?

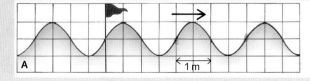

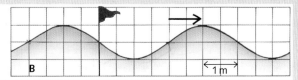

Related topics: SI units **1.02**; speed **2.01**; speed of sound **6.04**; frequency **6.05**; period and frequency **10.08**

125

6.02 Wave effects

The properties of waves can be studied using a **ripple tank** like the one below. Ripples (tiny waves) are sent across the surface of water. Obstacles are put in their path to see what effects are produced.

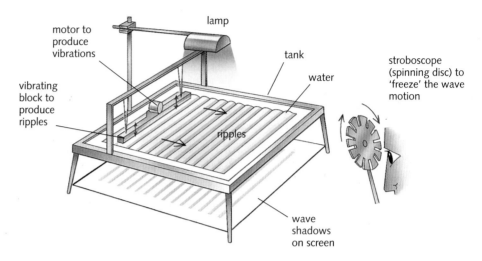

motor to produce vibrations

lamp

vibrating block to produce ripples

tank

water

stroboscope (spinning disc) to 'freeze' the wave motion

ripples

wave shadows on screen

Reflection

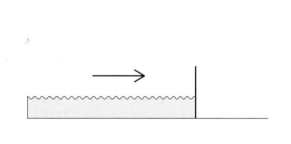

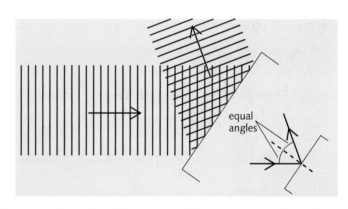

equal angles

A vertical surface is put in the path of the waves. The waves are reflected from the surface at the same angle as they strike it.

Refraction

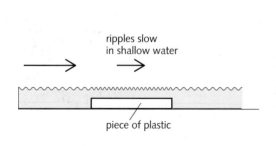

ripples slow in shallow water

piece of plastic

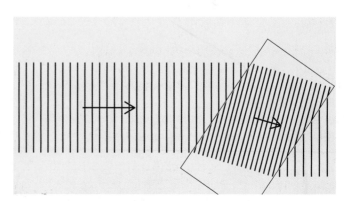

A flat piece of plastic makes the water more shallow, which slows the waves down. When the waves slow, they change direction. The effect is called **refraction**.

Refraction can be explained as follows.

The waves keep oscillating up and down at the same rate (frequency), so when they slow, the wavefronts close up on each other. That follows from the wave equation on the right. As the frequency is unchanged, a decrease in speed must cause a decrease in wavelength. From the last diagram on the opposite page, you can see that if the wavefronts close up on each other, their direction of travel must change, unless they are travelling at right-angles to the boundary.

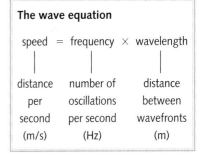

The wave equation

speed = frequency × wavelength

| distance per second (m/s) | number of oscillations per second (Hz) | distance between wavefronts (m) |

Diffraction

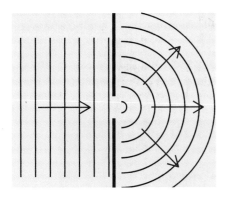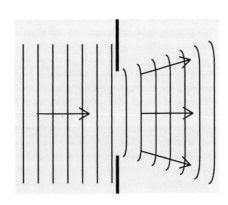

The waves bend round the sides of an obstacle, or spread out as they pass through a gap. The effect is called **diffraction**.

Diffraction is only significant if the size of the gap is about the same as the wavelength. Wider gaps produce less diffraction.

Wave evidence

Sound, light, and radio signals all undergo reflection, refraction, and diffraction. This suggests that they travel as waves. For example:

a Light reflects from mirrors; sound reflects from hard surfaces.
b Light bends when it passes from air into glass or water.
c Sound bends round obstacles such as walls and buildings, which is why you can hear round corners.
d Light spreads when it passes through tiny holes and slits. This suggests that light waves must have much shorter wavelengths than sound.
e Some radio signals can bend round very large obstacles such as hills. This suggests that radio waves must have long wavelengths.

Q

1 Say whether each of the effects b to e above is an example of *reflection*, *refraction*, or *diffraction*.
2 On the right, waves are moving towards a harbour.
   a) What will happen to waves striking the harbour wall at A?
   b) What will happen to waves slowed by the submerged sandbank at B?
   c) What will happen to waves passing through the harbour entrance at C?
   d) If the harbour entrance were wider, what difference would this make?

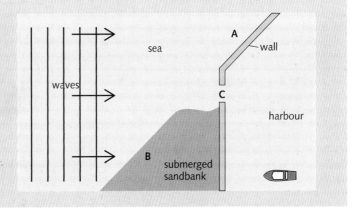

Related topics: waves and the wave equation **6.01**; reflection of sound **6.04**; light waves **7.01**; reflection of light **7.01–7.03**; refraction of light **7.04**; and **7.06**; radio waves **7.12**

Sound waves

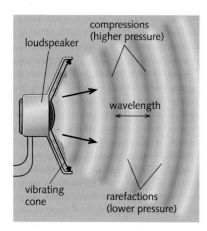

compressions (higher pressure)
loudspeaker
wavelength
vibrating cone
rarefactions (lower pressure)

When a loudspeaker cone vibrates, it moves forwards and backwards very fast. This squashes and stretches the air in front. As a result, a series of compressions ('squashes') and rarefactions ('stretches') travel out through the air. These are **sound waves**. When they reach your ears, they make your ear-drums vibrate and you hear a sound.

The nature of sound waves

Sound waves are caused by vibrations Any vibrating object can be a source of sound waves. As well as loudspeaker cones, examples include vibrating guitar strings, the vibrating air inside a trumpet, and the vibrating prongs of a tuning fork. Also, when hard objects (such as cymbals and steel drums) are struck, they vibrate and produce sound waves.

Wavefront essentials

For convenience, waves are often drawn using lines called **wavefronts**. In the case of sound waves, you can think of each wavefront as a compression.

Sound waves are longitudinal waves The air oscillates backwards and forwards as the compressions and rarefactions pass through it. When a compression passes, the air pressure rises. When a rarefaction passes, the pressure falls. The distance from one compression to the next is the **wavelength**.

Sound waves need a material to travel through This material is called a **medium**. Without it, there is nothing to pass on any oscillations. Sound cannot travel through a vacuum (completely empty space).

Sound waves can travel through solids, liquids, and gases Most sound waves reaching your ear have travelled through air. But you can also hear when swimming underwater, and walls, windows, doors, and ceilings can all transmit (pass on) sound.

Diffraction of sound waves

You can hear someone through an open window even if you cannot see them. That is because sound waves spread through gaps, or bend round obstacles, of a similar size to their wavelength. The effect is called **diffraction**. Most sound waves have wavelengths between a few centimetres and a few metres, so they are diffracted by everyday objects.

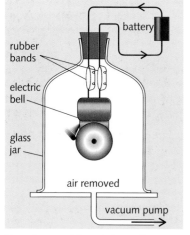

battery
rubber bands
electric bell
glass jar
air removed
vacuum pump

Sound cannot travel through a vacuum. When the air is removed from this jar, the bell goes quiet, even though the hammer is still striking the metal. (The rubber bands reduce the sound transmitted by the connecting wires.)

Displaying sounds

Sound waves can be displayed graphically using a microphone and an **oscilloscope** as on the right. When sound waves enter the microphone, they make a crystal or a metal plate inside it vibrate. The vibrations are changed into electrical oscillations, and the oscilloscope uses these to make a spot oscillate up and down on the screen. It moves the spot steadily sideways at the same time, producing a wave shape called a **waveform**. The waveform is really a graph showing how the air pressure at the microphone varies with time. It is *not* a picture of the sound waves themselves: sound waves are *not* transverse (up-and-down).

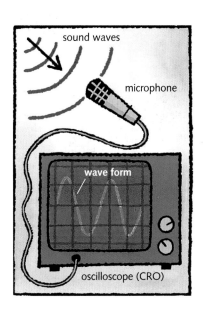

Wave effects explained

Reflection, refraction, and diffraction occur with all types of waves, including sound (for more about the reflection and refraction of sound, see the next spread). The **wave theory** can be used to explain these effects. It starts with the idea shown below left – that each point on a wavefront is the source of a new, circular wave which radiates from that point. Together, the new waves combine to create a new wavefront. Waves spread from all points on this wavefront... and so on. In this way, the wavefront travels along.

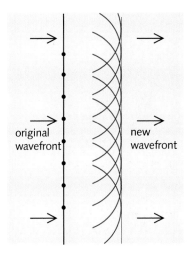

▲ A new wave radiates from each point on the original wavefront. Together, the new waves combine to create a new wavefront.

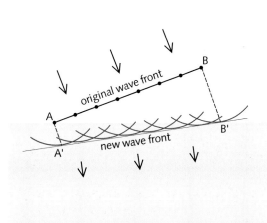

▲ Refraction (bending) as waves pass from one medium to another. A new wave travels from from A to A' in the same time as another travels from B to B'. But AA' is shorter than BB' because waves travel more slowly in the different medium. So the angle of the wavefront changes.

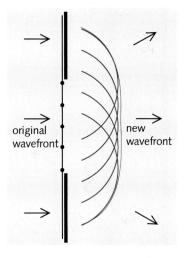

▲ Diffraction (spreading) as waves pass through a gap. Only part of the original wavefront can pass through. New waves coming from it combine to create a new, curved wavefront.

1 Give an example which demonstrates each of the following:
   a) Sound can travel through a gas.
   b) Sound can travel through a liquid.
   c) Sound can travel through a solid.
2 Explain each of the following:
   a) Sound cannot travel though a vacuum.
   b) It is possible to hear round corners.

3 a) Sound waves are *longitudinal* waves. Explain what this means.
   b) If sound waves are longitudinal, why are transverse (up-and-down) 'waves' seen on the screen of the oscilloscope above when someone whistles into the microphone?
4 Waves are diffracted when they pass through a narrow gap. How does the wave theory explain this?

Related topics: air pressure **3.08**; longitudinal waves **6.01**; diffraction **6.02**; loudspeaker **9.05**; oscilloscope **10.08**

129

6.04 Speed of sound and echoes

▶ Sound is much slower than light, so you hear lightning after you see it. Sound takes about 3 seconds to travel one kilometre. Light does it in almost an instant, so a 3 second gap between the flash and the crash means that the lightning is about a kilometre away.

Sound wave essentials

Sound waves are a series of compressions ('squashes') and rarefactions ('stretches') that travel through the air or other material.

Speed of sound

through...

air (dry) at 0 °C	330 m/s
air (dry) at 30 °C	350 m/s
water (pure) at 0 °C	1400 m/s
concrete	5000 m/s

The speed of sound

In air, the speed of sound is about 330 metres per second (m/s), or 760 mph. That is slower than Concorde but about four times faster than a racing car.

The speed of sound depends on the temperature of the air
Sound waves travel faster through hot air than through cold air.

The speed of sound does not depend on the pressure of the air
If atmospheric pressure changes, the speed of sound waves stays the same.

The speed of sound is different through different materials
Sound waves travel faster through liquids than through gases, and fastest of all through solids. There are some examples on the left.

Measuring the speed of sound

The speed of sound in air can be measured as shown below. A sound is made by hitting a metal block or plate with a hammer. When the control unit receives a pulse of sound from microphone A, it starts the clock. When it receives a pulse from microphone B, it stops it.

If B is 1.00 metre further away from the source of sound than A, and the clock records a time of 3.0 milliseconds (0.003 s):

$$\text{speed of sound} = \frac{\text{distance travelled}}{\text{time taken}} = \frac{1.00\,\text{m}}{0.003\,\text{s}} = 330\,\text{m/s}$$

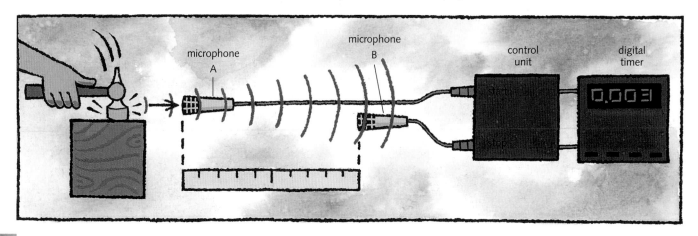

Refraction of sound★

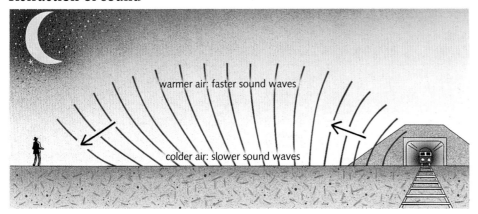

Distant trains and traffic often sound louder (and closer) at night. The reason is this. During the night time, when the ground cools quickly, air layers near the ground become colder than those above. Sound waves travel more slowly through this colder air. As a result, waves leaving the ground tend to bend back towards it, instead of spreading upwards. A bending effect like this, caused by a change in speed, is called **refraction**.

Echoes

Hard surfaces such as walls reflect sound waves. When you hear an **echo**, you are hearing a reflected sound a short time after the original sound. In the diagram on the right, the sound has to travel to the wall *and back again*. The time it takes is the **echo time**. So:

$$\text{speed of sound} = \frac{\text{distance travelled}}{\text{time taken}} = \frac{2 \times \text{distance to wall}}{\text{echo time}}$$

If the speed of sound is known, and the echo time is measured accurately, the distance to the wall can be calculated from the above equation. The principle is used in several devices, including the following:

● **Echo-sounder** This measures the depth of water under a boat. It sends pulses of sound waves towards the sea-bed and measures the echo time. The longer the time, the deeper the water.

● **Electronic tape-measure** A surveyor can use this to measure the distance between two walls. It works like an echo-sounder.

● **Radar** This uses the echo-sounding principle, but with microwaves instead of sound waves. It detects the positions of aircraft by measuring the 'echo times' of microwave pulses reflected from them.

Assume that the speed of sound in air is 330 m/s.

1 a) Why do you hear lightning after you see it?
 b) If lightning strikes, and you hear it 4 seconds after you see it, how far away is it?
2 Does sound travel faster through
 a) *cold air* or *warm air*? **b)** a *solid* or a *gas*?
3 When sound waves change direction because their speed changes, what is this effect called?

4 A ship is 220 metres from a large cliff when it sounds its foghorn.
 a) When the echo is heard on the ship, how far has the sound travelled?
 b) What time delay is there before the echo is heard?
 c) The ship changes its distance from the cliff. When the echo time is 0.5 seconds, how far is the ship from the cliff?

6.05 Characteristics of sound waves

Sound wave essentials

Sound waves are a series of compressions ('squashes') and rarefactions ('stretches') that travel through the air or other material.

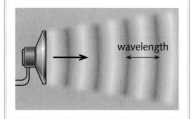

wavelength

Frequency and pitch

Sound waves are caused by vibrations – for example, the rapid, backwards-and-forwards oscillations of a loudspeaker cone.

The number of oscillations per second is called the **frequency**. It is measured in **hertz** (**Hz**). If a loudspeaker cone has a frequency of 100 Hz, it is oscillating 100 times per second and giving out 100 sound waves per second.

Different frequencies sound different to the ear. You hear *high* frequencies as *high* notes: musicians say that they have a **high pitch**. You hear *low* frequencies as *low* notes: they have a **low pitch**.

The human ear can detect frequencies ranging from about 20 Hz up to 20 000 Hz, although the ability to hear high frequencies decreases with age.

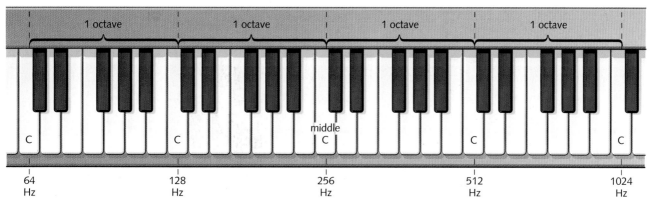

| 1 octave | 1 octave | 1 octave | 1 octave |

C   C   middle C   C   C

| 64 Hz | 128 Hz | 256 Hz | 512 Hz | 1024 Hz |

pitch		frequency
high	upper limit of hearing	20 000 Hz
	whistle	10 000 Hz
	high note (soprano)	1000 Hz
	low note (bass)	100 Hz
low	drum note	20 Hz

1000 Hz = 1 kilohertz (kHz)

Octaves★ Musical scales are based on these. If the pitch of a note increases by one octave, the frequency doubles, as shown on the keyboard above. This keyboard is tuned to **scientific pitch**. Bands and orchestras normally use frequencies that differ slightly from those shown.

The diagrams below show what happens if two steady notes, an octave apart, are picked up by a microphone and displayed on the screen of an oscilloscope. As the higher note has double the frequency of the lower note, the peaks occur twice as often and are only half as far apart.

▶ The **waveform** on each screen is a graph showing how the air pressure varies with time as the sound waves enter the microphone. The horizontal line is the time axis.

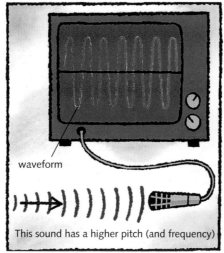

waveform

This sound has a higher pitch (and frequency)

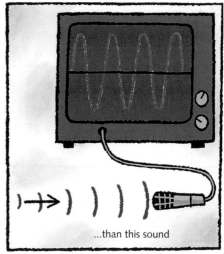

...than this sound

The wave equation

This equation applies to sound waves:

> speed = frequency × wavelength

In symbols: $v = f\lambda$ (λ = Greek letter *lambda*)

For example, if the speed of sound in air is 330 m/s:
sound waves of frequency 110 Hz have a wavelength of 3 m;
sound waves of frequency 330 Hz have a wavelength of 1 m;
so the *higher* the frequency, the *shorter* the wavelength.

Why the equation works

If 110 waves are sent out in one second, and each wave is 3 m long, then the waves must travel 330 metres in one second. In other words, if the frequency is 110 Hz and the wavelength is 3 m, the speed is 330 m/s.

Amplitude and loudness

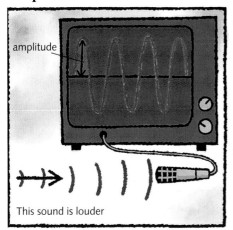
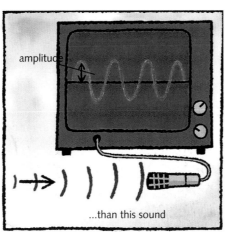

This sound is louder …than this sound

The sounds displayed on the oscilloscope screens above have the same frequency, but one is *louder* than the other. The oscillations in the air are bigger and the **amplitude** of the waveform is greater.

Sound waves carry energy. *Doubling* the amplitude means that *four* times as much energy is delivered per second.

Quality★

Middle C on a guitar does not sound quite the same as middle C on a piano, and its waveform looks different. The two sounds have a different **quality** or **timbre**. Each sound has a strong **fundamental frequency**, giving middle C. But other weaker frequencies are mixed in as well, as shown on the right. These are called **overtones**, and they differ from one instrument to another. With a synthesizer, you can select which frequencies you mix together, and produce the sound of a guitar, piano, or any other instrument.

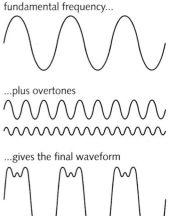

fundamental frequency…

…plus overtones

…gives the final waveform

Assume that the speed of sound in air is 330 m/s.
1 Here are the frequencies of four sounds:

   A: 400 Hz   B: 150 Hz   C: 500 Hz   D: 200 Hz

   a) Which sound has the highest pitch?
   b) Which two sounds are one octave apart?
   c) Which sound has the longest wavelength?
2 Why does a piano not sound quite like a guitar, even if both play the same note?

3 A sound is picked up by a microphone and displayed as a waveform on an oscilloscope. How would the waveform change if
   a) the sound had a higher pitch?
   b) the sound was louder?
4 The lower limit of human hearing is 20 Hz; the upper limit is 20 000 Hz.
   a) What is the upper limit in kHz?
   b) What is the wavelength at the lower limit?
   c) What is the wavelength at the upper limit?

Related topics: speed **2.01**; waves and the wave equation **6.01**; sound waves **6.03**; speed of sound **6.04**; oscilloscope **10.08**

133

6.06 Ultrasound

Sound wave essentials

Sound waves are a series of compressions ('squashes') and rarefactions ('stretches') that travel through the air or other material.

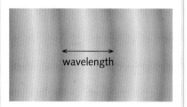

wavelength

The number of waves per second is called the **frequency**. It is measured in hertz (Hz).

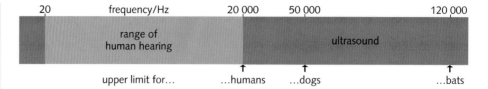

The human ear can detect sounds up to a frequency of about 20 000 Hz. Sounds above the range of human hearing are called **ultrasonic sounds**, or **ultrasound**. Here are some of the uses of ultrasound:

Cleaning and breaking★

Using ultrasound, delicate machinery can be cleaned without dismantling it. The machinery is immersed in a tank of liquid, then the vibrations of high-power ultrasound are used to dislodge the bits of dirt and grease.

In hospitals, concentrated beams of ultrasound can be used to break up kidney stones and gall stones without patients needing surgery.

Echo-sounding

Ships use **echo-sounders** to measure the depth of water beneath them. An echo-sounder sends pulses of ultrasound downwards towards the sea-bed, then measures the time taken for each echo (reflected sound) to return. The longer the time, the deeper the water. For example:

If a pulse of ultrasound takes 0.1 second to travel to the sea-bed and return, and the speed of sound in water is 1400 m/s:

$$\text{distance travelled} = \text{speed} \times \text{time} = 1400 \, \text{m/s} \times 0.1 \, \text{s} = 140 \, \text{m}$$

But the ultrasound has to travel down *and back*:
So: depth of water $= \frac{1}{2} \times 140 \, \text{m} = 70 \, \text{m}$

Most echo-sounders **scan** the area beneath them – they sweep their ultrasound beam backwards and forwards and from side to side. A computer displays the depth information as a picture on a screen.

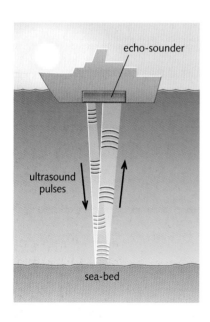

echo-sounder

ultrasound pulses

sea-bed

▶ This bat uses ultrasound to locate insects and other objects in front of it. It sends out a series of ultrasound pulses and uses its specially shaped ears to pick up the reflections. The process is called **echo-location**. It works like echo-sounding.

Metal testing★

The echo-sounding principle can be used to detect flaws in metals. A pulse of ultrasound is sent through the metal as on the right. If there is a flaw (tiny gap) in the metal, *two* reflected pulses are picked up by the detector. The pulse reflected from the flaw returns first, followed by the pulse reflected from the far end of the metal. The pulses can be displayed using an oscilloscope. The trace on the screen is a graph showing how the amplitude ('strength') of the ultrasound varies with time.

Scanning the womb★

The pregnant mother in the photograph below is having her womb scanned by ultrasound. Again, the echo-sounding principle is being used. A transmitter sends pulses of ultrasound into the mother's body. The transmitter also acts as a detector and picks up pulses reflected from the baby and different layers inside the body. The signals are processed by a computer, which puts an image on the screen.

Using ultrasound is much safer than using X-rays because X-rays can cause cell damage inside a growing baby. Also, ultrasound can distinguish between different layers of soft tissue, which an ordinary X-ray machine cannot.

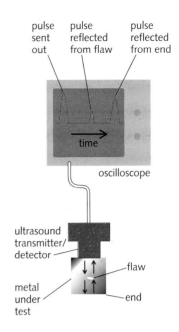

pulse sent out | pulse reflected from flaw | pulse reflected from end

time

oscilloscope

ultrasound transmitter/ detector

flaw

metal under test

end

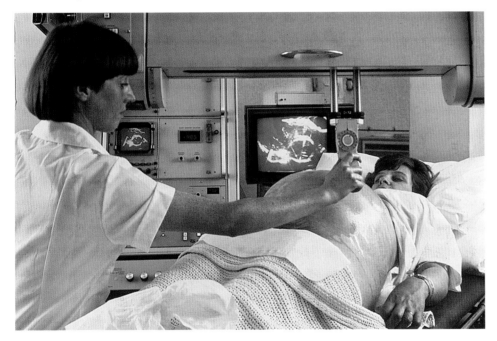

◀ An ultrasound scan of the womb. The nurse is moving an ultrasound transmitter/detector over the mother's body. A computer uses the reflected pulses to produce an image.

1 What is *ultrasound*?

2 Give *two* examples of the medical use of ultrasound.

3 a) What is an *echo-sounder* used for?
b) How does an echo-sounder work?

4 To answer this question, you will need the information on the right. A boat is fitted with an echo-sounder which uses ultrasound with a frequency of 40 kHz.
a) What is the frequency of the ultrasound in Hz?
b) If ultrasound pulses take 0.03 seconds to travel from the boat to the sea-bed and return, how deep is the water under the boat?
c) What is the wavelength of the ultrasound in water?

$$speed = \frac{distance\ travelled}{time\ taken}$$

speed of sound in water = 1400 m/s

speed = frequency × wavelength
(m/s)   (Hz)   (m)

1 kilohertz (kHz) = 1000 Hz

1 Kim and Sam are playing with a ball in the park. Unfortunately the ball finishes up in the middle of a pond, out of reach.

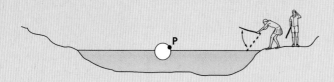

Kim thinks that hitting the water with a stick will make waves that will push the ball to the other side.

a) Which **two** of these words best describe the waves that are created on the water surface?

circular   longitudinal   plane   pressure   transverse   [2]

b) Kim hits the water surface regularly so that waves travel out to the ball and beyond it.

  (i) What happens to the ball?   [1]

Sam throws a stick which hits the ball at P.

  (ii) Sam is successful at moving the ball across the pond. Kim is not. Explain why.   [2]

c) (i) Kim hits the water surface regularly with the stick 20 times in 10 seconds. Calculate the frequency of the waves.   [2]

  (ii) The waves travel across the pond at 0.5 m/s. Calculate the wavelength.   [4]

MEG

2 a) The wave in the shallow tank of water shown in the figure moves at 0.08 m/s towards the left.

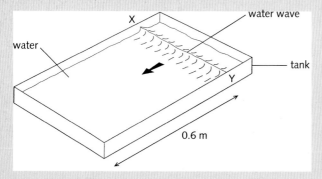

How long does it take for the wave to return to the position XY, but moving to the right?   [3]

b) A man is cutting down a tree with an axe. He hears the echo of the impact of the axe hitting the tree after 1.6 s.

  (i) What sort of obstacle could have caused the echo?   [1]

  (ii) The speed of sound is 330 m/s. How far is the tree from the obstacle?

c) Distinguish between the nature of the sound wave in (b) and the water wave in (a).   [2]

UCLES

3 The figure shows a cathode-ray oscilloscope trace for a sound wave produced by a loudspeaker.

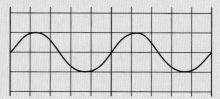

a) Copy the figure and draw the trace for a louder sound of the same pitch.   [2]

b) It takes 1/50th of a second (0.02 s) for the whole trace to be produced.

  (i) Show that the frequency of the sound produced by the loudspeaker is 100 Hz.

  (ii) Determine the wavelength in air of the sound produced by the loudspeaker. (The speed of sound in air is 330 m/s.)   [3]

UCLES

4 a) A sound wave travelling through air can be represented as shown in the diagram.

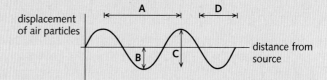

Which distance, **A**, **B**, **C**, or **D**, represents:

  (i) one wavelength?

  (ii) the amplitude of the wave?   [2]

b) The cone of a loudspeaker is vibrating. The diagram shows how the air particles are spread out in front of the cone at a certain time.

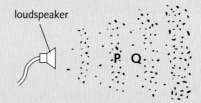

P is a compression, **Q** is a rarefaction.

  (i) Describe how the pressure in the air changes from **P** to **Q**.   [2]

  (ii) Describe the motion of the air particles as the sound wave passes.   [2]

  (iii) Copy the diagram of air particles above and mark and label a distance equal to one wavelength of the sound wave.   [1]

MEG

5 a) The first diagram on the next page shows a wave.

  (i) Copy the diagram and mark the amplitude, and label it A.   [1]

  (ii) State the number of cycles ('wavelengths') shown in the diagram.   [1]

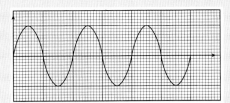

(iii) This complete wave was produced in 0.15 s. Calculate the period (time for one wave). [1]

(iv) Use the equation

$$frequency \ (Hz) = \frac{1}{period \ (s)}$$

to calculate the frequency of the wave. [1]

b) (i) Sound is a *longitudinal wave*. Explain what is meant by a longitudinal wave. [2]

(ii) If the amplitude of a sound wave is increased, what difference would you hear? [1]

WJEC

6

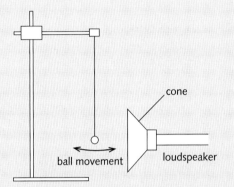

A light polystyrene ball is shown hanging very close to a loudspeaker. The loudspeaker gives out a sound of low frequency and the ball is seen to vibrate.

a) Explain how the sound from the loudspeaker causes the ball to move as described. [2]

b) Explain what will happen to the motion of the cone of the loudspeaker when:

(i) the sound is made louder; [1]

(ii) the pitch of the sound is increased. [1]

c) Calculate the frequency of a sound which has a wavelength of 0.5 m and travels at a speed of 340 m/s in air. Write down the formula that you use and show your working. [3]

WJEC

7 a) The figure shows a metal rod, 2.4 m long, being struck a sharp blow at one end using a light hammer. The time interval between the impact of the hammer and the arrival of the sound wave at the other end of the rod is measured electronically.

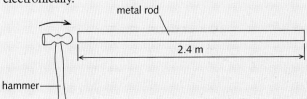

Four measurements of the time interval are 0.44 ms, 0.50 ms, 0.52 ms and 0.47 ms.

(i) Determine the average value of the four measurements.

(ii) Hence calculate a value for the speed of sound in the rod. [4]

UCLES

8 a) A microphone is connected to an oscilloscope (CRO). When different sounds, A, B, and C, are made, these are the waveforms seen on the screen:

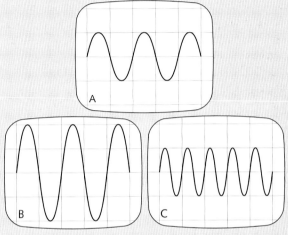

a) Comparing sounds A and B, how would they sound different? [2]

b) Comparing sounds A and C, how would they sound different? [2]

c) Which sound has the highest amplitude? [1]

d) Which sound has the highest frequency? [1]

e) The speed of sound is 330 m/s. If sound A has a frequency of 220 Hz, what is its wavelength? [2]

f) What is the frequency of sound C? [2]

9 Ultrasound waves are high frequency longitudinal waves. X-rays are high frequency transverse waves.

a) Explain the difference between transverse and longitudinal waves. [2]

b) The diagram shows an ultrasound probe used to obtain an image of an unborn baby.

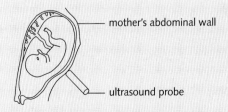

Give **two** reasons why ultrasound and not X-rays are used for this investigation. [2]

c) Describe **one** industrial use of ultrasonic waves. [2]

WJEC

Photocopy the list of topics below and tick the boxes of the ones that are included in your examination syllabus. (Your teacher should be able to tell you which they are.) Use your list when you revise. The spread number in brackets tells you where to find more information.

❏ 1 How waves transfer energy from one place to another without transferring matter. (6.01)

❏ 2 The difference between transverse and longitudinal waves, with examples of each. (6.01)

❏ 3 The meaning of wavelength. (6.01)

❏ 4 The meaning of amplitude. (6.01)

❏ 5 The meaning of frequency. (6.01)

❏ 6 The meaning of period. (6.01)

❏ 7 The link between period and frequency. (6.01)

❏ 8 The SI unit of frequency: the hertz. (6.01)

❏ 9 The equation linking speed, frequency, and wavelength. (6.01 and 6.05)

❏ 10 Demonstrating the following wave effects in a ripple tank:

– reflection
– refraction
– diffraction. (6.02)

❏ 11 How diffraction depends on the size of the gap through which the waves are passing. (6.02)

❏ 12 Evidence that sound is a form of wave motion. (6.02–6.03)

❏ 13 How sound waves are produced. (6.03)

❏ 14 What compressions and rarefactions are. (6.03)

❏ 15 Why sound waves need a material to travel through. (6.03)

❏ 16 Displaying waveforms on an oscilloscope screen. (6.03 and 6.05)

❏ 17 How the wave theory explains refraction and diffraction. (6.03)

❏ 18 The speed of sound in air. (6.04)

❏ 19 Factors affecting the speed of sound. (6.04)

❏ 20 Measuring the speed of sound. (6.04)

❏ 21 How the speed of sound is different in solids, liquids, and gases. (6.04)

❏ 22 The refraction of waves. (6.04)

❏ 23 How the reflection of sound causes echoes. (6.04)

❏ 24 Uses of the echo-timing principle. (6.04 and 6.06)

❏ 25 The frequency range of sound waves. (6.05)

❏ 26 The link between frequency and pitch. (6.05)

❏ 27 The link between amplitude and loudness. (6.05)

❏ 28 Interpreting the waveforms on an oscilloscope screen. (6.05)

❏ 29 Ultrasound frequencies. (6.06)

❏ 30 Uses of ultrasound, including

– cleaning and breaking
– echo-sounding
– metal testing
– scanning the womb. (6.06)

Rays and Waves

A rainbow forms as the Sun shines on raindrops. The raindrops, acting like tiny prisms, are splitting the white sunlight into its different spectral colours and reflecting them back. Because the Sun is behind the camera and the rainbow appears to extend to the ground, rain must be falling between the camera and the distant hills.

7.01 Light rays and waves

▶ If you can see a beam of light, this is because tiny particles of dust, smoke, or mist in the air are reflecting some of the light into your eyes.

For you to see something, light must enter your eyes. The Sun, lamps, lasers, and glowing TV screens all emit (send out) their own light. They are **luminous**. However, most objects are **non-luminous**. You see them only because daylight, or other light, bounces off them. They **reflect** light, and some of it goes into your eyes.

You can see this page because it reflects light. The white parts reflect most light and look bright. However, the black letters **absorb** nearly all the light striking them. They reflect very little and look dark.

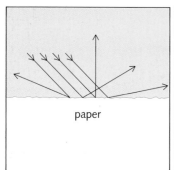

paper

Diffuse reflection

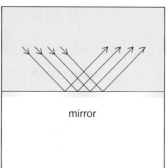

mirror

Regular reflection

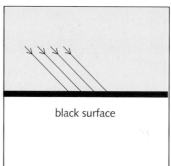

black surface

Absorption

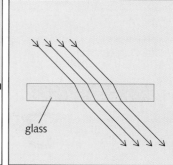

glass

Transmission

This solar-powered car uses the energy in sunlight to produce electricity for its motor.

Most surfaces are uneven, or contain particles that scatter light. As a result, they reflect light in all directions. The reflection is **diffuse**. However, mirrors are smooth and shiny. When they reflect light, the reflection is **regular**.

Transparent materials like glass and water let light pass right through them. They **transmit** light.

Features of light

Light is a form of radiation This means that light radiates (spreads out) from its source. In diagrams, lines called **rays** are used to show which way the light is going.

Light travels in straight lines You can see this if you look at the path of a sunbeam or a laser beam.

Light transfers energy Energy is needed to produce light. Materials gain energy when they absorb light. For example, solar cells use the energy in sunlight to produce electricity.

Light travels as waves Light radiates from its source rather as ripples spread across the surface of a pond. However, in the case of light, the 'ripples' are tiny, vibrating, electric and magnetic forces. Light waves have wavelengths of less than a thousandth of a millimetre (see below). Like other waves, they can be diffracted, but the effect is too small to notice unless the gaps are very narrow, for example, as in a fine mesh.

Some effects of light are best explained by thinking of light as a stream of tiny 'energy particles'. Scientists call these particles **photons**.

Light can travel through empty space Electric and magnetic ripples do not need a material to travel through. That is why light can reach us from the Sun and stars.

Light is the fastest thing there is In a vacuum (in space, for example), the speed of light is 300 000 kilometres *per second*. Nothing can travel faster than this. The speed of light seems to be a universal speed limit.

Wavelength and colour

When light enters the eye, the brain senses different wavelengths as different colours. The wavelengths range from 0.000 4 mm (violet light) to 0.000 7 mm (red light), and white light is made up of all the wavelengths in this range. Most sources emit a mixture of wavelengths. However, **lasers** emit light of a single wavelength and colour. Light like this is called **monochromatic** light.

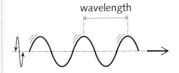

Wave essentials

wavelength

With **transverse** waves, like light, the oscillations (vibrations) are at right angles to the direction of travel.

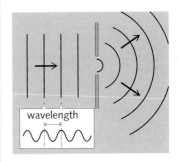

wavelength

Waves spread out as they pass through a gap. The effect is called **diffraction**. It is only significant if the size of the gap is about the same as the wavelength. Wider gaps cause less diffraction.

◀ Light from a laser is monochromatic (single wavelength and colour). Here, laser light is being used to measure the deflection of the rotating blades on an experimental jet engine.

1 Give *two* examples each of objects which
   a) emit their own light
   b) are only visible because they reflect light from another source.
2 What evidence is there that light travels in straight lines?
3 What happens to light when it strikes
   a) white paper b) black paper?
4 If the Moon is 384 000 km from Earth, the Sun is 150 000 000 km from Earth, and the speed of light is 300 000 km/s, calculate the time taken for light to travel from
   a) the Moon to the Earth b) the Sun to the Earth.
5 How do waves of violet light differ from waves of red light?
6 What is meant by *monochromatic* light?

7.02 Reflection in plane mirrors (1)

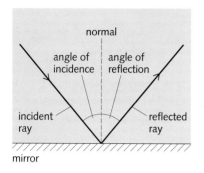

normal

angle of incidence | angle of reflection

incident ray | reflected ray

mirror

Definitions

Angle of incidence: this is the angle between the incident ray and the normal.

Angle of reflection: this is the angle between the reflected ray and the normal.

The laws of reflection

When a ray of light strikes a mirror, it is reflected as shown on the left. The incoming ray is the **incident ray**, the outgoing ray is the **reflected ray**, and the line at right-angles to the mirror's surface is called a **normal**. The mirror in this case is a **plane mirror**. This just means that it is a flat mirror, rather than a curved one.

There are two **laws of reflection**. They apply to all types of mirror:

1 The angle of incidence is equal to the angle of reflection.
2 The incident ray, the reflected ray, and the normal all lie in the same plane.

Put another way, light is reflected at the same angle as it arrives, and the two rays and the normal can all be drawn on one flat piece of paper.

Image in a plane mirror

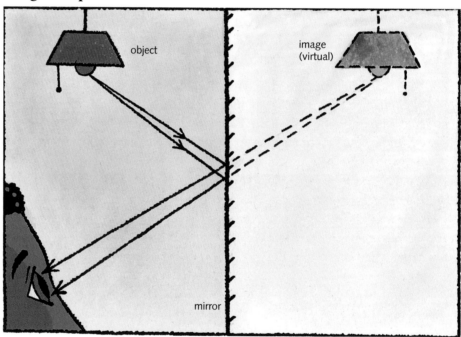

object

image (virtual)

mirror

In the diagram above, light rays are coming from an **object** (a lamp) in front of a plane mirror. Thousands of rays could have been drawn but, for simplicity, only two have been shown. After reflection, some of the rays enter the girl's eye. To the girl, they seem to come from a position behind the mirror, so that is where she sees an **image** of the lamp. Dotted lines have been drawn to show the point where two of the reflected rays appear to come from. The dotted lines are *not* rays.

The image seen in the mirror looks exactly the same as the object, apart from one important difference. The image is **laterally inverted** (back to front).

Real and virtual images In a cinema, the image on the screen is called a **real image** because rays from the projector focus (meet) to form it. The image in a plane mirror is not like this. Although the rays *appear* to come from behind the mirror, no rays actually pass through the image and it cannot be formed on a screen. An image like this is called a **virtual image**.

The word AMBULANCE is laterally inverted so that it reads correctly when seen in a driving mirror.

Finding the position of an image in a mirror

The position of an image in a plane mirror can be found by experiment:

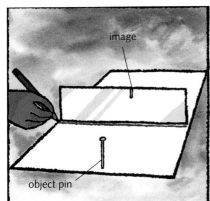

Put a mirror upright on a piece of paper. Put a pin (the object) in front of it. Mark the positions of the pin and the mirror.

Line up one edge of a ruler with the image of the pin. Draw a line along the edge to mark its position. Then repeat with the ruler in a different position.

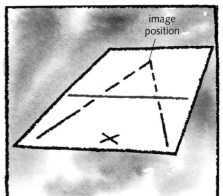

Take away the mirror, pin, and ruler. Extend the two lines to find out where they meet. This is the position of the image.

The result of the experiment can be checked like this. If a second pin is put *behind* the mirror, in the position found for the image, the pin should be in line with the image, as shown on the right. And it should stay in line when you move your head from side to side. Scientifically speaking, there should be **no parallax** (no relative movement) between the second pin and the image when you change your viewing position. If there is relative movement (parallax), then the two are not in the same position.

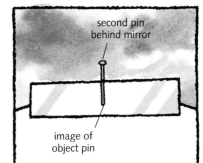

If a second pin is put in exactly the same position as the image of the first pin, it should stay in line with the image, wherever you view it from.

Rules for image size and position

When a plane mirror forms an image:

- The image is the same size as the object.
- The image is as far behind the mirror as the object is in front.
- A line joining equivalent points on the object and image passes through the mirror at right-angles.

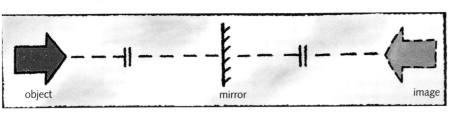

1 **a)** Copy the diagram on the right. Draw in the image in its correct position.
   b) From the object arrow's tip, A, draw two rays which reflect from the mirror and go into the person's eye.
   c) The image cannot be formed on a screen. What name is given to this type of image?
   d) Can the person see an image of the arrow's tail, B? If not, why not?
2 A man stands 10 m in front of a large, plane mirror. How far must he walk before he is 5 m away from his image?

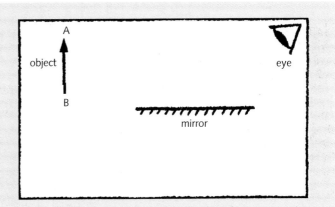

7.03 Reflection in plane mirrors (2)

Finding an image position by construction

In the diagrams below, O is a point object in front of a plane (flat) mirror. Here are two methods of finding the position of the image by geometric construction using a protractor. In Method 1, you deduce the position from the paths of two rays, but Method 2 is simpler!

Method 1

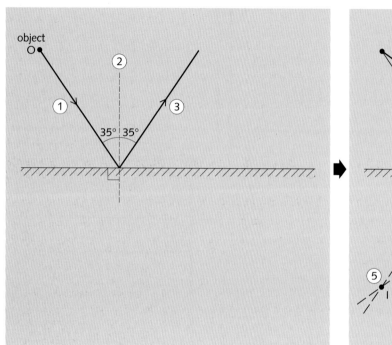

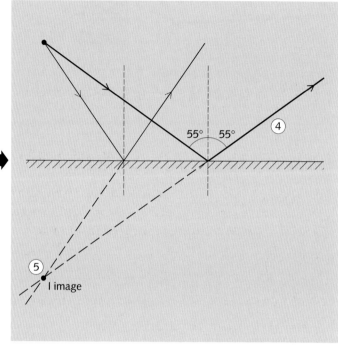

1 From the object, O, draw a ray which strikes the mirror at an angle of incidence of 35° (or value of your own choosing close to this).

2 Construct a normal (a line at right-angles to the mirror's surface) at the point where the ray strikes the mirror.

3 Draw the reflected ray from this point, so that the angle of reflection is equal to the angle of incidence.

4 Repeat steps 1 to 3 for a second ray with an angle of incidence of 55° (or value of your own choosing close to this).

5 Extend the two reflected ray backwards until they intersect (meet). The point of intersection, I, is the image position.

Method 2

This method is illustrated on the left. It uses the fact that the position of the image behind the mirror matches that of the object in front.

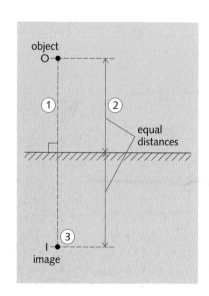

1 From the object, O, draw a line which passes through the mirror's surface at right angles. Extend this line well beyond the mirror.

2 Measure the distance from the object to the mirror.

3 At an equal distance behind the mirror, mark a point on the extended line. This point, I, is the image position.

Reflection problem

Example A horizontal ray of light strikes a plane mirror whose surface is angled at 55° to the ground, as shown below left.
a) What is the angle between the reflected ray and the ground?
b) If the mirror is re-angled to reflect the ray vertically upwards, what is the new angle between the surface of the mirror and the ground?

a)

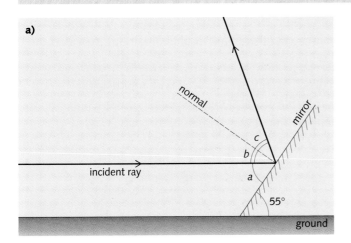

b)

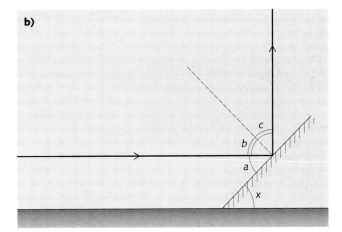

a) In the diagram above left, angles *a*, *b*, and *c* have also been labelled to help with the calculation. The incident ray is parallel to the ground, so the angle between the reflected ray and the ground is equal to *b* + *c*.

As the incident ray is parallel to the ground: $a = 55°$
But: $a + b = 90°$ So: $b = 35°$

As the angle of reflection = angle of incidence: $c = b$
So: $c = 35°$ Therefore: $b + c = 70°$

So, the angle between the reflected ray and the ground is 70°.

b) The situation is shown above right, where angles *a*, *b*, and *c* all now have new values. As before: $a + b = 90°$ and $c = b$. *x* is the unknown angle between the surface of the mirror and the ground. It is equal to *a*.

As the ray is reflected vertically: $b + c = 90°$ So *b* and *c* are both 45°
But: $a + b = 90°$ So: $a = 45°$ Therefore: $x = 45°$

So, the angle between the surface of the mirror and the ground must be changed to 45°.

Reflection essentials

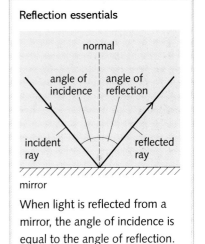

When light is reflected from a mirror, the angle of incidence is equal to the angle of reflection.

You will need a ruler, protractor, and sharp pencil.

1 In the diagram on the right, two rays leave a point object O and strike a plane mirror.
   a) Make an exact copy of the diagram.
   b) Measure the angle of incidence of each ray.
   c) Draw in the two reflected rays at the correct angles.
   d) Find where the image is formed and label it.
   e) By drawing or calculation, work out what angle the mirror would need to be turned through so that ray B is reflected back the way it came.

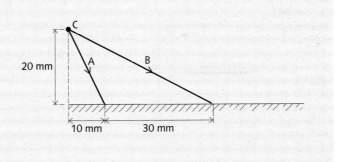

7.04 Refraction of light

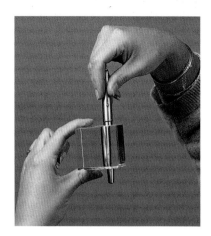

The 'broken pen' illusion on the left occurs because light is bent by the glass block. The bending effect is called **refraction**.

The diagram below shows how a ray of light passes through a glass block. The line at right-angles to the side of the block is called a **normal**. The ray is refracted towards the normal when it enters the block, and away from the normal when it leaves it. The ray emerges parallel to its original direction (provided the block has parallel sides).

Refraction would also occur if the glass were replaced with another transparent material, such as water or acrylic plastic, although the angle of refraction would be slightly different. The material that light is travelling through is called a **medium**.

Definitions

Angle of incidence: this is the angle between the incident ray and the normal.

Angle of refraction: this is the angle between the refracted ray and the normal.

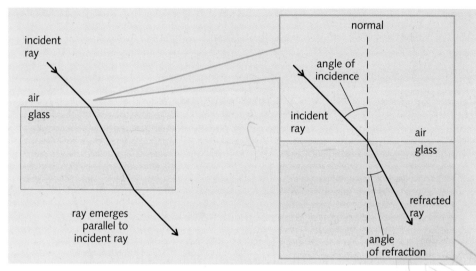

Real and apparent depth

Because of refraction, water (or glass) looks less deep than it really is. Its apparent depth is less than its real depth. This diagram shows why:

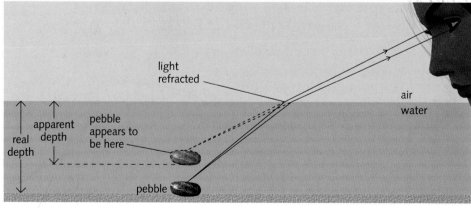

Why light is refracted

Scientists explain refraction as follows. Light is made up of tiny waves. These travel more slowly in glass (or water) than in air. When a light beam passes from air into glass, as shown on the left, one side of the beam is slowed before the other. This makes the beam 'bend'.

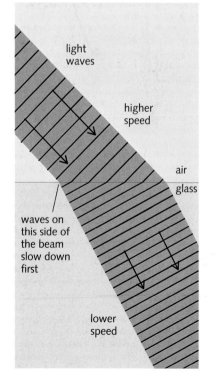

Refractive index

In a vacuum (empty space), the speed of light is 300 000 km/s. In air, it is effectively the same. However, in glass, light slows to 200 000 km/s.

The **refractive index** of a medium is defined like this:

$$\text{refractive index} = \frac{\text{speed of light in vacuum}}{\text{speed of light in medium}}$$

So, in the case of glass:

$$\text{refractive index} = \frac{300\,000\,\text{km/s}}{200\,000\,\text{km/s}} = 1.5$$

Some refractive index values are given on the right. The medium with the *highest* refractive index has the *greatest* bending effect on light because it slows the light the *most*.

medium	refractive index
diamond	2.42
glass (crown)	1.52
acrylic plastic (Perspex)	1.49
water	1.33

The above figures are based on more accurate values of the speed of light than those used on the left.

The refractive index of glass varies depending on the type of glass. Refractive index also varies slightly depending on the colour of the light.

Refraction by a prism

A **prism** is a triangular block of glass or plastic. The sides of a prism are not parallel. So, when light is refracted by a prism, it comes out in a different direction. It is **deviated**.

If a narrow beam of white light is passed through a prism, it splits into a range of colours called a **spectrum**, as shown below. The effect is called **dispersion**. It occurs because white is not a single colour but a mixture of all the colours of the rainbow. The prism refracts each colour by a different amount.

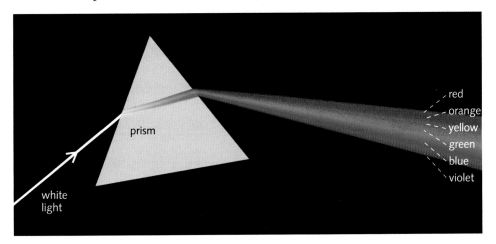

◀ Most people think that they can see about six colours in the spectrum of white light. However, the spectrum is really a continuous change of colour from beginning to end.

Red light is deviated (bent off-course) least by a prism. Violet light is deviated most. However, in this diagram, the difference has been exaggerated.

For questions 1b and 3, you will need to refer to the table at the top of the page. Assume that the speed of light in a vacuum is 300 000 km/s.

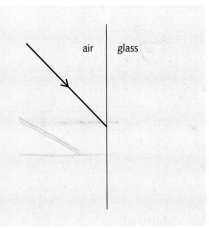

1 a) Copy the diagram on the right. Draw in and label the *normal*, the *refracted ray*, the *angle of incidence*, and the *angle of refraction*.
b) How would your diagram be different if the ray was passing into water rather than glass?
2 a) When white light passes through a prism, it spreads into a spectrum of colours. What is the spreading effect called?
b) Which colour is deviated most by a prism?
c) Which colour is deviated least?
3 Calculate the speed of light in water.

Related topics: refraction of waves **6.02**; colour and wave length **7.01**; light waves **7.01**; refraction calculations **7.06**

7.05 Total internal reflection

The inside surface of water, glass, or other transparent material can act like a perfect mirror, depending on the angle at which the light strikes it.

The diagrams below show what happens to three rays leaving an underwater lamp at different angles. Angle c is called the **critical angle**. For angles of incidence greater than this, there is no refracted ray. All the light is reflected. The effect is called **total internal reflection**.

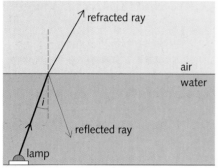

i = angle of incidence

The ray splits into a refracted ray and a weaker reflected ray.

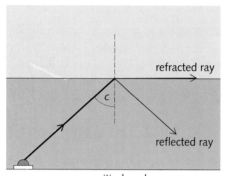

c = critical angle

The rays splits, but the refracted ray only just leaves the surface.

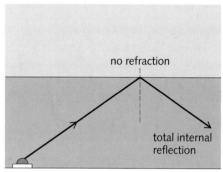

angle of incidence greater than c

There is no refracted ray. The surface of the water acts like a perfect mirror.

The value of the critical angle depends on the material. For example:

critical angle			
water 49°	acrylic plastic 42°	glass (crown) 41°	diamond 24°

Reflecting prisms

In the diagrams below, inside faces of prisms are being used as mirrors. Total internal reflection occurs because the angle of incidence on the face (45°) is greater than the critical angle for glass or acrylic plastic.

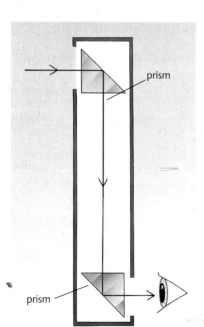

prism

prism

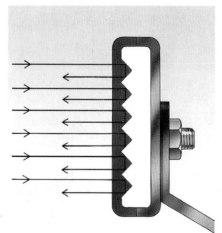

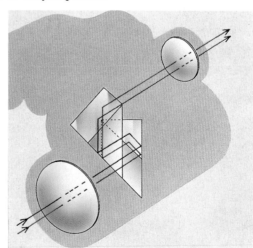

Periscope This is an instrument for looking over obstacles. Prisms reflect the light, although they can be be replaced with mirrors.

Rear reflectors (on cars and cycles) The direction of the incoming light is reversed by two total internal reflections.

Binoculars The lens system in each 'barrel' produces an upside-down image. Reflecting prisms are used to turn it the right way up.

Optical fibres

Optical fibres are very thin, flexible rods made of special glass or transparent plastic. Light put in at one end is total internally reflected until it comes out of the other end, as shown below. Although some light is absorbed by the fibre, it comes out almost as bright as it goes in – even if the fibre is several kilometres long.

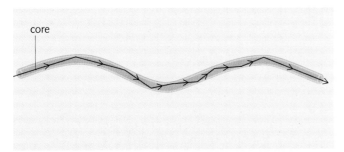

Single optical fibre In the type shown above, the inner glass core is coated with glass of a lower refractive index.

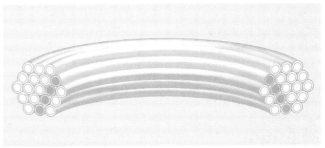

Bundle of optical fibres Provided the fibres are in the same positions at both ends, a picture can be seen through them.

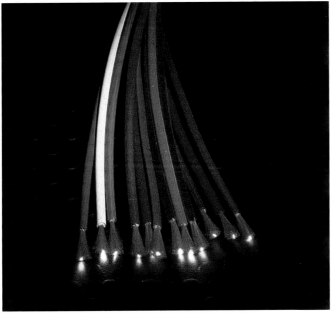

▲ Optical fibres can carry telephone calls. The signals are coded and sent along the fibre as pulses of laser light. Fewer booster stations are needed than with electrical cables.

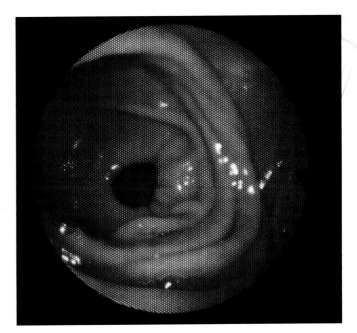

▲ This photograph was taken through an **endoscope**, an instrument used by surgeons for looking inside the body. An endoscope contains a long, thin bundle of optical fibres.

1 Glass has a critical angle of 41°. Explain what this means.

2 **a)** Copy and complete the diagrams on the right to show where each ray will go after it strikes the prism.
   b) If the prisms on the right were transparent triangular tanks filled with water, would total internal reflection still occur? If not, why not?

3 **a)** Give *two* examples of the practical use of optical fibres.
   b) Give *two* other examples of the practical use of total internal reflection.

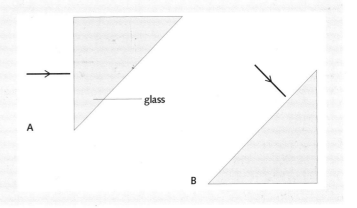

A

glass

B

Related topics: refraction **7.04**; calculating the critical angle **7.06**; optical fibres in communications **7.13**

149

7.06 Refraction calculations

Refraction essentials

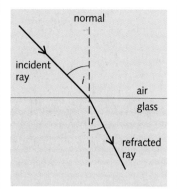

normal

incident ray

i

air

glass

r

refracted ray

A light ray bends as it enters a glass block. The bending effect is called **refraction**. It occurs because light waves slow down when they pass from air into glass or other **medium** (see 7.04). Passing from glass back in to air, they would speed up again. So, if the ray in the diagram were reversed, it would pass back into the air along the same path as it came in.

Measuring refractive index

To find the refractive index of, say, glass, you could direct a ray (from a ray box) at a glass block, mark the positions of the incident and refracted rays, measure their angles, then use the equation on the right. A semi-circular block is useful for experiments like this. If the ray passes through point O below, no bending occurs at the circular face, so it is easier to vary and measure the angles.

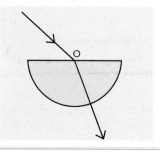

O

Snell's law

When light is refracted, an increase in the angle of incidence i produces an increase in the angle of refraction r. In 1620, the Dutch scientist Willebrord Snell discovered the link between the two angles: their *sines* are always in proportion.

When light passes from one medium into another:

$$\frac{\sin i}{\sin r} \text{ is constant}$$

This is known as **Snell's law**. It is illustrated by these examples:

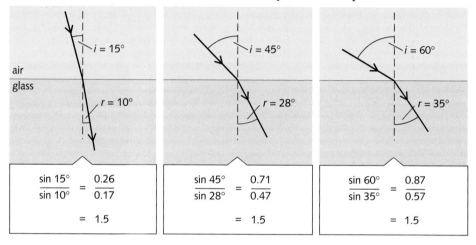

air
glass

$i = 15°$ $r = 10°$

$i = 45°$ $r = 28°$

$i = 60°$ $r = 35°$

$\dfrac{\sin 15°}{\sin 10°} = \dfrac{0.26}{0.17}$	$\dfrac{\sin 45°}{\sin 28°} = \dfrac{0.71}{0.47}$	$\dfrac{\sin 60°}{\sin 35°} = \dfrac{0.87}{0.57}$
$= 1.5$	$= 1.5$	$= 1.5$

Refractive index

The refractive index of a medium is defined like this:

$$\text{refractive index} = \frac{\text{speed of light in vacuum}}{\text{speed of light in medium}}$$

In a vacuum, the speed of light is 300 000 km/s – and effectively the same in air. In glass, it drops to 200 000 km/s. So, the refractive index of glass is 300 000 km/s ÷ 200 000 km/s, which is 1.5. This is the same as the value of $\sin i \div \sin r$ in the diagrams above.

Here is an alternative definition of refractive index:

$$\text{refractive index} = \frac{\sin i}{\sin r}$$

Example Light (in air) strikes water at an angle of incidence of 45°. If the refractive index of water is 1.33, what is the angle of refraction?

Applying the above equation: $1.33 = \dfrac{\sin 45°}{\sin r}$

Rearranged, this gives $\sin r = \sin 45°/1.33$. When calculated, this gives $\sin r = 0.532$. So the angle of refraction r is 32°.

Calculating the critical angle★

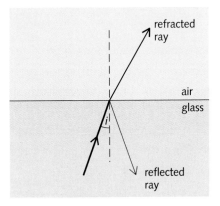

i = angle of incidence

refracted ray

air
glass

reflected ray

c = critical angle

refracted ray

reflected ray

angle of incidence greater than c

no refraction

total internal reflection

In the diagrams above, rays are travelling from glass towards air at different angles. When the angle of incidence is greater than the **critical angle**, there is no refracted ray. All the light is reflected. There is **total internal reflection**.

Knowing the refractive index of a material, the critical angle can be calculated. For example:

On the right, the middle diagram above has been redrawn with the ray direction reversed. This time, the angle of *incidence* is 90°, and angle c is now the angle of *refraction*. If the refractive index of glass is 1.5:

$$\text{refractive index} = \frac{\sin 90°}{\sin c} = \frac{1}{\sin c} \qquad (\text{as } \sin 90° = 1)$$

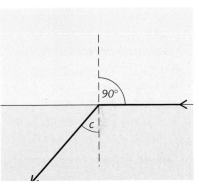

▲ Compare this with the middle diagram at the top of the page.

rearranging: $\quad \sin c = \dfrac{1}{1.5} = 0.67$

so c, the critical angle of glass, = 42°.

Note: this figure differs slightly from that in Spread 7.03 because a simplified value for the refractive index of glass has been used in the calculation.

From the above calculation, it follows that the critical angle c of any medium can be calculated using this equation:

For a medium of refractive index n: $\quad \sin c = \dfrac{1}{n}$

 Q

To answer these questions, you will need a calculator (or set of tables) containing sine values.

1 The refractive index of water is 1.33. Calculate the angle of refraction if light (in air) strikes water at an angle of incidence of **a)** 24° **b)** 53°.
2 A transparent material has a refractive index of 2.0.
  a) Calculate the critical angle.
  b) If the refractive index were less than 2.0, would the critical angle be *greater* or *less* than before?
3 Diamond has a refractive index of 2.42. The speed of light in a vacuum (or in air) is 300 000 km/s. Calculate:
  a) the speed of light in diamond **b)** the critical angle for diamond.

▲ When a diamond is cut, the facets (faces) are angled so that they produce total internal reflection. Reflected light gives the diamond its 'sparkle'.

7.07 Lenses (1)

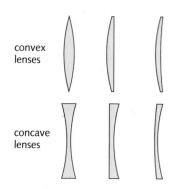

convex lenses

concave lenses

Lenses bend light and form images. There are two main types of lens. The diagram on the left shows some examples of each.

Convex lenses These are thickest in the middle and thin round the edge. When rays parallel to the principal axis pass through a convex lens, they are bent inwards. The point F where they converge (meet) is called the **principal focus**. Its distance from the centre of the lens is the **focal length**. A convex lens is known as a **converging lens**.

Rays can pass through the lens in either direction, so there is another principal focus F′ on the opposite side of the lens and the same distance from it.

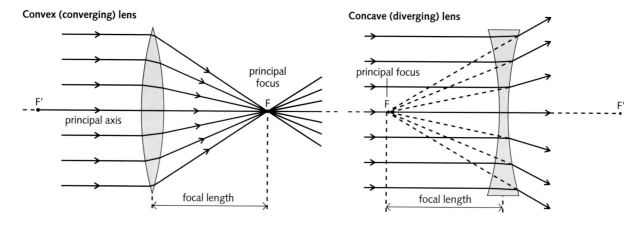

Convex (converging) lens

principal axis

F′

principal focus

F

focal length

Concave (diverging) lens

principal focus

F

F′

focal length

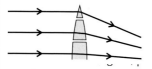

How lenses bend light

or other transparent material. Each section of a lens acts like a tiny prism, refracting (bending) light as it goes in and again as it comes out. Expensive lenses have special coatings to reduce the colour-spreading of the prisms.

Concave lenses★ These are thin in the middle and thickest round the edge. When rays parallel to the principal axis pass through a concave lens, they are bent outwards. The principal focus is the point from which the rays appear to diverge (spread out). A concave lens is a **diverging lens**.

Real images formed by convex lenses

In the diagram below, rays from a very distant object are being brought to a focus by a convex lens. Rays come from all points on the object. However, for simplicity, only a few rays from one point have been shown. Together, the rays form an image which can be picked up on a screen. An image like this is called a **real image**. It is formed in the **focal plane**.

In a camera, a convex lens is used as below to form a real image on a piece of film or CCD. The image in the eye is formed in the same way.

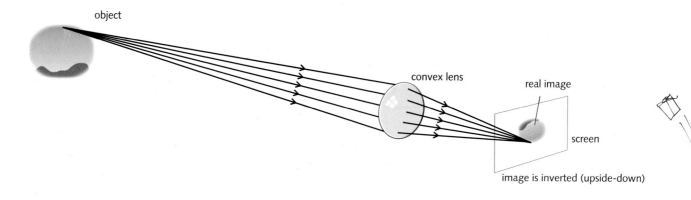

object

convex lens

real image

screen

image is inverted (upside-down)

The rays from a point on a very distant object are effectively parallel, so the image passes through the principal focus. However, for an object at any other distance, the image is in a different position.

You can predict where a convex lens will form an image by drawing a **ray diagram**. There are two examples below. Each has these features:

- For simplicity, rays are drawn from just one point on the object.
- The rays used are the **standard rays** described on the right. These are chosen because it is easy to work out where they go. Only two of them are needed to find where the image is.
- For simplicity, rays are shown bending at the line through the middle of the lens. In reality, bending takes place at each surface.

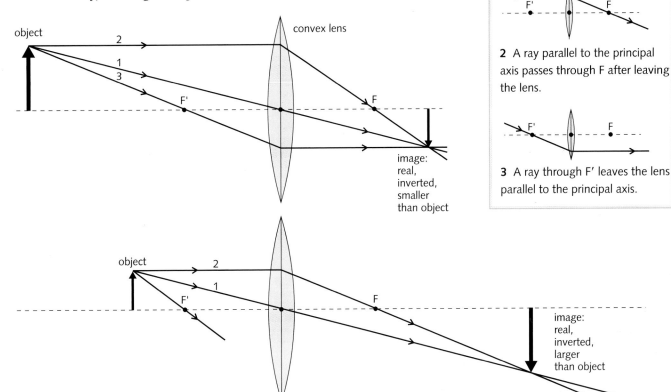

The ray diagrams above show that as the object is moved towards the lens, the image becomes bigger and further away.

A film projector uses a convex lens to form a magnified, real image on a screen a long way away from it, as in the lower diagram.

Standard rays

In ray diagrams, any two of the following rays are needed to fix the image position and size:

1 A ray through the centre passes straight through the lens.

2 A ray parallel to the principal axis passes through F after leaving the lens.

3 A ray through F′ leaves the lens parallel to the principal axis.

Q

1. **a)** Which of the lenses on the right is a convex lens?
   b) Which one is a converging lens?
   c) What is meant by the *principal focus* of the convex lens?
   d) What is meant by the *focal length* of the convex lens?
2. **a)** If a convex lens picks up rays from a very distant object, where is the image formed?
   b) If the object is moved towards the lens, what happens to the position and size of the image?
3. Draw a ray diagram like one of those above, but with the object exactly 2 × focal length away from the lens. Draw in and describe the image.

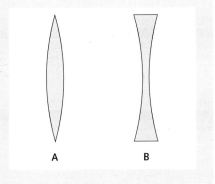

Related topics: mirrors **7.04**; refraction by a prism **7.04**; camera, projector, and enlarger **7.09**; the human eye **7.10**

153

7.08 Lenses (2)

Convex lens as a magnifying glass

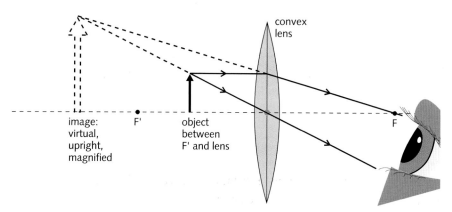

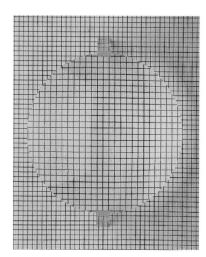

▲ Thick, bulging convex lenses have the shortest focal lengths and make the most powerful magnifying glasses.

Thin convex lenses have longer focal lengths and are much less powerful.

If an object is closer to a convex lens than the principal focus, the rays never converge. Instead, they appear to come from a position behind the lens. The image is upright and magnified. It is called a **virtual image** because no rays actually meet to form it and it cannot be picked up on a screen. Used like this, a convex lens is often called a **magnifying glass**.

Drawing accurate ray diagrams

Problems like the one below can be solved by doing a ray diagram as an accurate scale drawing on graph paper:

Example An object 2 cm high stands on the principal axis at a distance of 9 cm from a convex lens. If the focal length of the lens is 6 cm, what is the image's position, height, and type?

For accuracy, you need to choose a scale that makes the diagram as large as possible. In the drawing below, 1 cm on the paper represents 2 cm of actual distance. When the final measurements are scaled up, they show that the image is 18 cm from the lens, 4 cm high, and real.

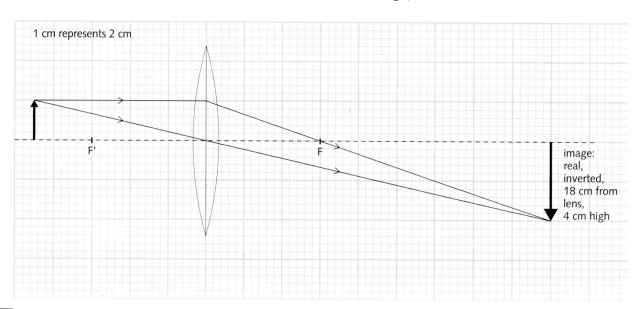

1 cm represents 2 cm

image: real, inverted, 18 cm from lens, 4 cm high

Measuring the focal length of a convex lens★

You can find an approximate value for the focal length of a convex lens by forming an image of a distant window (or other distant bright object) on a screen. Rays from the window are almost parallel, so the image is close to the principal focus of the lens. Therefore the distance from the image to the lens is approximately the same as the focal length.

For a more accurate measurement, you can use the method shown on the right. Here, the object is a set of illuminated crosswires. Its position is adjusted until a clear image is picked up alongside it on the card. As the rays are being reflected back to the position they came from, they must be parallel when they strike the mirror. So the crosswires must be exactly at the principal focus of the lens.

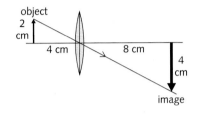

Linear magnification★

In the diagram on the right, the image distance is *twice* the object distance. As the two triangles formed are similar, the sizes of the image and object are in the same proportion as their distances. So, the image height is *twice* the object height. The **linear magnification** in this case is 2:

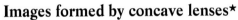

$$\text{linear magnification} = \frac{\text{image height}}{\text{object height}} = \frac{\text{image distance}}{\text{object distance}}$$

This equation applies whatever the type of lens or image.

Images formed by concave lenses★

In the diagram below, two standard rays have been used to show how a concave lens forms an image. Wherever the object is positioned, the image is always small, upright, and virtual.

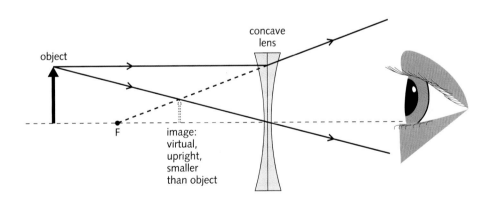

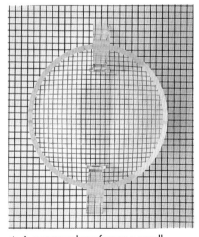

▲ A concave lens forms a small, upright, virtual image.

Q

1 a) An object 2 cm high is placed 12 cm away from a convex lens of focal length 6 cm. By doing an accurate drawing on graph paper, find the position, height, and type of image.
b) The object is moved so that it is only 10 cm away from the lens. Use another drawing to find the new position, height, and type of image.
c) What is the linear magnification in part b?

2 Where should the object be placed if the image formed by a convex lens is to be
a) virtual, and larger than the object?
b) real, and the same size as the object?
c) real, and larger than the object?

3 Draw a labelled diagram to show how you could quickly find an approximate value for the focal length of a convex lens.

Related topics: virtual image **7.02**; focal length and ray diagrams **7.07**

7.09 Camera, projector, and enlarger★

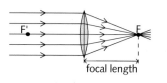
The camera

A camera uses a convex lens to form a small, inverted, real image at the back.

The lens is moved in or out to make focusing adjustments. In many cameras, this happens automatically. In cheaper cameras, the lens is fixed.

The shutter opens and shuts quickly to let a small amount of light into the camera. On some cameras, the speed of the shutter can be adjusted.

The film is kept in darkness until the shutter opens. It is coated with light-sensitive chemicals that are changed by the different shades and colours in the image. When the film is processed, these changes are 'fixed', and the film used in printing the photograph. Digital cameras have a **CCD** (charge-coupled device) instead of film. This is a light-sensitive microchip that captures the image electronically.

The diaphragm is a set of sliding plates between the lens and the film. It controls the **aperture** (diameter) of the hole through which the light passes. In bright sunshine, a small aperture might be used to cut down the amount of light reaching the film or CCD. Many cameras have automatic aperture adjustment. Cheaper cameras usually have a fixed aperture.

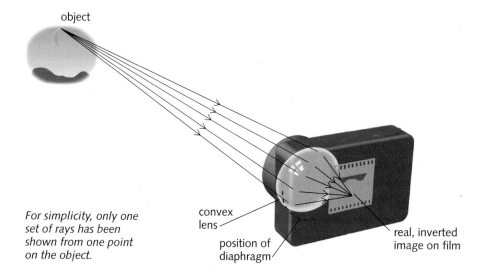

For simplicity, only one set of rays has been shown from one point on the object.

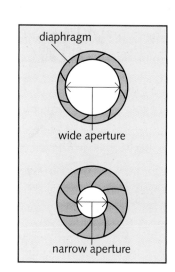

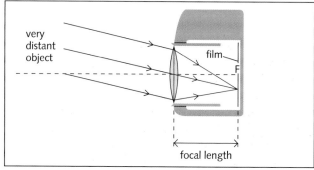

▲ For very distant objects, the film needs to be at the principal focus F of the lens.

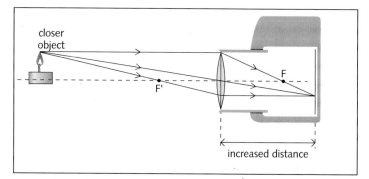

▲ For closer objects, the lens must be moved further away from the film.

The projector

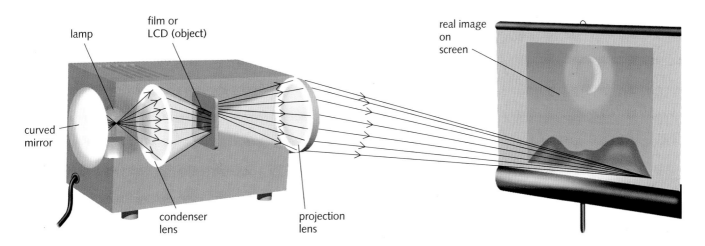

A projector uses a convex lens to form a large, inverted, real image on a screen. The object is a brightly lit piece of film or LCD with a picture on it.

The projection lens forms the image on the screen. To get a large image, the lens has to be a long way from the screen. The lens is moved backwards or forwards in its holder to make focusing adjustments.

The film or LCD must be upside down to get an upright picture on the screen. As the image is large and distant, the film or LCD must be positioned just outside the principal focus F′ of the projection lens.

The condenser lens concentrates light on the film or LCD so that it is very bright and evenly lit.

The enlarger

An enlarger is used when photographic prints are being made. Its job is to produce a printed image that is much larger than the one on the film. It magnifies in a similar way to a projector.

You can see the basic layout of an enlarger on the right. The piece of film is evenly illuminated by the lamp above it. The projection lens forms a magnified image on the photographic paper at the bottom.

As the paper is light sensitive, it must only be exposed to light for a short time to capture the image. So the enlarger must be used in a darkroom. Some enlargers have a timed shutter to control the amount of light passing through the film.

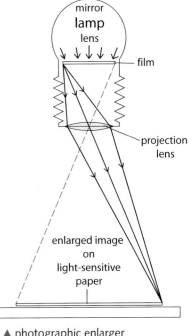

▲ photographic enlarger

1 Give *two* ways in which the amount of light entering a camera can be controlled.
2 An object moves closer to a camera. Which way must the lens be moved to keep the image in focus?
3 In a projector, what job is done by
   a) the condenser lens
   b) the projection lens?
4 Why must the film or transparency be put into a projector upside-down?

5 Someone is using a projector when they decide to move the screen closer and re-focus the image.
   a) How will the size of the image be affected?
   b) Which way must the lens be moved?
6 Someone is using an enlarger. If they want to print a *bigger* picture on photographic paper, say whether they should increase or decrease each of these:
   a) The distance between the paper and lens
   b) The distance between the film and lens.

7.10 The human eye★

Like a camera, the human eye uses a convex lens system to form a small, inverted, real image of an object in front of it.

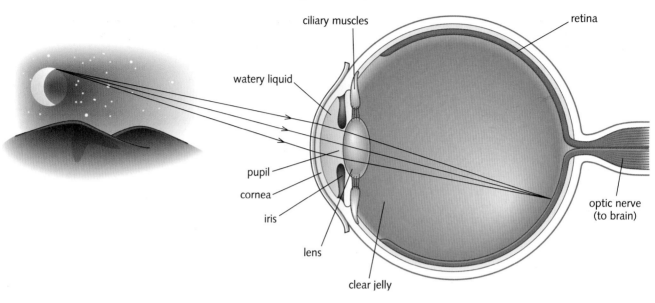

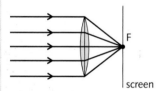

The cornea is a curved 'window' over the front of the eye. The cornea and the watery liquid behind do most of the converging of the light.

The lens is used to make focusing adjustments: the process is called **accommodation**. The lens does not move backwards and forwards as in a camera. Instead, it is flexible, and its shape is changed by the ring of **ciliary muscles** around it. When an object moves closer to the eye, the lens thickens so that the image stays in focus on the back.

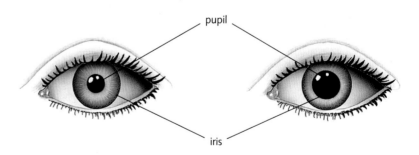

The iris is the bit that makes your eyes brown or blue. Its job is to control the amount of light entering the eye. The light passes through a gap in the middle, called the **pupil**. If you walk into a dark room, the pupil automatically becomes bigger to let in more light.

The retina is the 'screen' on the back of the eye where the image is formed. It contains over 100 million light-sensitive cells. These react to light by sending nerve impulses (electrical signals) along the **optic nerve** to the brain. The brain uses the nerve impulses to form a view of the outside world.

The image on the retina is upside-down. However, the brain is so used to this view, that it thinks of it as the right way up!

Convex lens essentials

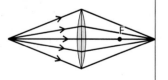

Rays from a distant point are effectively parallel. A convex lens makes them converge at its **principal focus** F.

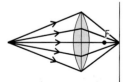

If the rays come from a closer point, the lens makes them converge further away from it.

If a thicker lens is used, its focal length is shorter, so the rays converge nearer the lens.

The image formed when rays converge is called a **real image**.

Correcting defects in vision

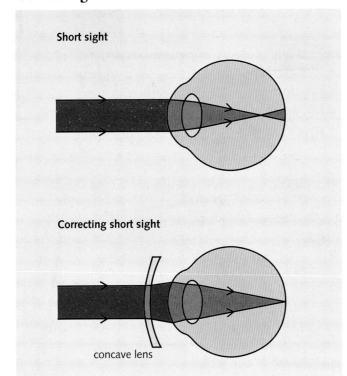

Short sight

Correcting short sight

concave lens

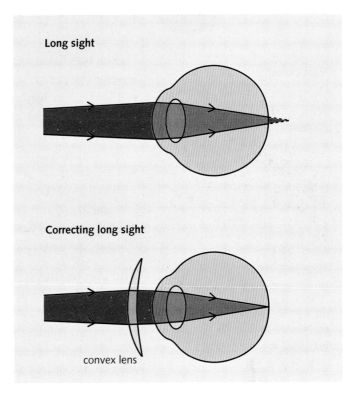

Long sight

Correcting long sight

convex lens

With many people, changes in the shape of the eye lens are not enough to produce sharp focusing on the retina. To overcome the problem, spectacles or contact lenses have to be worn.

Short sight In a short-sighted eye, the lens cannot be made thin enough for looking at distant objects. So the rays are bent inwards too much. They converge before they reach the retina. To correct the fault, a *concave* (diverging) lens is placed in front of the eye.

Long sight In a long-sighted eye, the lens cannot be made thick enough for looking at close objects. So the rays are not bent inwards enough. When they reach the retina, they have still not met. To correct the fault, a *convex* (converging) lens is placed in front of the eye.

From middle age onwards, the eye lens becomes less flexible and loses its ability to accommodate for objects at different distances. To overcome this difficulty, some people wear **bifocals** – spectacles whose lenses have a top part for looking at distant objects and a bottom part for close ones. If the spectacles have **progressive** lenses, the change is continuous from top to bottom.

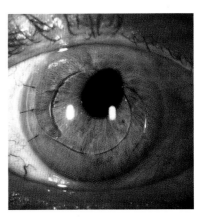

▲ This person's eye has been fitted with a plastic lens because the natural lens had developed too many cloudy patches, called cataracts.

1 Which part of the eye
   a) is the 'screen' on which the image is formed
   b) does most of the converging of the light
   c) controls the amount of light entering the eye?
2 If an object moves closer to the eye, the lens has to adjust to refocus the image.
   a) What is this process called?
   b) How does the lens make focusing adjustments?

3 A short-sighted person cannot see distant objects clearly. Why not?
4 A long-sighted person cannot see close objects clearly. Why not?
5 What type of spectacle lens or contact lens is needed to correct for
   a) short sight
   b) long sight?

Related topics: convex and concave lenses, real images, ray diagrams **7.07–7.08**, camera **7.09**

7.11 # Electromagnetic waves (1)

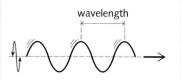
Atom

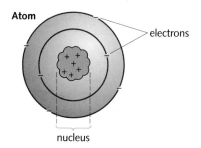
electrons
nucleus

In an atom, the electrons have negative (−) charge and the nucleus has positive (+) charge. Electromagnetic waves are emitted whenever charged particles oscillate or lose energy.

Wave equation

For any set of moving waves:

speed = frequency × wavelength
(m/s)   (Hz)   (m)

If the speed of the waves is unchanged, an increase in frequency means a decrease in wavelength, and vice versa.

1000 Hz = 1 kilohertz (kHz)

1 000 000 Hz = 1 megahertz (MHz)

Light waves belong to a whole family of **electromagnetic waves**. These have several features in common. For example:

- They can travel through a vacuum (for example, space).
- They travel through a vacuum at a speed of 300 000 kilometres per second. This is usually called the **speed of light**, although it is the speed of all electromagnetic waves.
- They are transverse waves – their oscillations are at right-angles to the direction of travel. It is electric and magnetic fields that are oscillating, not material.
- They transfer energy. A source loses energy when it radiates electromagnetic waves. A material gains energy when it absorbs them.

The electromagnetic spectrum

The full range of electromagnetic waves is called the **electromagnetic spectrum**. It is shown in the chart on the opposite page. The range of wavelengths is huge. At one end are the longest radio waves with wavelengths of several kilometres. At the other end are the shortest gamma rays with wavelengths of less than one-billionth of a millimetre.

Where electromagnetic waves come from

All matter is made of atoms. Atoms are themselves made up of a central **nucleus** with tiny particles called **electrons** orbiting around it. The nucleus and the electrons are electrically charged. Sometimes, electrons can escape from their atoms. For example, when an electric current passes through a wire, the current is a flow of free electrons.

Electromagnetic waves are emitted (sent out) whenever charged particles oscillate or lose energy in some way. For example, the vibrating atoms in a hot, glowing bulb filament emit infrared and light, and an oscillating electric current emits radio waves. The higher the frequency of oscillation, or the greater the energy change, the shorter the wavelength of the electromagnetic waves produced.

You may need information from the next spread, 7.12.

1 Give *three* properties (features) common to all electromagnetic waves.
2 Put the following in order of wavelength, starting with the longest:
    ultraviolet   X-rays   red light   violet light   microwaves   infrared
3 Name a type of electromagnetic radiation that
    a) is visible to the eye   **b)** is emitted by hot objects
    c) is diffracted by hills   **d)** can cause fluorescence
    e) is used for radar   **f)** can pass through dense metals.
4 A VHF radio station emits radio waves at a frequency of 100 MHz.
    a) What is the frequency in Hz?
    b) What is the wavelength? (speed of radio waves = 3×10^8 m/s)
    c) What is the wavelength of radio waves from a long-wave transmitter, broadcasting at a frequency of 200 kHz?

The electromagnetic spectrum

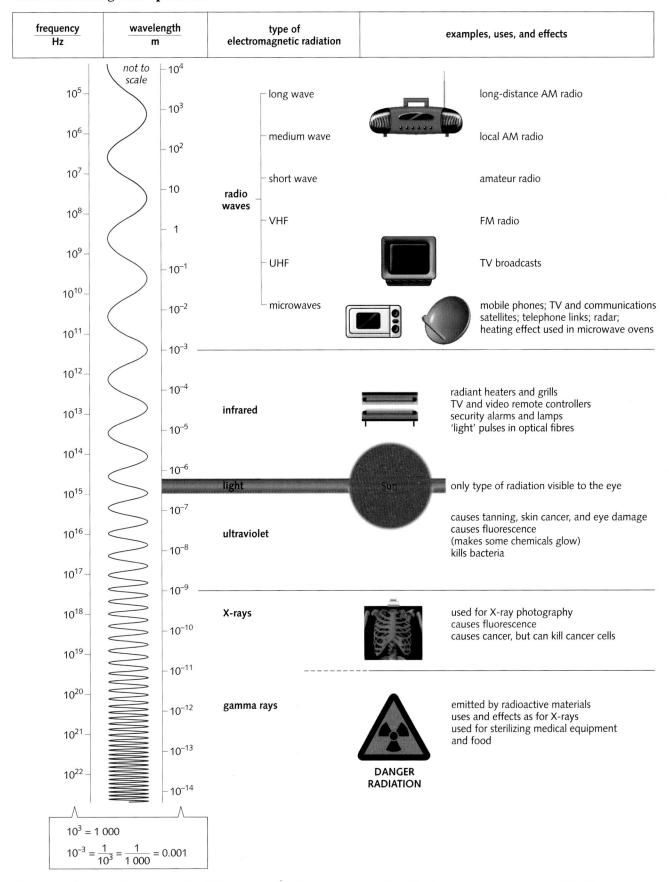

frequency Hz	wavelength m	type of electromagnetic radiation	examples, uses, and effects

$10^3 = 1\ 000$

$10^{-3} = \dfrac{1}{10^3} = \dfrac{1}{1\ 000} = 0.001$

For more information about the different types of electromagnetic radiation, see the next spread, 7.12.

Related topics: thermal radiation **5.08**; transverse waves, frequency and wavelength **6.01**; radar **6.04**; light waves **7.01**; light spectrum **7.04**; gamma rays **7.12** and **11.02**; atoms and electric charge **8.01**

Electromagnetic waves (2)

7.12

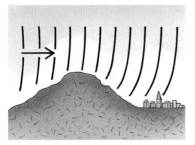

Radio waves of long and medium wavelengths diffract (bend) round hills.

This dish receives microwaves from a satellite.

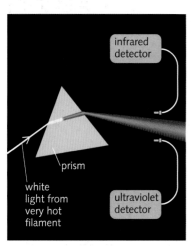

▲ Infrared and ultraviolet can be detected just beyond the two ends of the visible part of the spectrum.

Radio waves

Stars are natural emitters of radio waves. However, radio waves can be produced artificially by making a current oscillate in a transmitting aerial (antenna). In a simple radio system, a microphone controls the current to the aerial so that the radio waves 'pulsate'. In the radio receiver, the incoming pulsations control a loudspeaker so that it produces a copy of the original sound. Radio waves are also used to transmit TV pictures.

Long and medium waves will diffract (bend) around hills, so a radio can still receive signals even if a hill blocks the direct route from the transmitting aerial. Long waves will also diffract round the curved surface of the Earth.

VHF and UHF waves have shorter wavelengths. VHF (very high frequency) is used for stereo radio and UHF (ultra high frequency) for TV broadcasts. These waves do not diffract round hills. So, for good reception, there needs to be a straight path between the transmitting and receiving aerials.

Microwaves have the shortest wavelengths (and highest frequencies) of all radio waves. They are used by mobile phones, and for beaming TV and telephone signals to and from satellites and across country.

Like all electromagnetic waves, microwaves produce a heating effect when absorbed. Water absorbs microwaves of one particular frequency. This principle is used in microwave ovens, where the waves penetrate deep into food and heat up the water in it. However, if the body is exposed to microwaves, they can cause internal heating of body tissues.

Infrared radiation and light

When a radiant heater or grill is switched on, you can detect the infrared radiation coming from it by the heating effect it produces in your skin. In fact, *all* objects emit some infrared because of the motion of their atoms or molecules. Most radiate a wide range of wavelengths.

As an object heats up, it radiates more and more infrared, and shorter wavelengths. At about 700 °C, the shortest wavelengths radiated can be detected by the eye, so the object glows 'red hot'. Above about 1000 °C, the whole of the visible spectrum is covered, so the object is 'white hot'.

Short-wavelength infrared is often called 'infrared light', even though it is invisible. However, strictly speaking, light is just the part of the electromagnetic spectrum that is visible to the eye.

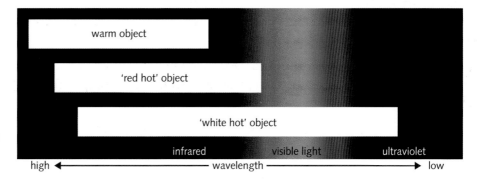

Security alarms and lamps can be switched on by motion sensors that pick up the changing pattern of infrared caused by an approaching person. At night, photographs can be taken using infrared. In telephone networks, signals are sent along optical fibres as pulses of infrared 'light'. And remote controllers for TVs and video recorders work by transmitting infrared pulses.

Ultraviolet radiation

Very hot objects, such as the Sun, emit some of their radiation beyond the violet end of the visible spectrum. This is ultraviolet radiation. It is sometimes called 'ultraviolet light', even though it is invisible.

The Sun's ultraviolet is harmful to living cells. If too much penetrates the skin, it can cause skin cancer. If you have a black or dark skin, the ultraviolet is absorbed before it can penetrate too far. But with a fair skin, the ultraviolet can go deeper. Skin develops a tan to try to protect itself against ultraviolet. Ultraviolet can also damage the retina in the eye and cause blindness.

As ultraviolet is harmful to living cells, it is used in some types of sterilizing equipment to kill bacteria (germs).

Fluorescence Some materials fluoresce when they absorb ultraviolet: they convert its energy into visible light and glow. In fluorescent lamps, the inside of the tube is coated with a white powder which gives off light when it absorbs ultraviolet. The ultraviolet is produced by passing an electric current through the gas (mercury vapour) in the tube.

X-rays

X-rays are given off when fast-moving electrons lose energy very quickly. For example, in an X-ray tube, the radiation is emitted when a beam of electrons hits a metal target. Short-wavelength X-rays are extremely penetrating. A dense metal like lead can reduce their strength, but not stop them. Long-wavelength X-rays are less penetrating. For example, they can pass through flesh but not bone, so bones will show up on an X-ray photograph. In engineering, X-rays can be used to take photographs that reveal flaws inside metals – for example faulty welds in pipe joints. Airport security systems also use them to detect any weapons hidden in luggage.

All X-rays are dangerous because they damage living cells deep in the body and can cause cancer or mutations (genetic change). However, concentrated beams of X-rays can be used to *treat* cancer by destroying abnormal cells.

Gamma rays

Gamma rays come from radioactive materials. They are produced when the nuclei of unstable atoms break up or lose energy. They tend to have shorter wavelengths than X-rays because the energy changes that produce them are greater. However, there is no difference between X-rays and gamma rays of the same wavelength.

Like X-rays, gamma rays can be used in the treatment of cancer, and for taking X-ray-type photographs. As they kill harmful bacteria, they are also used for sterilizing food and medical equipment.

For questions, see the previous spread, 7.11.

Sunbeds use ultraviolet to cause tanning in some types of skin.

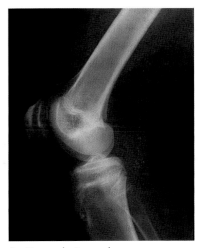

An X-ray photograph

Ionizing radiations

Ultraviolet, X-rays, and gamma rays cause **ionization** – they strip electrons from atoms in their path. The atoms are left with an electric charge, and are then known as **ions**.

Ionization is harmful because it can kill or damage living cells, or make them grow abnormally as cancers.

Related topics: infrared and thermal radiation **5.08**; diffraction **6.02**; light spectrum **7.04**; optical fibres **7.05** and **7.13**; X-ray tube **10.08**; radioactivity, gamma rays, and ionization **11.02**

7.13 Communications★

Telephone, radio, and TV are all forms of telecommunication – ways of transmitting information over long distances. The information may be sounds, pictures, or computer data. The diagram below left shows a simple telephone system. An **encoder** (the microphone) turns the incoming information (speech) into a form which can be transmitted (electrical signals). The signals pass along the **transmission path** (wires) to a **decoder** (the earphone). This turns the signals back into useful information (speech).

Other telecommunication systems use different types of signal and transmission path. The signals may be changes in voltage, changes in the intensity of a beam of light, or changes in the strength or frequency of radio waves. They may be transmitted using wires, optical fibres, or radio waves.

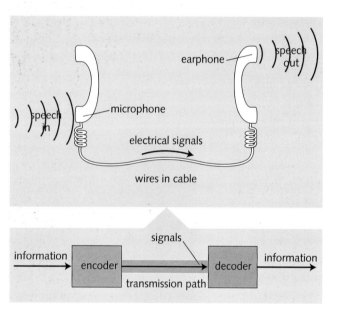

▲ Like all telecommunications systems, a simple telephone system sends signals from a coder to a decoder.

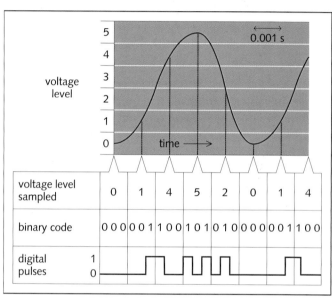

▲ How an analogue signal is converted into digital pulses. Real systems use hundreds of levels and a much faster sampling rate.

Analogue and digital transmission

The sound waves entering a microphone make the voltage across it vary – as shown in the graph above right. A continuous variation like this is called an **analogue signal**. The table shows how it can be converted into **digital signals** – signals represented by numbers. The original signal is **sampled** electronically many times per second. In effect, the height of the graph is measured repeatedly, and the measurements changed into **binary codes** (numbers using only 0's and 1's). These are transmitted as a series of pulses and turned back into an analogue signal at the receiving end.

Advantages of digital transmission Signals lose power as they travel along. This is called **attenuation**. They are also spoilt by **noise** (electrical interference). To restore their power and quality, digital pulses can be 'cleaned up' and amplified at different stages by **regenerators**. Analogue signals can also be amplified, but the noise is amplified as well, so the signals are of lower quality when they reach their destination.

Optical fibres

For long-distance transmission, telephone networks often use **optical fibres**. These are long, thin strands of glass which can carry digital signals in the form of pulses of light. At the transmitting end, electrical signals are encoded into light signals by an **LED** (light-emitting diode) or a **laser diode**. At the receiving end, the light signals are decoded by a **photodiode** which turns them back into electrical signals. Optical fibre cables are thinner and lighter than electric cables. They carry more signals and with less attenuation. They are not affected by electrical interference, and cannot be 'tapped'.

Optical fibres

Storing and retrieving information

When you listen to a recording, the music is being recreated electronically from stored information. Here are some of the methods used for storing and retrieving (getting back) information of this type:

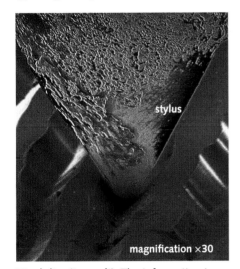

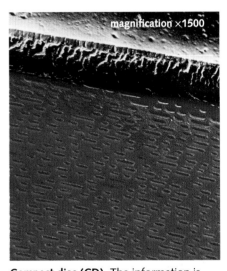

Vinyl disc ('record') The information is recorded as a long, wavy-sided groove on the surface of the disc. To retrieve it, the disc is rotated so that a stylus travels along the groove. The wavy sides make the stylus vibrate and the vibrations are turned into electrical signals. They are analogue signals.

Magnetic tape This is usually on spools in a cassette. The information is recorded as a pattern of varying magnetism along the tape. To retrieve it, the tape is pulled past a tiny coil, and the motion generates electrical signals in the coil. The recording is normally analogue, though it can be digital.

Compact disc (CD) The information is recorded digitally as a sequence of microscopic bumps on a metal layer inside the disc. To retrieve it, the disc is rotated and laser light is reflected from the bumps. The reflected pulses are picked up by a photodiode and turned into electrical signals.

Q

1 The diagram on the right shows part of a telephone system.
   a) In what form do the signals travel along the fibre?
   b) What does the laser diode do?
   c) What does the photodiode do?
   d) What does the regenerator do?
   e) Give *two* advantages of sending digital signals rather than analogue ones.
   f) Give *two* advantages of using an optical fibre link rather than a cable with wires in it.
2 Give an example of information being stored
   a) in digital form b) in analogue form.

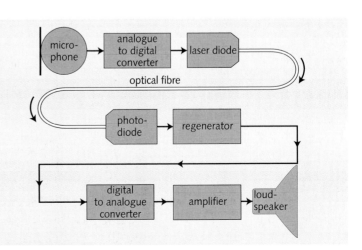

Related topics: sound waves **6.03**; optical fibres **7.04**; magnetic storage **9.04**; signals **10.01**; analogue and digital **10.01**; LEDs **10.01**

1 The diagram shows a light signal travelling through an optical fibre made of glass.

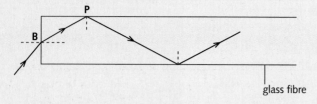

glass fibre

a) State **two** changes that happen to the light when it passes from air into the glass fibre at **B**. [2]
b) Explain why the light follows the path shown after hitting the wall of the fibre at **P**. [2]

WJEC

2 Lenses are used in many optical devices.
a) Copy and complete the table below about the images formed by some optical devices. [6]

optical device	nature of image	size of image	position of image
eye	real		
projector		magnified	
magnifying glass			further from lens than the object

b) An object is placed closer to a converging (convex) lens than its principal focus. The figure shows an incomplete ray diagram for the formation of the image.

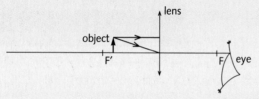

(i) Copy and complete the ray diagram and draw the image formed.
(ii) Use the ray diagram to help you describe **three** properties of this image formed by the lens. [7]

NEAB

3 a) Copy the diagram and draw the path of the ray of yellow light as it passes through and comes out of the glass prism. [2]

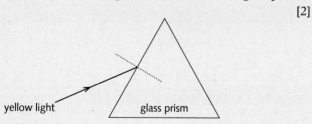

yellow light   glass prism

b) What do we call this effect? [1]
c) State why light changes direction when it enters a glass prism. [1]

WJEC

4 The figure shows an object OB in front of a converging lens. The principal foci of the lens are labelled F and F'. An image of OB will be formed to the right of the lens.

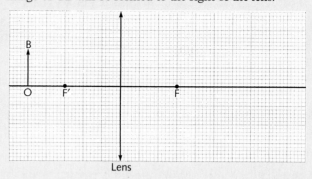

Lens

a) Copy the figure and draw two rays from the top of the object B which pass through the lens and go to the image. [2]
b) Draw the image formed. Label this image I. [1]
c) Calculate the linear magnification produced by the lens. [3]

NEAB

5

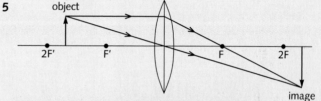

The diagram shows a converging lens forming a real image of an illuminated object.
State **two** things that happen to the image when the object is moved towards F'. [2]

WJEC

6

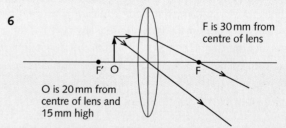

F is 30 mm from centre of lens

O is 20 mm from centre of lens and 15 mm high

The diagram shows an object **O** placed in front of a convex (converging) lens and the passage of two rays from the top of the object through the lens.
a) Copy and complete the diagram (using the dimensions given) to show where the image is formed. [1]
b) State **two** properties of the image. [2]
c) Use the information from the completed diagram and the equation

$$linear\ magnification = \frac{height\ of\ image}{height\ of\ object}$$

to calculate the magnification produced by the lens. [3]

WJEC

7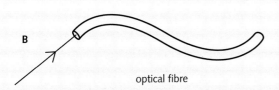

In the diagrams above, rays of light strike a mirror and one end of an optical fibre.

a) Copy and complete the diagrams to show what will happen to each of the rays. [2]

b) Which diagram shows an example of total internal reflection? [1]

c) Give two practical uses of optical fibres. [2]

d) The light in each ray is **monochromatic**. What does this mean? [1]

8 A ray of light, in air, strikes one side of a rectangular glass block. The refractive index of the glass is 1.5.

a) Draw a diagram to show the direction the ray will take in the glass if the angle of incidence is 0°. [2]

b) Draw a diagram to show the approximate direction the ray will take in the glass if the angle of incidence is 45°, and calculate the angle of refraction. [4]

c) If the speed of light in air is 3×10^8 m/s, calculate the speed of light in the glass. [2]

9 Light and gamma rays are both examples of electromagnetic radiation.

a) Name three other types of electromagnetic radiation. [3]

b) State two differences between light and gamma rays. [2]

c) The speed of light is 3×10^8 m/s. Calculate the frequency of yellow light of wavelength 6×10^{-7} m. [2]

10

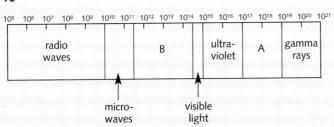

The diagram shows the main regions of the electromagnetic spectrum. The numbers show the frequencies of the waves measured in hertz (Hz).

a) Name the regions
  (i) **A**, [1]
  (ii) **B**. [1]

b) (i) Write down, **in words**, the equation connecting wave speed, wavelength and wave frequency. [1]
  (ii) Calculate the frequency of the radiation with a wavelength of 0.001 m (10^{-3} m), given that all electromagnetic waves travel at a speed of 300 000 000 m/s (3×10^8 m/s) in space. [2]
  (iii) State to which part of the electromagnetic spectrum the radiation in part (ii) belongs. [1]

c) Explain how and why microwaves can cause damage to or even kill living cells. [2]

WJEC

11 The figure shows a square block of glass JKLM with a ray of light incident on side JK at an angle of incidence of 60°. The refractive index of the glass is 1.50.

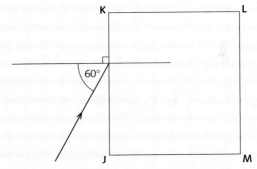

a) Calculate the angle of refraction of the ray. [2]

b) Calculate the critical angle for a ray of light in this glass. [2]

c) Explain why the ray shown cannot emerge from side KL but will emerge from side LM. [3]

UCLES

12 a)

less than	the same as	greater than

Copy the sentences below and use **one** of the three phrases above to complete each sentence. Each phrase may be used once, more than once or not at all.

  (i) The wavelength of radio waves is _____ the wavelength of ultraviolet radiation. [1]
  (ii) In a vacuum the speed of ultraviolet radiation is _____ the speed of light. [1]
  (iii) The frequency of ultraviolet radiation is _____ the frequency of infrared radiation. [1]

b) Name the part of the electromagnetic spectrum that is used to:
  (i) send information to and from satellites; [1]
  (ii) kill harmful bacteria in food. [1]

WJEC

Photocopy the list of topics below and tick the boxes of the ones that are included in your examination syllabus. (Your teacher should be able to tell you which they are.) Use your list when you revise. The spread number in brackets tells you where to find more information.

❏ 1 The characteristics (features) of light. (7.01)

❏ 2 The speed of light. (7.01 and 7.11)

❏ 3 Defining a normal, the angle of incidence, and the angle of reflection. (7.02)

❏ 4 The laws of reflection. (7.02)

❏ 5 How an image is formed in a plane mirror. (7.02)

❏ 6 The difference between a real and virtual image. (7.02 and 7.07)

❏ 7 Finding the position of an image in a plane mirror. (7.02 and 7.03)

❏ 8 Using the link between the angle of incidence and the angle of reflection. (7.03)

❏ 9 The refraction of light. (7.04)

❏ 10 Defining the angle of refraction. (7.04)

❏ 11 How light changes speed when it passes from one medium to another. (7.04)

❏ 12 How light is refracted by a prism. (7.04)

❏ 13 Why water (or a glass block) looks less deep than it really is. (7.04)

❏ 14 Why light is refracted. (7.04)

❏ 15 The meaning of refractive index. (7.04 and 7.06)

❏ 16 How a prism forms a spectrum. (7.04)

❏ 17 Total internal reflection. (7.05)

❏ 18 The meaning of critical angle. (7.05)

❏ 19 Uses of reflecting prisms. (7.05)

❏ 20 Optical fibres and their uses. (7.05 and 7.13)

❏ 21 Snell's law for refraction. (7.06)

❏ 22 Calculating the critical angle. (7.06)

❏ 23 The properties of convex lenses. (7.07)

❏ 24 The difference between a convex and a concave lens. (7.07)

❏ 25 Converging and diverging lenses. (7.07)

❏ 26 The meanings of principal focus and focal length. (7.07)

❏ 27 Using a convex lens to form a real image. (7.07)

❏ 28 Using a convex lens as a magnifying glass. (7.08)

❏ 29 Drawing accurate ray diagrams for a convex lens. (7.08)

❏ 30 Measuring the focal length of a convex lens. (7.08)

❏ 31 Linear magnification. (7.08)

❏ 32 How a concave lens forms an image. (7.08)

❏ 33 The camera and its parts. (7.09)

❏ 34 The projector and its parts. (7.09)

❏ 35 How a photographic enlarger works. (7.09)

❏ 36 The human eye and its parts. (7.10)

❏ 37 Short and long sight and their correction. (7.10)

❏ 38 Electromagnetic waves. (7.11)

❏ 39 The electromagnetic spectrum. (7.11)

❏ 40 The characteristics and properties of:
   – radio waves
   – microwaves
   – infrared rays
   – ultraviolet rays
   – X-rays.
   – gamma rays. (7.12)

❏ 41 Communications and signals. (7.13)

❏ 42 The difference between analogue and digital signals. (7.13)

❏ 43 The advantages of digital transmission. (7.13)

❏ 44 Using optical fibres in communication systems. (7.13)

❏ 45 Methods of recording: vinyl, tape, and CD. (7.13)

8 Electricity

- ELECTRIC CHARGE
- CONDUCTORS AND INSULATORS
- ELECTRIC FIELDS
- CURRENT, VOLTAGE, AND RESISTANCE
- SERIES AND PARALLEL CIRCUITS
- ELECTRICAL POWER
- MAINS ELECTRICITY
- CALCULATING ELECTRICAL ENERGY

The city of Bogotá, Colombia, at night. Like other cities, it is so bright that it can even be seen from space. Modern industrial societies rely heavily on the use of electricity – not only for lighting, as shown here, but also for running factory machinery, information and communications systems, and heating. Typically, electricity accounts for about one sixth of an industrialized country's energy use.

8.01 Electric charge (1)

This person has been charged up. Her hairs all carry the same type of charge, so they repel each other.

Electric charge, or 'electricity', can come from batteries and generators. But some materials become charged when they are rubbed. Their charge is sometimes called **electrostatic charge** or 'static electricity'. It causes sparks and crackles when you take off a pullover, and if you slide out of a car seat and touch the door, it may even give you a shock.

Negative and positive charges

Polythene and Perspex can be charged by rubbing them with a dry, woollen cloth.

When two charged polythene rods are brought close together, as shown below, they *repel* (try to push each other apart). The same thing happens with two charged Perspex rods. However, a charged polythene rod and a charged Perspex rod *attract* each other. Experiments like this suggest that there are two different and opposite types of electric charge. These are called **positive** (+) charge and **negative** (−) charge:

> Like charges repel; unlike charges attract.
> The closer the charges, the greater the force between them.

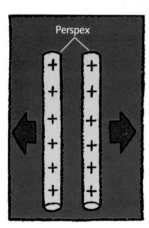

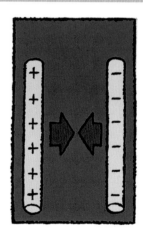

Where charges come from

Everything is made of tiny particles called atoms. These have electric charges inside them. A simple model of the atom is shown on the left. There is a central **nucleus** made up of **protons** and **neutrons**. Orbiting the nucleus are much lighter electrons:

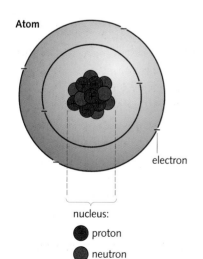

Atom

electron

nucleus:
- proton
- neutron

Electrons have a negative (−) charge.
Protons have an equal positive (+) charge.
Neutrons have no charge.

Normally, atoms have equal numbers of electrons and protons, so the *net* (overall) charge on a material is zero. However, when two materials are rubbed together, electrons may be transferred from one to the other. One material ends up with more electrons than normal and the other with less.

So one has a net negative charge, while the other is left with a net positive charge. Rubbing materials together does not *make* electric charge. It just *separates* charges that are already there.

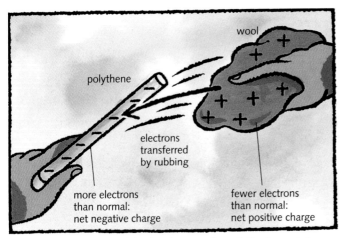

▲ When polythene is rubbed with a woollen cloth, the polythene pulls electrons from the wool.

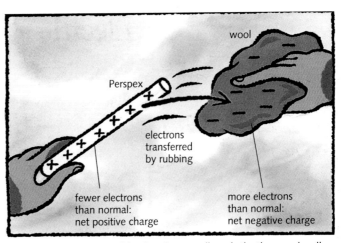

▲ When Perspex is rubbed with a woollen cloth, the wool pulls electrons from the Perspex.

Conductors and insulators

When some materials gain charge, they lose it almost immediately. This is because electrons flow through them or the surrounding material until the balance of negative and positive charge is restored.

Conductors are materials that let electrons pass through them. Metals are the best electrical conductors. Some of their electrons are so loosely held to their atoms that they can pass freely between them. These **free electrons** also make metals good thermal conductors.

Most non-metals conduct charge poorly or not at all, although carbon is an exception.

Insulators are materials that hardly conduct at all. Their electrons are tightly held to atoms and are not free to move – although they can be transferred by rubbing. Insulators are easy to charge by rubbing because any electrons that get transferred tend to stay where they are.

Semiconductors★ These are 'in-between' materials. They are poor conductors when cold, but much better conductors when warm.

Conductors	
Good	Poor
metals	water
especially:	human body
silver	earth
copper	
aluminium	
carbon	

Semiconductors	
silicon	germanium

Insulators	
plastics	glass
e.g:	rubber
PVC	dry air
polythene	
Perspex	

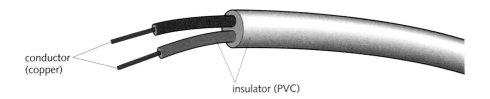

conductor (copper)

insulator (PVC)

◄ The 'electricity' in a cable is a flow of electrons. Most cables have copper conducting wires with PVC plastic around them as insulation.

1 Say whether the following *attract* or *repel*:
   a) two negative charges
   c) a negative charge and a positive charge
   b) two positive charges.

2 In an atom, what kind of charge is carried by
   a) protons b) electrons c) neutrons?

3 What makes copper a better electrical conductor than polythene?

4 Why is it easy to charge polythene by rubbing, but not copper?

5 Name one non-metal that is a good conductor.

6 When someone pulls a plastic comb through their hair, the comb becomes negatively charged.
   a) Which ends up with more electrons than normal, the comb or the hair?
   b) Why does the hair become positively charged?

Related topics: thermal conduction **5.06**; atoms **11.01**

Electric charge (2)

8.02

Attraction of uncharged objects

A charged object will attract any uncharged object close to it. For example, the charged screen of a TV will attract dust.

The diagram on the left shows what happens if a positively charged rod is brought near a small piece of aluminium foil. Electrons in the foil are pulled towards the rod, which leaves the bottom of the foil with a net positive charge. As a result, the top of the foil is attracted to the rod, while the bottom is repelled. However, the attraction is stronger because the attracting charges are closer than the repelling ones.

A charged object attracts an uncharged one.

Earthing★

If enough charge builds up on something, electrons may be pulled through the air and cause sparks – which can be dangerous. To prevent charge building up, objects can be **earthed**: they can be connected to the ground by a conducting material so that the unwanted charge flows away.

▶ An aircraft and its tanker must be earthed during refuelling, otherwise charge might build up as the fuel 'rubs' along the pipe. One spark could be enough to ignite the fuel vapour.

Detecting charge

Electrostatic charge can be detected using a **leaf electroscope** as above. If a charged object is placed near the cap, charges are induced in the electroscope. Those in the gold leaf and metal plate repel, so the leaf rises.

Induced charges

Charges that 'appear' on an uncharged object because of a charged object nearby are called **induced charges**. In the diagram below, a metal sphere is being charged by induction. The sphere ends up with an *opposite* charge to that on the rod, which never actually touches the sphere.

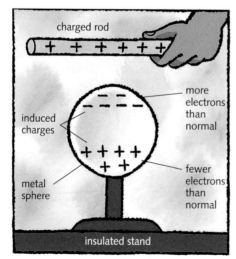

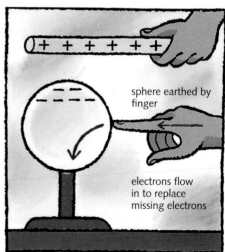

Unit of charge

The SI unit of charge is the **coulomb** (**C**). It is equal to the charge on about 6 million million million electrons, although it is not defined in this way. One coulomb is a relatively large quantity of charge, and it is often more convenient to measure charge in **microcoulombs**:

1 microcoulomb (μC) = 10^{-6} C   (one millionth of a coulomb)

The charge on a rubbed polythene rod is, typically, only about $0.005\,\mu$C.

Using electrostatic charge★

In the following examples, the charge comes from an electricity supply rather than from rubbing.

Electrostatic precipitators are fitted to the chimneys of some power stations and factories. They reduce pollution by removing tiny bits of ash from the waste gases. Inside the chamber of a precipitator (see right), the ash is charged by wires, and then attracted to the metal plates by an opposite charge. When shaken from the plates, the ash collects in the tray at the bottom.

Industrial inkjet printers, used for printing lettering on boxes, use the force between charges to control where the ink drops go.

Photocopiers work using the principle shown below:

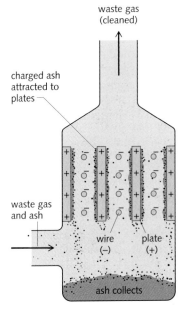

▲ An electrostatic precipitator uses charge to remove bits of ash from the waste gases produced by a factory or power station.

1	2	3	4	5

 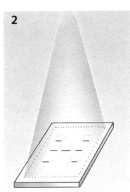 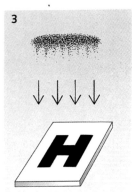 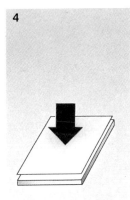 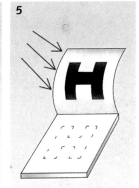

| Inside the photocopier, a light-sensitive plate (or drum) is given a negative charge. | An image of the original document is projected onto the plate. The bright areas lose their charge but the dark areas keep it. | Powdered ink (called toner) is attracted to the charged (dark) areas. | A blank sheet of paper is pressed against the plate and picks up powdered ink. | The paper is heated so that the powdered ink melts and sticks to it. The result is a copy of the original document. |

Q

1 **a)** Give an example of where electrostatic charge might be a hazard.
   b) How can the build-up of electrostatic charge be prevented?
2 How many microcoulombs are there in one coulomb?
3 On the right, a charged rod is held close to a metal can. The can is on an insulated stand.
   a) Copy the diagram. Draw in any induced charges on the can.
   b) Why is the can attracted to the rod even though the net (overall) charge on the can is zero?
   c) If you touch the can with your finger, electrons flow through it. In which direction is the flow?
   d) What type of charge is left on the can after it has been touched?

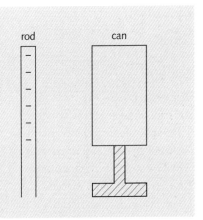

8.03 Electric fields

Atom and charge essentials

Electric charge can be positive (+) or negative (−). Like charges repel. Unlike charges attract.

Charges come from atoms. In an atom, the charged particles are electrons (−) and protons (+). Normally, an atom has equal amounts of − and + charge, so it is uncharged. However, if an atom gains or loses electrons, it is left with a net (overall) negative or positive charge.

Most materials are made up of groups of atoms, called molecules.

A charged object will cause a redistribution of the + and − charges in uncharged objects nearby. Concentrations of + or − charge which occur because of this are called induced charges.

An electric current is a flow of charge. When a metal conducts, there is a flow of electrons.

The girl on the right has given herself an electric charge by touching the dome of a Van de Graaff generator. The dome can reach over 100 000 volts, although this is reduced when she touches it. However, the current that flows into her body (0.000 02 amperes or less) is far too small to be dangerous.

The force of repulsion between the charges on the girl's head and hairs is strong enough to make her hairs stand up. If electric charges feel a force, then, scientifically speaking, they are in an **electric field**. So there is an electric field around the dome and the girl.

Electric field patterns

In diagrams, lines with arrows on them are used to represent electric fields. There are some examples of field patterns below. In each case, the arrows show the direction in which the force on a *positive* (+) charge would act. As like charges repel, the field lines always point *away* from positive (+) charge and *towards* negative (−) charge.

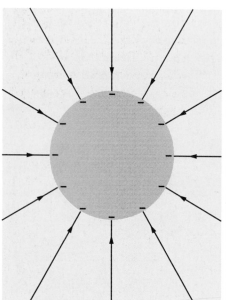

▲ Electric field close to a negatively charged sphere. The field around a Van de Graaff dome is similar to this.

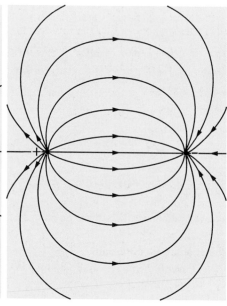

▲ Electric field between two opposite, point charges.

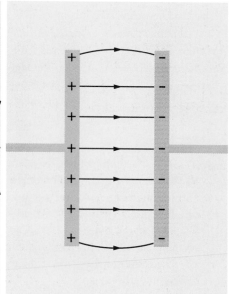

▲ Electric field between two parallel plates with opposite charges on them.

Curves, points, and ions★

When a conductor is charged up, the charges repel each other, so they collect on the outside. The charges are most concentrated near the sharpest curve. This is where the electric field is strongest and the field lines closest together.

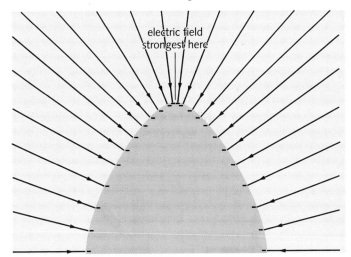

▲ The electric field is strongest where the charges are most concentrated and the field lines are closest together.

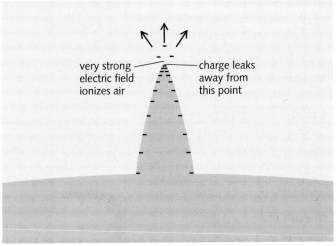

▲ At a sharp point, the electric field may be strong enough to ionize the air so that it will conduct charge away.

If a sharp spike is put on the dome of a Van de Graaff generator, any charge on the dome immediately leaks away from the point. At the point, the metal is very sharply curved. Here, the charge is so concentrated that the electric field is strong enough to ionize the air (see below). Ionized air conducts, so the dome loses its charge through the air.

Ions are electrically charged atoms (or groups of atoms). Atoms become ions if they lose (or gain) electrons. A stream of ions is a flow of charge, so it is another example of a current.

Most of the molecules in air are uncharged, but not all, as shown on the right. Flames, air movements, and natural radiation from space or rocks can all remove electrons from molecules in air so that ions are formed. Although these soon recombine with any free electrons around, more are being formed all the time. With no ions in it, air is a good electrical insulator. But with ions present, it has charges that are free to move, so the air becomes a conductor.

In a thunderstorm, the concentrations of different ions may be so great that a very high current may flow through the air, causing a flash of lightning.

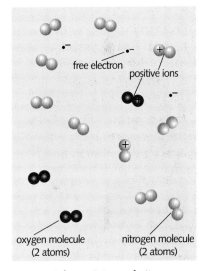

Air is mainly a mixture of nitrogen and oxygen molecules. The charged ones are called ions.

1 The diagram on the right shows electric field lines round a charged metal sphere (in air).
a) Copy the diagram. Draw in the direction of the electric field on each field line.
b) If a positive charge were placed at X, in which direction would it move?
c) If a negative charge were placed at X, in which direction would it move?
d) If a sharp spike were placed on top of the sphere, what would happen to the charge on the sphere?

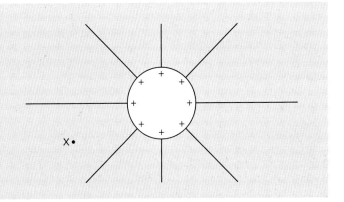

Related topics: atoms and molecules **5.01**; charges and conductors **8.01**; induced charges **8.02**; Ionizing radiation **11.02**

175

8.04

Current in a simple circuit

Charge essentials

Electric charge can be positive (+) or negative (−). Like charges repel, unlike charges attract.

Charges come from atoms. In atoms, the charged particles are protons (+) and **electrons** (−).

Electrons can move through some materials, called **conductors**. Copper is the most commonly used conductor.

The unit of charge is the **coulomb (C)**.

An electric **cell** (commonly called a **battery**) can make electrons move, but only if there is a conductor connecting its two terminals. Then, chemical reactions inside the cell push electrons from the negative (−) terminal round to the positive (+) terminal.

The cell below is being used to light a bulb. As electrons flow through the bulb, they make a filament (thin wire) heat up so that it glows. The conducting path through the bulb, wires, switch, and battery is called a **circuit**. There must be a *complete* circuit for the electrons to flow. Turning the switch OFF breaks the circuit and stops the flow.

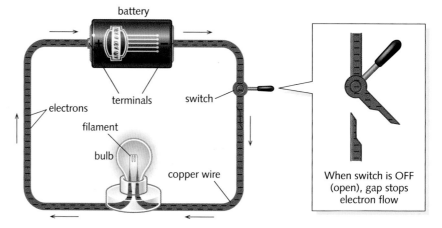

The above circuit can be drawn using **circuit symbols**:

Ammeter

To measure a current, you need to choose a meter with a suitable range on its scale. This ammeter cannot measure currents above 1 A. Also to measure, say, 0.1 A accurately, it would be better to use a meter with a lower range.

When connecting up a meter, the red (+) terminal should be on the same side of the circuit as the + terminal of the battery.

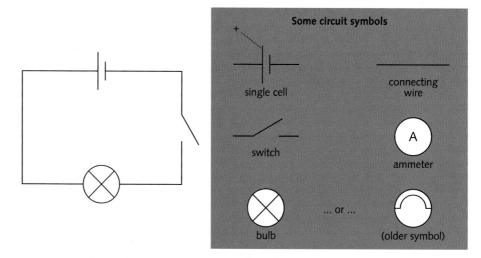

Measuring current

A flow of charge is called an electric **current**. The higher the current, the greater the flow of charge.

The SI unit of current is the **ampere (A)**. About 6 million million million electrons flowing round a circuit every second would give a current of 1 A. However, the ampere is not defined in this way.

Currents of about an ampere or so can be measured by connecting an **ammeter** into the circuit. For smaller currents, a **milliammeter** is used. The unit in this case is the **milliampere (mA)**. 1000 mA = 1 A

Some typical current values

current through a small torch bulb   0.2 A   (200 mA)
current through a car headlight bulb   4 A
current through an electric kettle element   10 A

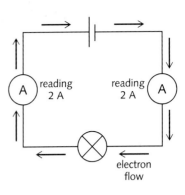

Putting ammeters (or milliammeters) into a circuit has almost no effect on the current. As far as the circuit is concerned, the meters act just like pieces of connecting wire.

The circuit on the right has two ammeters in it. Any electrons leaving the battery must flow through both, so both give the same reading:

The current is the same at all points round a simple circuit.

Charge and current

There is a link between charge and current:

If charge flows at this rate...   then the current is...
1 coulomb per second   1 ampere
2 coulombs per second   2 amperes   ...and so on.

The link can also be expressed as an equation:

$$\text{charge} = \text{current} \times \text{time}$$
$$\text{(C)} \quad\quad \text{(A)} \quad\quad \text{(s)}$$

For example, if a current of 2 amperes flows for 3 seconds, the charge delivered is 3 coulombs.

Current direction

Some circuit diagrams have arrowheads marked on them. These show the **conventional current direction**: the direction from + to − round the circuit. Electrons actually flow the other way. Being negatively charged, they are repelled by negative charge, so are pushed out of the negative terminal of the battery.

The conventional current direction is equivalent to the direction of transfer of positive charge. It was defined before the electron was discovered and scientists realised that positive charge did not flow through wires. However, it isn't 'wrong'. Mathematically, a transfer of positive charge is the same as a transfer of negative charge in the opposite direction.

Definitions

Although it is convenient to think of 1 ampere as 1 coulomb per second, the coulomb is actually defined in terms of the ampere:

1 coulomb is the charge that passes when a current of 1 ampere flows for 1 second.

The ampere is one of the SI base units. It is defined in terms of the magnetic force produced by a current.

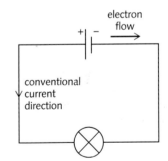

Q

1 Convert these currents into amperes: **a)** 500 mA **b)** 2500 mA
2 Convert these currents into milliamperes: **a)** 2.0 A **b)** 0.1 A
3 **a)** Draw the circuit on the right using circuit symbols.
   b) On your diagram, mark in and label the conventional current direction and the direction of electron flow.
   c) The current reading on one of the ammeters is shown. What is the reading on the other one?
   d) Which bulb(s) will go out if the switch contacts are moved apart? Give a reason for your answer.
4 What charge is delivered if
   a) a current of 10 A flows for 5 seconds
   b) a current of 250 mA flows for 40 seconds?

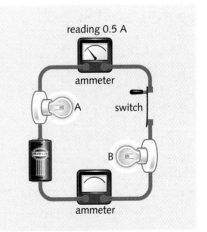

8.05 Potential difference

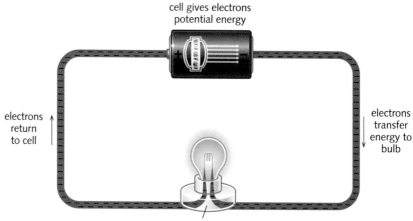

cell gives electrons
potential energy

electrons
return
to cell

electrons
transfer
energy to
bulb

electrons lose potential
energy: energy radiated

Circuit essentials

A cell can make electrons flow round a circuit. The flow of electrons is called a current. Electrons carry a negative (−) charge. As like charges repel, electrons are pushed out of the negative (−) terminal of the cell.

Charge is measured in coulombs (C).

Energy and work essentials

Energy is measured in joules (J). Potential energy is the energy that something has because of its state or position.

Work is also measured in joules (J). If something loses energy, it does work; if it gains energy, then work is done on it. The gain or loss of energy is equal to the work done.

The cell above is pushing out electrons. The electrons repel each other, so, like the coils of a compressed spring, they have potential energy. As the electrons slowly flow round the circuit, they transfer energy from the cell to the bulb. The energy is radiated by the hot filament.

PD (voltage) across a cell

A cell normally has a **voltage** marked on it. The higher its voltage, the more energy it gives to the electrons pushed out. The scientific name for voltage is **potential difference** (**PD**). PD can be measured by connecting a **voltmeter** across the terminals of the cell. The SI unit of PD is the volt (**V**):

If the PD across a cell is 1 volt, then 1 joule of potential energy is given to each coulomb of charge. In other words, 1 volt means 1 joule per coulomb (J/C).

If the PD across a cell is 2 volts, then 2 joules of potential energy are given to each coulomb of charge, ...and so on.

A cell produces its highest PD when not in a circuit and not supplying current. This maximum PD is called the **electromotive force** (**EMF**) of the cell. When a current is being supplied, the PD drops because of energy wastage inside the cell. For example, a car battery labelled '12 V' might only deliver 9 V when being used to turn a starter motor.

Cells in series

To produce a higher PD, several cells can be connected in **series** (in line) as shown below. The word 'battery' really means a collection of joined cells, although it is commonly used for a single cell as well.

Voltmeter and symbol. (For information about range and connection, see note under ammeter in previous spread.)

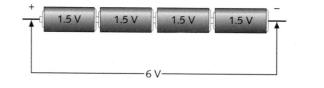

+   1.5 V   1.5 V   1.5 V   1.5 V   −

—6 V—

battery made up of
several cells
(symbol)

PDs around a circuit

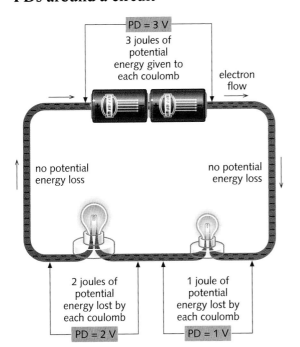

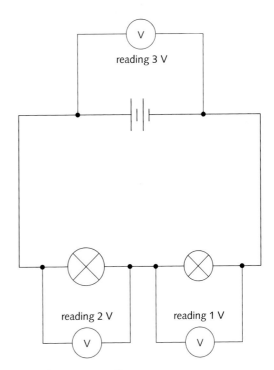

In the circuit above, the electrons flow through two bulbs. They lose some of their potential energy in the first bulb and the rest in the second. In total, all the energy supplied by the battery is radiated by the bulbs. Almost none is spent in the connecting wires.

Like the battery, each bulb has a PD across it:
If a bulb (or other component) has a PD of 1 volt across it, then 1 joule of potential energy is spent by each coloumb of charge passing through it.

The second diagram shows the same circuit with voltmeters connected across different sections (the voltmeters do not affect how the circuit works). The readings illustrate a principle which applies in any circuit:

> Moving round a circuit, from one battery terminal to the other, the sum of the PDs across the components is equal to the PD across the battery.

Definitions

The electromotive force (EMF) of a cell (or other source) is the work done per unit of charge by the cell in driving charge round a complete circuit (including the cell itself).

The potential difference (PD) across a component is the work done per unit of charge in driving charge through the component.

Q

1 In what unit is each of these measured?
   a) PD   b) EMF   c) charge   d) current   e) energy
2 In the circuit on the right, the two bulbs are of different sizes and brightnesses.
   a) What type of meter is meter X?
   b) What type of meter is meter Y?
   c) What is the reading on meter Y?
   d) How much potential energy does each coulomb have as it leaves the battery?
   e) How much potential energy is lost by each coulomb passing through bulb A?
   f) How much charge passes through A every second?
   g) How much energy is radiated from A every second?

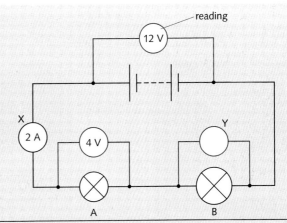

Current is measured in amperes (A) using an ammeter. If 1 ampere flows for 1 second, the charge passing is 1 coulomb.

Related topics: SI units 1.02; energy 4.01–4.02; electrons and charge 8.01–8.02; charge and current 8.04; cell arrangements 8.09

Resistance (1)

To make a current flow through a conductor, there must be a potential difference (voltage) across it. Copper connecting wire is a good conductor and a current passes through it easily. However, a similar piece of nichrome wire is not so good and less current flows for the same PD. The nichrome wire has more **resistance** than the copper.

Resistance is calculated using the equation below. The SI unit of resistance is the **ohm** (Ω). (The symbol Ω is the Greek letter *omega*.)

$$\text{resistance } (\Omega) = \frac{\text{PD across conductor (V)}}{\text{current through conductor (A)}}$$

For example, if a PD of 6 V is needed to make a current of 3 A flow through a wire: resistance = 6 V/3 A = 2 Ω.

With a *lower* resistance, a *lower* PD would be needed to give the same current. Even copper connecting wire has some resistance. However, it is normally so low that only a very small PD is needed to make a current flow through it, and this can be neglected in calculations.

Some factors affecting resistance

The resistance of a conductor depends on several factors:

- **Length** Doubling the length of a wire doubles its resistance.
- **Cross-sectional area** Halving the 'end on' area of a wire doubles its resistance. So a thin wire has more resistance than a thick one.
- **Material** A nichrome wire has more resistance than a copper wire of the same size.
- **Temperature** For metal conductors, resistance increases with temperature. For semiconductors, it decreases with temperature.

Resistance and heating effect

There is a heating effect whenever a current flows through a resistance. This principle is used in heating elements, and also in the filaments of bulbs. The heating effect occurs because electrons collide with atoms as they pass through a conductor. The electrons lose energy. The atoms gain energy and vibrate faster. Faster vibrations mean a higher temperature.

high resistance

↑

━━━ ═ ═ ━━━
long, thin, nichrome wire

━━━━━━━━━━
long, thin, copper wire

▬▬▬━━▬▬▬
long, thick, copper wire

▬▬▬▬▬
short, thick, copper wire

low resistance

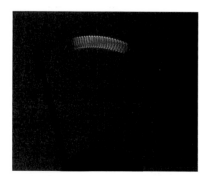

▲ The filament of a bulb is made of very thin tungsten wire. Tungsten has a high melting point.

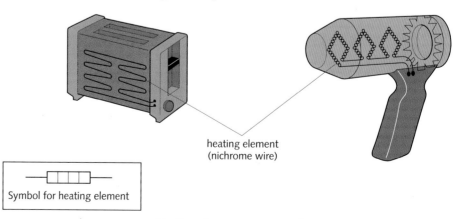

─┤▢▢▢▢├─

Symbol for heating element

heating element
(nichrome wire)

Heating elements are normally made of nichrome.

Resistance components

Resistors are specially made to provide resistance. In simple circuits, they reduce the current. In more complicated circuits, such as those in radios, TVs, and computers, they keep currents and PDs at the levels needed for other components (parts) to work properly.

Resistors can have values ranging from a few ohms to several million ohms. For measuring higher resistances, these units are useful:

1 kilohm (kΩ) = 1000 Ω   1 megohm (MΩ) = 1 000 000 Ω

Like all resistances, resistors heat up when a current flows through them. However, if the current is small, the heating effect is slight.

Variable resistors (rheostats) are used for varying current. The one on the right is controlling the brightness of a bulb. In hi-fi equipment, rotary (circular) variable resistors are used as volume controls.

Thermistors have a high resistance when cold but a much lower resistance when hot. They contain semiconductor materials. Some electrical thermometers use a thermistor to detect temperature change.

Light-dependent resistors (LDRs) have a high resistance in the dark but a low resistance in the light. They can be used in electronic circuits which switch lights on and off automatically.

Diodes have an extremely high resistance in one direction but a low resistance in the other. In effect, they allow current to flow in one direction only. They are used in electronic circuits.

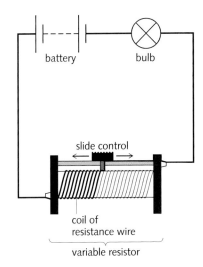

Moving the slide control of the variable resistor to the right increases the length of resistance wire in the circuit. This reduces the current and dims the bulb.

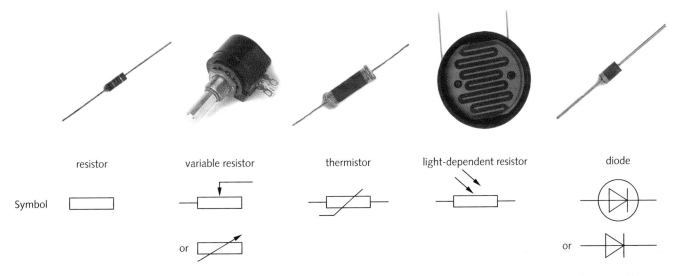

| Symbol | resistor | variable resistor | thermistor | light-dependent resistor | diode |

Q

1 When a kettle is plugged into the 230 V mains, the current through its element is 10 A.
   a) What is the resistance of its element?
   b) Why does the element need to have resistance?

2 In the diagram at the top of this page, a variable resistor is controlling the brightness of a bulb. What happens if the slide control is moved to the left? Give a reason for your answer.

3 Which of the components in the photographs above has each of these properties?
   a) A high resistance in the dark but a low resistance in the light.
   b) A resistance that falls sharply when the temperature rises.
   c) A very low resistance in one direction, but an extremely high resistance in the other.

Related topics: SI units **1.02**; temperature, vibrating atoms, and thermometers **5.02**; conductors and semiconductors **8.01**; current, circuits and symbols **8.04**; potential difference **8.05**; diodes **10.02**; LDRs and thermistors **10.04**; resistor colour code **page 325**

8.07 Resistance (2)

Resistance equation

$$\text{resistance} = \frac{\text{potential difference}}{\text{current}}$$

Units:

resistance: ohm (Ω)

potential difference (PD): volt (V)

current: ampere (A)

This triangle gives the V, I, and R equations. To find the equation for I, cover up the I, ...and so on.

V, I, R equations

The resistance equation can be written using symbols:

$$R = \frac{V}{I}$$

where R = resistance, V = PD (voltage), and I = current

(Note the difference between the symbol V for PD and the symbol V for volt.)

The above equation can be rearranged in two ways:

$$V = IR \qquad \text{and} \qquad I = \frac{V}{R}$$

These are useful if the PD across a known resistance, or the current through it, is to be calculated.

> *Example* A 12 Ω resistor has a PD of 6 V across it. What is the current through the resistor?

In this case: $V = 6\,\text{V}$, $R = 12\,\Omega$, and I is to be found. So:

$$I = \frac{V}{R} = \frac{6}{12} = 0.5 \qquad \text{(omitting units for simplicity)}$$

So the current is 0.5 A.

How current varies with PD for a metal conductor

The circuit below left can be used to investigate how the current through a conductor depends on the PD across it. The conductor in this case is a coiled-up length of nichrome wire, kept at a steady temperature by immersing it in a large amount of water. The PD across the nichrome can be varied by adjusting the variable resistor.

Typical results are shown in the table and graph below. The experiment is also one method of measuring resistance.

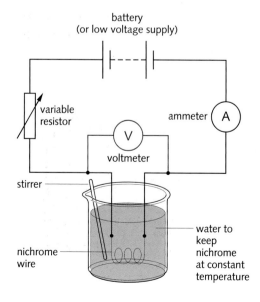

PD	current	PD/current
1.0 V	0.2 A	5.0 Ω
2.0 V	0.4 A	5.0 Ω
3.0 V	0.6 A	5.0 Ω
4.0 V	0.8 A	5.0 Ω
5.0 V	1.0 A	5.0 Ω

resistance

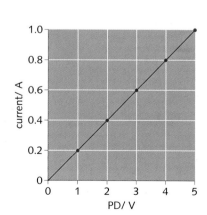

Ohm's law

In the experiment on the opposite page, the results have these features:

- A graph of current against PD is a straight line through the origin.
- If the PD doubles, the current doubles, ...and so on.
- PD ÷ current always has the same value ($5\,\Omega$ in this case).

Mathematically, these can be summed up as follows:

> The current is proportional to the PD.

This is known as **Ohm's law**, after George Ohm, the 19th century scientist who first investigated the electrical properties of wires.

Metal conductors obey Ohm's law, provided their temperature does not change. Put another way, a metal conductor has a constant resistance, provided its temperature is constant. This is not always the case with other types of conductor.

Current – PD graphs

Here are two more examples of current–PD graphs. In both, the resistance varies depending on the PD. In the case of the diode, the negative part of the graph is for readings obtained when the PD is reversed (i.e. when the diode is connected into the test circuit the opposite way round).

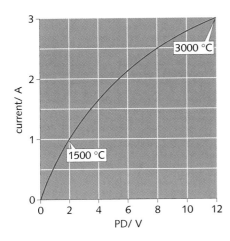

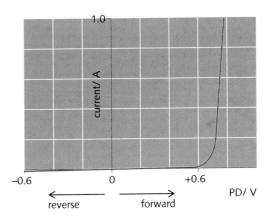

Tungsten filament As the current increases, the temperature rises and the resistance goes up. So the current is not proportional to the PD.

Semiconductor diode The current is not proportional to the PD. And if the PD is reversed, the current is almost zero. In effect, the diode 'blocks' current in the reverse direction.

1 The graph lines A and B on the right are for two different conductors. Which conductor has the higher resistance?

2 Using the left-hand graph above, calculate the resistance of the tungsten filament when its temperature is **a)** 1500 °C **b)** 3000 °C.

3 In the right-hand graph above, does the diode have its highest resistance in the forward direction or the reverse? Explain your answer.

4 A resistor has a steady resistance of $8\,\Omega$.
   a) If the current through the resistor is 2 A, what is the PD across it?
   b) What PD is needed to produce a current of 4 A?
   c) If the PD falls to 6 V, what is the current?

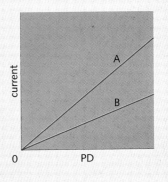

8.08 More about resistance factors

Resistance essentials

To make a current flow through a conductor, there must be a PD (voltage) across it. The resistance of the conductor is calculated like this:

$$resistance = \frac{PD}{current}$$

Units:
resistance: ohms (Ω)
PD: volts (V)
current: amperes (A)

Even copper connecting wires have some resistance, although this is usually very small. Resistors and heating elements are designed to have resistance.

The resistance of a wire depends on its length and cross-sectional area. It also depends on the material and its temperature (although for metals, the change of resistance with temperature is small).

The effects of length and area

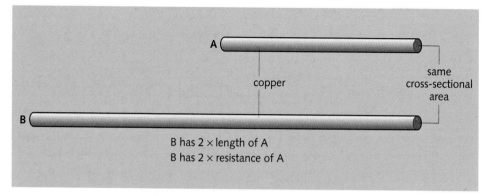

B has 2 × length of A
B has 2 × resistance of A

The copper wires above have the same cross-sectional area and temperature. But B is *twice* as long as A. As a result, it has *twice* the resistance of A. If B were *three* times as long as A, it would have three *times* the resistance, and so on. Results like this can be summed up as follows:

Provided other factors do not change:

resistance $\propto$ length   (the symbol $\propto$ means 'directly proportional to')

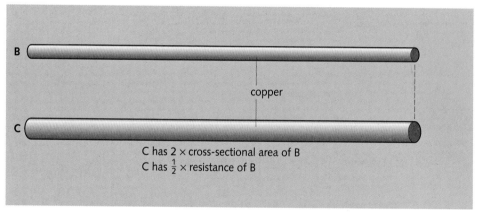

C has 2 × cross-sectional area of B
C has $\frac{1}{2}$ × resistance of B

The copper wires above have the same length and temperature. But C has *twice* the cross-sectional area of B. As a result, it has *half* the resistance of B. If C had *three* times the cross-sectional area of B, it would have *one third* of the resistance, and so on. Results like this can be summed up as follows:

Provided other factors do not change:

resistance $\propto \dfrac{1}{area}$ ('area' means 'cross-sectional area')

The above proportionalities are true for other types of wire, although the resistances will differ. For example, nichrome has much more resistance than copper of the same length, cross-sectional area, and temperature.

The results can be combined as follows:

For any given conducting material at constant temperature:

$$resistance \propto \frac{length}{area}$$

Proportionality problems

When there are mathematical problems to solve, equations are much more useful than proportionalities. Fortunately, the proportionality linking resistance (R), length (l), and area (A) can be converted into an equation like this:

$$R = \rho \times \frac{l}{A}$$

(ρ = Greek letter 'rho')

where ρ is a constant for the material at a particular temperature. ρ is called the **resistivity** of the material (Table 1). Rearranging the above equation gives:

$$\rho = \frac{R \times A}{l}$$

This is useful when comparing different wires, A and B, made from the same material. As ρ is the same for each wire (at a particular temperature):

$$\frac{\text{resistance}_A \times \text{area}_A}{\text{length}_A} = \frac{\text{resistance}_B \times \text{area}_B}{\text{length}_B}$$

Example Wire A has a resistance of 12 Ω. If wire B is twice the length of A and twice the diameter, what is its resistance? (Assume that both wires are at the same temperature.)

As wire B has twice the diameter of A, it has four times the cross-sectional area (see the box above right).

The resistance of wire B is to be found: call it R_B. As no measurements are given, use letters to represent these as well, as in the diagram on the right.

If length$_A$ = x, then length$_B$ = $2x$

If area$_A$ = A, then area$_B$ = $4A$

Also, resistance$_A$ = 12 Ω and resistance$_B$ = R_B

Substituting the above values in the previous equation gives:

$$\frac{12 \times A}{x} = \frac{R_B \times 4A}{2x}$$

(omitting units for simplicity)

Rearranging and cancelling gives: $R_B = 6$

So, the resistance of wire B is 6 Ω.

Diameter and area

| area A | area 4A |

If one wire has *twice* the diameter of another, as above, then it has *four times* the cross-sectional area. That follows from the equation for the area of a circle: $A = \pi r^2$. Doubling the diameter doubles the radius. So, replacing r in the equation with $2r$ gives:

new area
$= \pi(2r)^2 = 4\pi r^2 = 4A$.

Similarly *three* times the diameter gives *nine times* the area, and so on. So:

area $\propto$ diameter^2

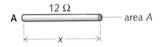

A ⎯ 12 Ω ⎯ area A

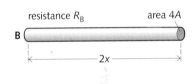

B — resistance R_B — area 4A

Typical resistivity values/ Ω m	
Constantan	49×10^{-8}
Manganin	44×10^{-8}
Nichrome	100×10^{-8}
Tungsten	55×10^{-8}

Table 1

1 Wire X has a resistance of 18 Ω. Wire Y is made of the same material and is at the same temperature. If Y is the same length as X, but 3 times the diameter, what is its resistance?

2 Wires A and B are made of the same material and are at the same temperature. The chart on the right gives some information about them.

a) If you were to use part of wire A to make an 18 Ω resistor, what length would you need?

b) What is the resistance of wire B?

c) What length of wire B would you need to make a 20 Ω resistor?

	wire A	wire B
length	1000 mm	2500 mm
area	2.0 mm²	0.5 mm²
resistance	25 Ω	

Series and parallel circuits (1)

Circuit essentials

Potential difference (PD), or voltage, is measured in volts (V). The greater the PD aross a bulb or other component, the greater the current flowing through it. Current is measured in amperes (A).

Bulbs, resistors, and other components have resistance to a flow of current. Resistance is measured in ohms (Ω).

The bulbs above have to get their power from the same supply. There are two basic methods of connecting bulbs, resistors, or other components together. The circuits below demonstrate the differences between them.

Bulbs in series and parallel

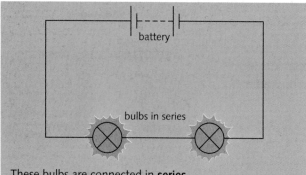

These bulbs are connected in **series**.
● The bulbs share the PD (voltage) from the battery, so each glows dimly.
● If one bulb is removed, the other goes out because the circuit is broken.

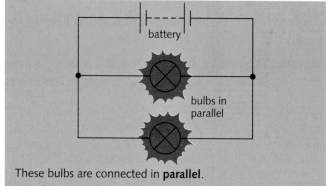

These bulbs are connected in **parallel**.
● Each gets the full PD from the battery because each is connected directly to it. So each glows brightly.
● If one bulb is removed, the other keeps working because it is still part of an unbroken circuit.

Circuits and switches

If two or more bulbs have to be powered by one battery, as in a car lighting system, they are normally connected in parallel. Each bulb gets the full battery PD. Also, each can be switched on and off independently:

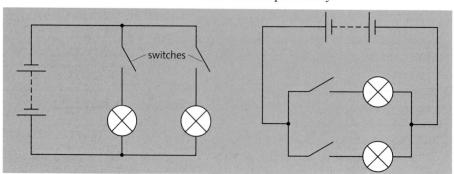

These diagrams show two different ways of drawing the same circuit for independently switched bulbs.

Basic circuit rules

There are some basic rules for all series and parallel circuits. They are illustrated by the examples below. The particular current values depend on the resistances and PDs. However, the equation on the right *always* applies to *every* resistor.

PD $=$ current $\times$ resistance
(V)   (A)   (Ω)

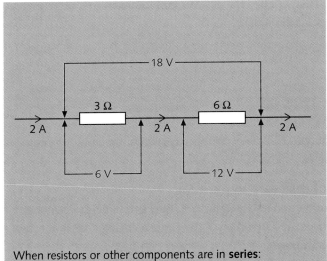

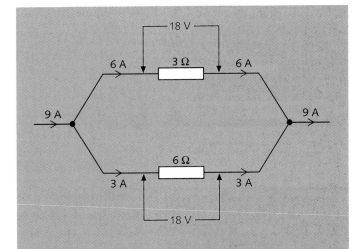

When resistors or other components are in **series**:
● the current through each of the components is the same
● the total PD (voltage) across all the components is the sum of the PDs across each of them.

When resistors or other components are in **parallel**:
● the PD (voltage) across each of the component is the same
● the total current in the main circuit is the sum of the currents in the branches.

Cell arrangements

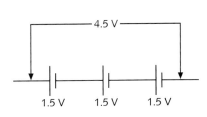

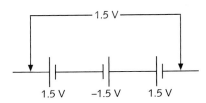

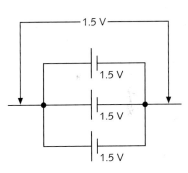

These cells are connected in series. The total PD (voltage) across them is the sum of the individual PDs.

Here, a mistake has occurred. One of the cells is the wrong way round, so it cancels out one of the others.

The PD across parallel cells is only the same as from one cell. But together, the cells can deliver a higher current.

1   When one of the lights on a Christmas tree breaks, the others go out as well. What does this tell you about the way the lights are connected?
2   Give *two* advantages of connecting bulbs to a battery in parallel.
3   Redraw either of the circuits on the left so that it has a single switch which turns both bulbs on and off together.
4   This question is about the circuit on the right:
    a) The readings on two of the ammeters are shown. What are the readings on ammeters X and Y?
    b) If the PD across the battery is 6 V, what is the PD across each of the bulbs? (Note: you can neglect the PD across an ammeter.)

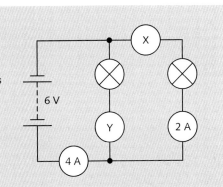

8.10 Series and parallel circuits (2)

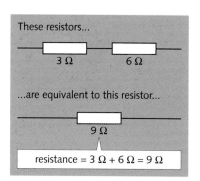

These resistors...

3 Ω   6 Ω

...are equivalent to this resistor...

9 Ω

resistance = 3 Ω + 6 Ω = 9 Ω

Combined resistance of resistors in series

If two (or more) resistors are connected in series, they give a *higher* resistance than any of the resistors by itself. The effect is the same as joining several lengths of resistance wire to form a longer length.

If resistors R_1 and R_2 are in series, their combined resistance R is given by this equation:

$$R = R_1 + R_2$$

There is an example on the left. For three or more resistors, the above equation can be extended by adding R_3 ... and so on.

Combined resistance of resistors in parallel

If two (or more) resistors are connected in parallel, they give a *lower* resistance than any of the resistors by itself. The effect is the same as using a thick piece of resistance wire instead of a thin one. There is a wider conducting path than before.

If two resistors R_1 and R_2 are in parallel, their combined resistance R is given by this equation (there is a proof at the bottom of the page):

$$\frac{1}{R} = \frac{1}{R_1} + \frac{1}{R_2}$$

For three or more resistors, the equation can be extended by adding $1/R_3$, ... and so on.

If the above equation for two resistors is rearranged, it becomes: $R = \dfrac{R_1 \times R_2}{R_1 + R_2}$

In words: combined resistance = $\dfrac{\text{resistances multiplied}}{\text{resistances added}}$

For example, if $3\,\Omega$ and $6\,\Omega$ resistors are in parallel:

$$\text{combined resistance} = \frac{6 \times 3}{6 + 3} = 2\,\Omega$$

Note: this method of calculation works only for *two* resistors in parallel.

These resistors...

3 Ω

6 Ω

...are equivalent to this resistor...

2 Ω

Omitting units for simplicity:

$$\frac{1}{\text{resistance}} = \frac{1}{3} + \frac{1}{6}$$
$$= \frac{2}{6} + \frac{1}{6}$$
$$= \frac{3}{6}$$
$$= \frac{1}{2}$$

So: resistance = 2 Ω

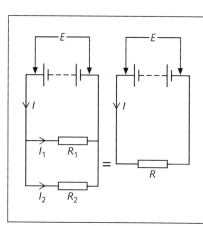

Proving the parallel resistor equation

In the circuit on the left, R_1 has the full battery PD of E across it. So does R_2.

As current = $\dfrac{\text{PD}}{\text{resistance}}$: $I_1 = \dfrac{E}{R_1}$ and $I_2 = \dfrac{E}{R_2}$

But $I = I_1 + I_2$ so $I = \dfrac{E}{R_1} + \dfrac{E}{R_2}$

If resistor R is equivalent to R_1 and R_2 in parallel, it must take the same current I from the battery:

$I = \dfrac{E}{R}$ Therefore: $\dfrac{E}{R} = \dfrac{E}{R_1} + \dfrac{E}{R_2}$ so $\dfrac{1}{R_2} = \dfrac{1}{R_1} + \dfrac{1}{R_2}$

Solving circuit problems

To solve problems about circuits, you need to know the basic circuit rules on the previous spread. You also need to know the link between PD (voltage), current, and resistance. This is given on the right.

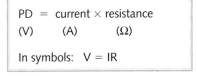

PD = current × resistance
(V)   (A)   (Ω)

In symbols: V = IR

Example 1 Calculate the PDs across the 3 Ω resistor and the 6 Ω resistor in the circuit on the right.

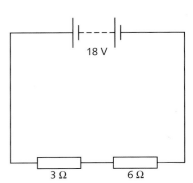

The first stage is to calculate the total resistance in the circuit, and then use this information to find the current:

total resistance $= 3\,\Omega + 6\,\Omega = 9\,\Omega$

so: current $I = \dfrac{PD}{resistance} = \dfrac{18\,V}{9\,\Omega} = 2\,A$

Knowing that the 3 Ω resistor has a current of 2 A through it, you can calculate the PD across it:

$PD = current \times resistance = 2\,A \times 3\,\Omega = 6\,V$

The PD across the 6 Ω resistor can be worked out in the same way. However, it can also be deduced from the fact that the PDs across the two resistors must add up to 18 V, the PD across the battery. By either method, the PD across the 6 Ω resistor is 12 V.

Example 2 Calculate the currents I, I_1, and I_2 in the circuit on the right.

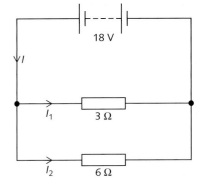

The 3 Ω resistor has the full battery PD of 18 V across it. So:

$I_1 = \dfrac{PD}{resistance} = \dfrac{18\,V}{3\,\Omega} = 6\,A$

Using the same method: $I_2 = 3\,A$

The current I is the total of the currents in the two branches. So:

$I = I_1 + I_2 = 6\,A + 3\,A = 9\,A$

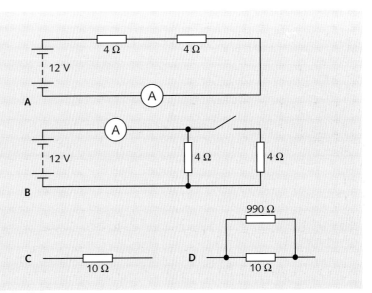

1 In circuit A on the right:
   a) What does the ammeter read?
   b) What is the PD across each of the resistors?
2 In circuit B on the right:
   a) What does the ammeter read when the switch is open (OFF)?
   b) What is the current through each of the 4 Ω resistors when the switch is closed (ON)?
   c) What does the ammeter read when the switch is closed?
   d) What is the combined resistance of the two resistors when the switch is closed?
3 Which resistor arrangement, C or D, on the right has the lower resistance? Check your answer by calculation.

8.11 Electrical power

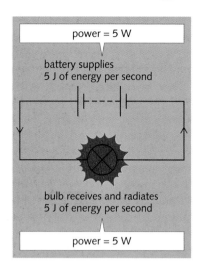

power = 5 W

battery supplies
5 J of energy per second

bulb receives and radiates
5 J of energy per second

power = 5 W

Circuit essentials

In a circuit like the one above, the charge is carried by electrons. Charge is measured in coulombs (C).

The flow of electrons is called a current. Current is measured in amperes (A).

Potential difference (PD), or voltage, is measured in volts (V). The greater the PD across a battery, the more potential energy each electron is given. The greater the PD aross a bulb or other component, the more energy each electron loses as it passes through.

Energy is measured in joules (J).

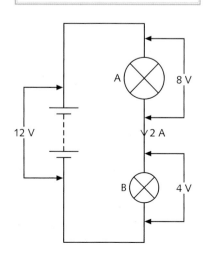

In the circuit on the left, the battery gives electrons potential energy. In the bulb, this is changed into thermal energy (heat) and then radiated.

Power is the rate at which energy is transformed (changed from one form to another). The SI unit of power is the **watt** (**W**):

$$\text{power} = \frac{\text{energy transformed}}{\text{time taken}}$$

The battery on the left is supplying 5 joules of energy every second, so its power is 5 watts. The bulb is taking energy at the same rate, so its power is also 5 watts.

Appliances such as toasters, irons, and TVs have a **power rating** marked on them, either in watts or in kilowatts:

1 kilowatt (kW) = 1000 watts

Some typical power ratings are shown below. Each figure tells you the power the appliance will take *if connected to a supply of the correct voltage*. For any other voltage, the actual power would be different.

Electrical power equation

For circuits, there is a more useful version of the power equation. If a battery, bulb, or other component has a PD (voltage) across it and a current through it, the power is given by this equation:

$$\begin{array}{ccc} \text{power} = & \text{PD} \times & \text{current} \\ \text{(W)} & \text{(V)} & \text{(A)} \end{array}$$

In symbols:

$$P = VI$$

Example In the circuit on the left, what is the power of the battery and each of the bulbs?

For the battery: power = PD × current = 12 V × 2 A = 24 W
For bulb A: power = PD × current = 8 V × 2 A = 16 W
For bulb B: power = PD × current = 4 V × 2 A = 8 W

The bulbs are the only items getting power from the battery, so their total power (16 W + 8 W) is the same as that supplied by the battery (24 W).

Why the electrical power equation works

The equation power = PD × current is a result of how the volt, ampere, coulomb, joule, and watt are related. The following example should explain why.

Here are two ways of describing what is happening on the right:

General description	Scientific description
Each coulomb of charge gains 12 joules of energy from the battery	PD = 12 volts
2 coulombs of charge leave the battery every second	current = 2 amperes
So 12 x 2 joules of energy leave the battery every second	power = 24 watts

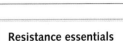

Power dissipated in a resistor★

Although the following section is about resistors, it also applies to heating elements and any other components that have resistance.

When a current flows through a resistor, it has a heating effect. Electrons lose potential energy, which is changed into thermal energy. Scientifically speaking, energy is **dissipated** in the resistor.

For calculating the rate of energy dissipation, there is another useful version of the electrical power equation. It is found like this:

power = PD × current

But: PD = current × resistance

So: power = current × resistance × current

So:

power = current² × resistance

In symbols:

$$P = I^2R$$

> *Example* What power is dissipated in a 5 Ω resistor when the current through it is **a)** 2 A **b)** 4 A?

Omitting some of the units for simplicity:

a) Power = current² × resistance = 2^2 × 5 = 20 W

b) Power = current² × resistance = 4^2 × 5 = 80 W

Note that *doubling* the current produces *four times* the power dissipation.

Resistance essentials

Resistors have a resistance measured in ohms (Ω). The higher the resistance, the less current flows through the resistor for each volt of PD across it.

$$resistance = \frac{PD}{current}$$

1 In 5 seconds, a hairdrier takes 10 000 joules of energy from the mains supply. What is its power **a)** in watts **b)** in kilowatts?

2 If an electric heater takes a current of 4 A when connected to a 230 V supply, what is its power?

3 If a light bulb has a power of 36 W when connected to a 12 V supply, what is the current through it?

4 In the circuit on the right, both resistors receive the full battery PD.
   a) Use the resistance equation (in the box above right) to calculate the current through each of the resistors.
   b) Calculate the power dissipated in each of the resistors.
   c) Calculate the power of the battery.
   d) If a battery of twice the PD was used, what would its power be?

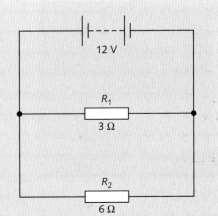

Related topics: SI units **1.02**; energy **4.01**; power **4.04**; current **8.04**; PD **8.05**; resistance **8.061–8.10**

191

8.12 Mains electricity (1)

Circuit essentials

A PD (potential difference) is needed to make a current flow round a circuit. PD is measured in volts (V) and is more commonly called voltage. Current is measured in amperes (A).

When you plug a kettle into a mains socket, you are connecting it into a circuit, as shown below. The power comes from a generator in a power station. The supply voltage depends on the country. For household circuits, some countries use a voltage in the range 220–240 V, others in the range 110–130 V.

Mains current is **alternating current** (AC). It flows backwards and forwards, backwards and forwards..... 50 times per second, in some countries. The **mains frequency** is 50 hertz (Hz). In other countries, the mains frequency is 60 Hz. AC is easier to generate than one-way direct current (DC) like that from a battery.

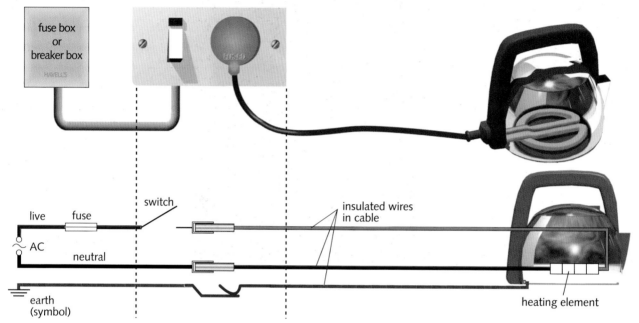

Live (or hot, or active) wire This goes alternately negative and positive, making the current flow backwards and forwards through the circuit.

Neutral (or cold) wire This completes the circuit. In many systems, it is kept at zero voltage by the electricity supply company.

Switch This is fitted in the live wire. It would work equally well in the neutral, but wire in the cable would still be live with the switch OFF. This would be dangerous if, for example, the cable was accidentally cut.

Fuse This is a thin piece of wire which overheats and melts if the current is too high. Like the switch, it is placed in the live wire, often as a cartridge. If a fault develops, and the current gets too high, the fuse 'blows' and breaks the circuit before the cable can overheat and catch fire. Many circuits use a **circuit breaker** instead of a fuse (see the next spread).

Earth (grounded) wire This is a safety wire. It connects the metal body of the kettle to earth and stops it becoming live. For example, if the live wire comes loose and touches the metal body, a current immediately flows to earth and blows the fuse. This means that the kettle is then safe to touch.

Double insulation Some appliances – radios for example – do not have an earth wire. This is because their outer case is made of plastic rather than metal. The plastic acts as an extra layer of insulation around the wires.

This table lamp has an insulating body and does not need an earth wire.

Plugs★

Plugs are a safe and simple way of connecting appliances to the mains. At least 13 different types of plug are in use around the world. You can see an example on the right. In this case, the plug has two metal pins (live and neutral), with an earth connection made by two metal contacts at the edge. Some plugs have a third pin for the earth connection, and some lack the earth altogether.

A few countries use a three-pin plug with a fuse inside, as below. The fuse value might be 3 A, 5 A, or 13 A. This tells you the current needed to blow the fuse. It must be greater than the normal current through the appliance, but as close to it as possible, so that the fuse will blow as soon as the current gets too high.

▲ This two-pin plug has earth connections in grooves at the edge.

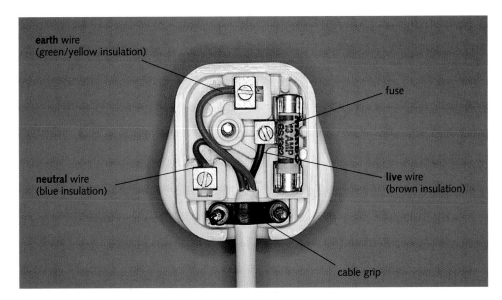

earth wire
(green/yellow insulation)

fuse

neutral wire
(blue insulation)

live wire
(brown insulation)

cable grip

◀ When wiring a plug like this, check the following:

● The wires are connected to the correct terminals, using the colour code shown.

● The cable is held firmly by the grip.

● The correct fuse is fitted.

If you know the power of an appliance, you can use the equation on the right to work out whether a 3 A, 5 A, or 13 A fuse is needed. Here are two examples:

Kettle: 2300 W 230 V

current = power / voltage = 2300 W/230 V = 10 A
So a 13 A fuse is needed.

TV: 115 W 230 V

current = power / voltage = 115 W/230 V = 0.5 A   So a 3 A fuse is needed.

The TV would still work with a 13 A fuse. But if a fault developed, its circuits might overheat and catch fire without the fuse blowing.

Electrical power equation

power	= voltage	× current
(watts)	(volts)	(amperes)
(W)	(V)	(A)

Q

1 In a mains plug, which wire, *live*, *neutral*, or *earth*,
  a) goes alternately + and −
  b) is a safety wire
  c) is cut off from the mains supply if a fuse blows?
2 What is a fuse, and how does it work?
3 Why should the switch always be in the live wire rather than the neutral?
4 Why do some appliances not have an earth wire?
5 Some countries use plugs with a fuse in. Work out whether the plug for each appliance on the right should be fitted with a 3 A, 5 A or a 13 A fuse.

Supply voltage: 230 V	
appliance	*power*
hairdrier	1500 W
drill	400 W
iron	1200 W
table lamp	60 W

Mains electricity (2)

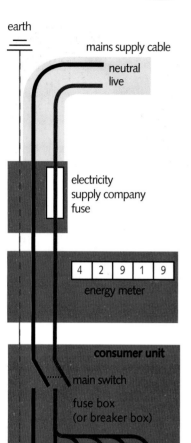

earth

mains supply cable

neutral
live

electricity supply company fuse

4	2	9	1	9

energy meter

consumer unit

main switch

fuse box (or breaker box)

15A 15A 15A 30A 5A

Circuits around the house★

The electricity supply company's cable into each house contains a live and a neutral wire. In the **consumer unit**, these wires branch into several parallel circuits for the lights, cooker, and mains sockets. The cable for most circuits also contains an earth wire.

In the consumer unit, each circuit passes through a fuse or, alternatively, a **circuit breaker**. A circuit breaker is an automatic switch which 'trips' (turns off) when the current rises above the specified value. It can be reset by turning the switch on or by pressing a button. Depending on what it contains, the consumer unit is more commonly called the **fuse box** or the **breaker box**.

For extra safety, some circuits may be fitted with a **residual current device** (**RCD**). This compares the currents in the live and neutral wires. These should be the same. If they are not, then current must be flowing to earth – perhaps through someone touching an exposed wire. The RCD senses the difference and switches off the current before any harm can be done.

Mains sockets★

In most countries, each group of mains sockets (or sometimes, each individual socket) is protected by its own fuse or circuit breaker in the consumer unit, as shown in the diagram on this page.

A few countries use a system in which each appliance is protected by a fuse in its plug (see previous spread). In such cases, as many as ten sockets might be connected to the same cable running from the consumer unit, with a single fuse or circuit breaker to protect the whole cable.

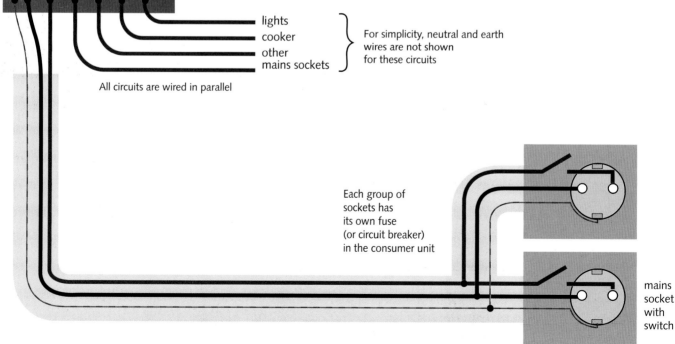

lights
cooker
other
mains sockets

} For simplicity, neutral and earth wires are not shown for these circuits

All circuits are wired in parallel

Each group of sockets has its own fuse (or circuit breaker) in the consumer unit

mains socket with switch

Using two-way switches★

In most houses, you can turn the landing lights on or off from upstairs or downstairs. For this, **two-way switches** are used:

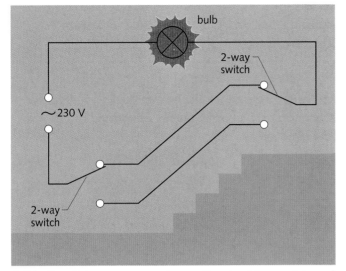

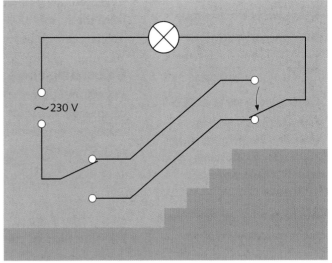

▲ If both switches are up – or down – the circuit is complete and a current flows through the bulb.

▲ But if one switch is up and the other is down, the circuit is broken. Each switch reverses the effect of the other one.

Safety first

Mains electricity can be dangerous. Here are some of the hazards:

- Old, frayed wiring. Broken strands mean that a wire will have a higher resistance at one point. When a current flows through it, the heating effect may be enough to melt the insulation and cause a fire.
- Long extension leads. These may overheat if used when coiled up. The current warms the wire, but the heat has less area to escape from a tight bundle.
- Water in sockets or plugs. Water will conduct a current, so if electrical equipment gets wet, there is a risk that someone might be electrocuted.
- Accidentally cutting cables. With lawnmowers and hedgetrimmers, a plug-in RCD should always be used to avoid the risk of electrocution.

If an accident happens, and someone is electrocuted, you must switch off at the socket and pull out the plug before giving any help.

A plug-in RCD gives protection against the risk of electrocution.

Q

1 If you use a hairdrier (or other appliance) on a mains system as on the left, the current flows through two fuses. Where are these fuses?
2 What is the purpose of a circuit breaker?
3 a) What is the purpose of an RCD?
   b) Why should an RCD always be used with a lawnmower?
4 If an accident occurs and someone is electrocuted, what *two* things must you do before giving help?
5 Copy and complete the circuit diagram on the right so that the bulb can be controlled by either switch.

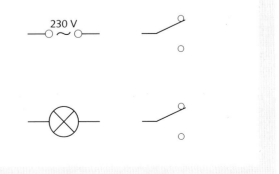

Related topics: resistance and heating effect **8.06**; parallel circuits **8.09**; how a circuit breaker works **9.04**

8.14 Electrical energy calculations

In a circuit, appliances such as kettles, toasters, and food mixers take energy from the supply and transform it (change it into other forms). For example, appliances with heating elements change it into thermal energy (heat).

Calculating energy

Energy and power are linked by the equation in the box on the left. If the power of an appliance is known, the energy transformed in any given time can be calculated by rearranging the equation like this:

energy transformed	= power	× time
(J)	(W)	(s)

For example, if a 1000 W heating element is switched on for 5 seconds (s): energy transformed = 1000 W × 5 s = 5000 J. So the heating element gives off 5000 J of thermal energy.

Electrical energy equation

As power = PD × current, the above equation can also be written like this:

energy transformed = PD	× current	× time
(J) (V)	(A)	(s)

In symbols: $E = VIt$

Example A 12 V water heater takes a current of 2 A. If it is switched on for 60 seconds, how much thermal energy does it produce?

Energy transformed = PD × current × time = 12 V × 2 A × 60 s = 1440 J

In this case, all the energy is transformed into thermal energy, so the heater produces 1440 J of thermal energy.

Measuring energy in kilowatt-hours★

Electricity supply companies use the **kilowatt-hour**, rather than the joule, as their unit of energy measurement:

One kilowatt-hour (kWh) is the energy supplied when an appliance whose power is 1 kW is used for 1 hour.

1 kW is 1000 W, and 1 hour is 3600 s. So, if a 1 kW appliance is used for 1 hour: energy supplied = power × time = 1000 W × 3600 s = 3 600 000 J. Therefore: 1 kWh = 3 600 000 J.

Energy in kilowatt-hours is calculated like this:

energy supplied	= power	× time
(kWh)	(kW)	(hours)

For example:
If a 2 kW heater is used for 3 hours, the energy supplied is 6 kWh.

Batteries are a very convenient, portable source of electricity, but their energy can cost over 200 times more per kilowatt-hour than energy from the mains.

Calculating the cost of electricity★

CENTRAL ELECTRICITY

CUSTOMER
ACCOUNT NO. **3742 463**

PRESENT METER READING	PREVIOUS METER READING	UNITS USED	COST PER UNIT (cu) INCL. TAX	COST (cu)
42935	41710	1225	10	12 250

As currencies differ from one country to another, the examples of prices on this page are given in 'cu', standing for currency unit.

◀ Part of an electricity bill, based on the meter readings below. On most bills, the cost is shown before tax is added, and there may be an additional standing charge to pay as well.

The 'electricity meter' in a house is an energy meter. The more energy you take, the more you have to pay. The reading on the meter gives the total energy supplied in **Units**. The Unit is another name for the kilowatt-hour.

The diagrams on the right show the meter readings at the beginning and end of a quarter (three-month period). In this case:

energy supplied = 42935 kWh − 41710 kWh = 1225 kWh = 1225 Units

If the electricity supply company charges 10cu per Unit:
cost of energy supplied = 1225 × 10cu = 12 250cu

The cost of running individual appliances can be calculated as follows:

Example 1 If energy is 10cu per unit, what is the cost of running a 2 kW heater for 3 hours?

Energy supplied = power × time = 2 kW × 3 h = 6 kWh = 6 Units

As the cost per Unit is 10cu:
total cost = 6 × 10cu = 60cu

Example 2 If energy costs 10cu per unit, what is the cost of running a 100 W lamp for 30 minutes?

To calculate the number of kWh, the power must be in *kilo*watts and the time in *hours*. In this case: 100 W is 0.1 kW, and 30 minutes is 0.5 h. So:

energy supplied = power × time = 0.1 kW × 0.5 h = 0.05 kWh = 0.05 Units

As the cost per Unit is 10cu:
total cost = 0.05 × 10cu = 0.5cu

meter reading

4 1 7 1 0
kWh

meter reading 3 months later

4 2 9 3 5
kWh

1 Calculate the energy supplied to a 60 W bulb
   a) in 1 second **b)** in 1 minute.
2 A bulb takes a current of 3 A from a 12 V battery.
   a) What is the power of the bulb?
   b) How much energy is supplied in 10 minutes?
3 A 2 kW heater is switched on for 4 hours. Calculate the thermal energy given off by the heater
   a) in kWh **b)** in joules.

4 If energy costs 10cu per Unit, calculate the cost of using
   a) a 3 kW electric fire for 5 hours
   b) five 60 W bulbs for 12 hours.
   c) a 1200 W hairdrier for 15 minutes.
5 Someone decides to replace six 60 W filament bulbs with six 15 W low-energy bulbs. If energy costs 10cu per Unit and the bulbs are used, on average, for 4 hours per day, what will the annual saving be?

Related topics: power **4.04**; thermal energy from heaters **5.10**; electrical power **8.11**; mains electricity **8.13**

197

1 a) When a balloon is rubbed in your hair, the balloon becomes negatively charged.
 (i) Explain how the balloon becomes negatively charged. [2]
 (ii) State what you know about the size and sign of the charge left on your hair. [2]
b) The negatively charged balloon is brought up to the surface of a ceiling. The balloon sticks to the ceiling. Explain how and why this happens. [3]

WJEC

2 Read the following passage carefully before answering the questions.

Spraying crops with chemical fertilizers or insecticides has become more efficient. A portable high voltage generator gives the drops of liquid insecticide a small positive charge. This makes the liquid break up into smaller drops and causes the spray to become finer and spread out more.
The plants, which are all reasonable conductors, are in contact with the earth. As the droplets of spray get near the plants, the plants themselves become slightly charged and attract the droplets.

a) (i) Explain why the positive charge on the droplets makes the spray spread out. [1]
 (ii) State what charge appears on the plants as the droplets come near to them. [1]
b) Explain fully how the plants themselves become fully charged. [3]
c) Suggest two reasons why it is an advantage to both the farmer and the environment to use very small charged droplets during insecticide spraying. [2]

WJEC

3

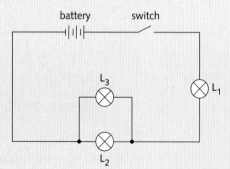

The circuit shows a battery connected to a switch and three identical lamps, L$_1$, L$_2$ and L$_3$.
a) Copy the diagram and add:
 (i) an arrow to show the current direction in the circuit when the switch is closed; [1]
 (ii) a voltmeter V , to measure the voltage across L$_1$; [1]
 (iii) a switch, labelled S, that controls L$_3$ only. [1]
b) State and explain what effect adding another cell to the battery would have on the lamps in the circuit. [2]

WJEC

4 The circuit diagram shows a battery connected to five lamps. The currents through lamps **A** and **B** are shown.

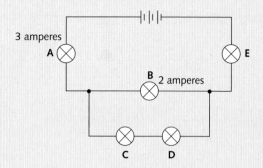

Write down the current flowing through
a) lamp **C**, [1]
b) lamp **E**. [1]

WJEC

5 (i) How much energy is transferred by a battery of EMF 4.5 V when 1.0 C of charge passes through it? [1]
 (ii) How much power is developed in a battery of EMF 4.5 V when a current of 1.0 A is passing through it? [1]

UCLES

6 The diagram shows a circuit which contains two resistors.

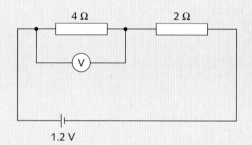

Calculate
a) the total resistance of the two resistors in series, (Ω) [1]
b) the current flowing through the cell, (A) [1]
c) the current flowing through the 4 Ω resistor, (A) [1]
d) the reading of the voltmeter, (V) [1]
e) the power produced in the 4 Ω resistor. (W) [1]

UCLES

7 A small electric hairdryer has an outer case made of plastic. The following information is printed on the case:

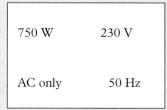

a) Explain the meaning of these terms:
  (i) AC only [1]
  (ii) 50 Hz [1]
b) The hairdrier does not have an earth wire. Instead, it is **double insulated**. Explain what this means. [2]
c) What current does the hairdryer take? [2]
d) The hairdryer is protected by its own fuse.
  (i) What is the purpose of the fuse? [1]
  (ii) Given a choice of a 5 A or a 13 A fuse for the hairdryer, which would you select, and why? [2]
e) If the hairdryer were used in a country where the mains voltage was only 110 V, what difference would this make, and why? [3]

8 A small generator is labelled as having an output of 5 kW, 240 V AC (at constant frequency). It is used to provide emergency lighting for a large building in the event of a breakdown of the mains supply. The circuit is shown below.

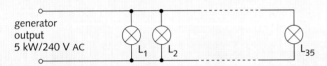

There are 35 light fittings on the circuit, each with a 240 V, 60 W lamp.
a) Calculate the maximum current which the generator is designed to supply. [2]
b) (i) Calculate the power needed when all the lamps are turned on at the same time.
  (ii) Explain why this generator is suitable for supplying the power required but would not be suitable if all the 60 W lamps were exchanged for 150 W lamps. [4]

c) Write down two reasons why all the lamps are connected in parallel rather than in series. In each answer, you should refer to both types of circuit. [4]
d) Calculate the resistance of the filament of each 60 W lamp. [4]
e) The figure below shows the current output of the generator when it is supplying all 35 of the 60 W lamps.

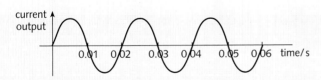

  (i) Calculate the frequency of the supply from the generator.
  (ii) Copy the diagram and sketch another graph to show the approximate current output of the generator when 17 lamps are removed from their fittings. [4]

9 A student investigates how the current through a lamp varies with the voltage (PD) across it. She uses the circuit shown below.

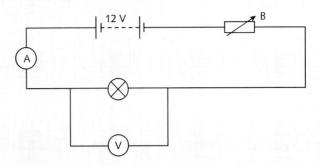

a) Three of the components are labelled, A, V, and B. Write down what each one is. [3]
b) Describe how the student should carry out the experiment. [3]
From her results, the student plots this graph:

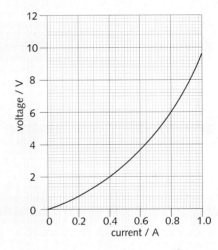

c) What is the current when the voltage across the lamp is 2.0 V? [1]
d) What is the resistance of the lamp when the voltage across it is 2.0 V? [2]
e) What is the resistance of the lamp when the voltage across it is 6.0 V? [2]
f) What happens to the resistance of the lamp as the voltage across it is increased? [1]

10 A small electric heater takes a power of 60 W from a 12 V supply.
  a) What is the current through the heater? [2]
  b) What is the resistance of the heater? [2]
  c) How much charge (in C) passes through the heater in 20 seconds? [2]
  d) How much energy (in J) is transformed by the heater in 20 seconds? [2]

Photocopy the list of topics below and tick the boxes of the ones that are included in your examination syllabus. (Your teacher should be able to tell you which they are.) Use your list when you revise. The spread number in brackets tells you where to find more information.

❑ 1 The two types of electric charge. (8.01)

❑ 2 Charges attracting and repelling. (8.01)

❑ 3 How charges come from the atom. (8.01)

❑ 4 Electrical conductors and insulators. (8.01)

❑ 5 Why charged objects attract uncharged ones. (8.02)

❑ 6 The need for earthing to prevent charge build-up. (8.02)

❑ 7 Detecting charge. (8.02)

❑ 8 Induced charges. (8.02)

❑ 9 The SI unit of charge: the coulomb. (8.02)

❑ 10 Uses of electrostatic charge, including the photocopier and electrostatic precipitator. (8.02)

❑ 11 Electric fields and their features. (8.03)

❑ 12 Ionization of air. (8.03)

❑ 13 The basic principles of a simple circuit. (8.04)

❑ 14 Measuring current. (8.04)

❑ 15 The SI unit of current: the ampere. (8.04)

❑ 16 The equation linking charge and current. (8.04)

❑ 17 The conventional current direction. (8.04)

❑ 18 Measuring PD (voltage). (8.05)

❑ 19 The meaning of EMF. (8.05)

❑ 20 The SI unit of PD (voltage): the volt. (8.05)

❑ 21 Rule linking the PDs round a circuit. (8.05)

❑ 22 The equation linking resistance, PD, and current. (8.06 and 8.07)

❑ 23 The SI unit of resistance: the ohm. (8.06)

❑ 24 Factors affecting the resistance of a wire. (8.06 and 8.08)

❑ 25 Using the heating effect of a current. (8.06)

❑ 26 Resistors, variable resistors, thermistors, light-dependent resistors, and diodes. (8.06)

❑ 27 Ohm's law. (8.07)

❑ 28 Interpreting current–PD graphs. (8.07)

❑ 29 The properties of circuits with components in series and in parallel. (8.07)

❑ 30 Using the relationship between the resistance, length, and cross-sectional area of a wire. (8.08)

❑ 31 The PD across cells in series, and in parallel. (8.09)

❑ 32 Equations for the combined resistance of resistors in series and in parallel. (8.10)

❑ 33 How to solve circuit problems. (8.10)

❑ 34 Calculating electrical power (in watts). (8.11)

❑ 35 The equation linking power, PD (voltage), and current. (8.11)

❑ 36 Calculating the power dissipated in a resistor (lost because of the heating effect). (8.10)

❑ 37 The difference between AC and DC. (8.12)

❑ 38 The features of a mains circuit. (8.12)

❑ 39 The function of a fuse. (8.12)

❑ 40 How an earth wire makes a mains circuit safer. (8.12)

❑ 41 Why switches, fuses, and circuit breakers should be in the live wire. (8.12)

❑ 42 How to wire a mains plug safely. (8.12)

❑ 43 Choosing fuses of the correct value. (8.12)

❑ 44 Household mains circuits. (8.13)

❑ 45 The hazards of mains electricity. (8.13)

❑ 46 The use of circuit breakers in mains circuits. (8.13)

❑ 47 Calculating electrical energy, and its cost. (8.14)

❑ 48 Measuring energy in kilowatt-hours. (8.14)

Magnets and Currents

- MAGNETS - MAGNETIC FIELDS - MAGNETIC EFFECT OF A CURRENT - ELECTROMAGNETS
- MAGNETIC FORCE ON A CURRENT - ELECTRIC MOTORS - ELECTROMAGNETIC INDUCTION
- GENERATORS - TRANSFORMERS - POWER TRANSMISSION AND DISTRIBUTION

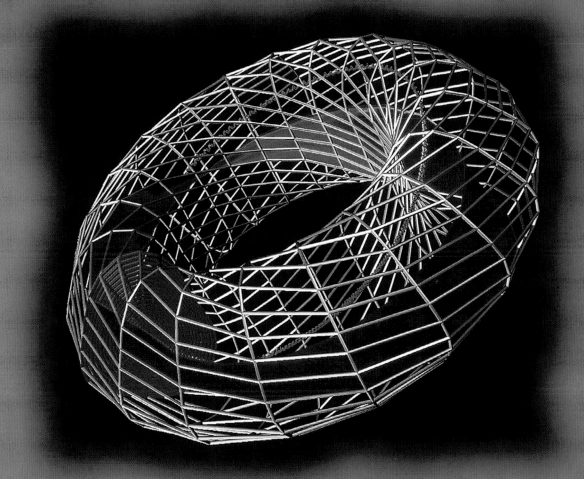

Computer model of the magnetic field inside the doughnut-shaped chamber of a nuclear fusion reactor. Like the Sun, fusion reactors release energy by smashing hydrogen atoms together to form helium. One day, they may provide the energy to run power stations on Earth.

In the reactor, the magnetic field is used to trap charged particles from hydrogen at a temperature of over 100 million °C.

9.01 Magnets

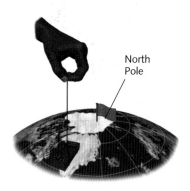

North Pole

Magnetic poles

If a small bar magnet is dipped into iron filings, the filings are attracted to its ends, as shown in the photograph on the opposite page. The magnetic force seems to come from two points, called the **poles** of the magnet.

The Earth exerts forces on the poles of a magnet. If a bar magnet is suspended as on the left, it swings round until it lies roughly north–south. This effect is used to name the two poles of a magnet. These are called:

- the **north-seeking pole** (or **N pole** for short)
- the **south-seeking pole** (or **S pole** for short).

If you bring the ends of two similar bar magnets together, there is a force between the poles as shown below:

Like poles repel; unlike poles attract.
The closer the poles, the greater the force between them.

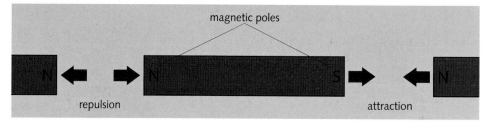

magnetic poles

repulsion

attraction

Induced magnetism

Materials such as iron and steel are attracted to magnets because they themselves become magnetized when there is a magnet nearby. The magnet **induces** magnetism in them, as shown below. In each case, the induced pole nearest the magnet is the *opposite* of the pole at the end of the magnet. The attraction between unlike poles holds each piece of metal to the magnet.

The steel and the iron behave differently when pulled right away from the magnet. The steel keeps some of its induced magnetism and becomes a **permanent magnet**. However, the iron loses virtually all of its induced magnetism. It was only a **temporary magnet**.

Properties of magnets

A magnet

- has a magnetic field around it (see the next spread).

- has two opposite poles (N and S) which exert forces on other magnets. Like poles repel; unlike poles attract.

- will attract magnetic materials by inducing magnetism in them. In some materials (e.g steel) the magnetism is permanent. In others (e.g. iron) it is temporary.

- will exert little or no force on a non-magnetic material.

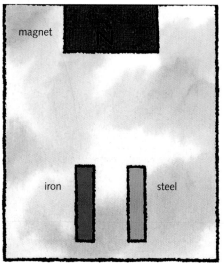

magnet

iron   steel

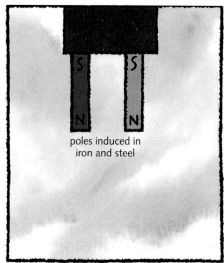

poles induced in iron and steel

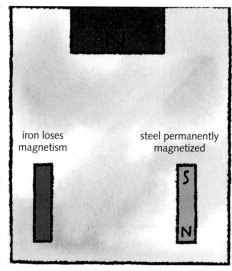

iron loses magnetism

steel permanently magnetized

Making a magnet

A piece of steel becomes permanently magnetized when placed near a magnet, but its magnetism is usually weak. It can be magnetized more strongly by stroking it with one end of a magnet, as on the right. However, the most effective method of magnetizing it is to place it in a long coil of wire and pass a large, direct (one-way) current through the coil. The current has a magnetic effect which magnetizes the steel.

Magnetic and non-magnetic materials

A **magnetic material** is one which which can be magnetized and is attracted to magnets. All strongly magnetic materials contain iron, nickel, or cobalt. For example, steel is mainly iron. Strongly magnetic metals like this are called **ferromagnetics**. They are described as *hard* or *soft* depending on how well they keep their magnetism when magnetized:

Hard magnetic materials such as steel, and alloys called Alcomax and Magnadur, are difficult to magnetize but do not readily lose their magnetism. They are used for permanent magnets.

Soft magnetic materials such as iron and Mumetal are relatively easy to magnetize, but their magnetism is only temporary. They are used in the cores of electromagnets and transformers because their magnetic effect can be 'switched' on or off or reversed easily.

Non-magnetic materials include metals such as brass, copper, zinc, tin, and aluminium, as well as non-metals.

Where magnetism comes from★

In an atom, tiny electrical particles called electrons move around a central nucleus. Each electron has a magnetic effect as it spins and orbits the nucleus.

In many types of atom, the magnetic effects of the electrons cancel, but in some they do not, so each atom acts as a tiny magnet. In an unmagnetized material, the atomic magnets point in random directions. But as the material becomes magnetized, more and more of its atomic magnets line up with each other.

Together, billions of tiny atomic magnets act as one big magnet.

If a magnet is hammered, its atomic magnets are thrown out of line: it becomes **demagnetized**. Heating it to a high temperature has the same effect.

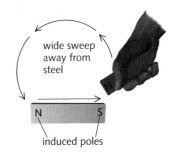

Magnetizing a piece of steel by stroking it with a magnet.

Ferrous and non-ferrous

Iron and alloys (mixtures) containing iron are called **ferrous** metals (*ferrum* is Latin for iron). Aluminium, copper, and the other non-magnetic metals are **non-ferrous**.

Magnetic materials are attracted to magnets and can be made into magnets.

1 What is meant by the *N pole* of a magnet?
2 Magnetic materials are sometimes described as *hard* or *soft*.
   a) What is the difference between the two types?
   b) Give one example of each type.
3 Name *three* ferromagnetic metals.
4 Name *three* non-magnetic metals.
5 The diagram on the right shows three metal bars. When different ends are brought together, it is found that A and B attract, A and C attract, but A and D repel. Decide whether each of the bars is a permanent magnet or not.

A   bar 1

B   C   bar 2

D   bar 3

Related topics: atoms and electrons 8.01; the Earth's magnetism 9.02; electromagnets 9.04; transformers 9.10–9.11

Magnetic fields

In the photograph below, iron filings have been sprinkled on paper over a bar magnet. The filings have become tiny magnets, pulled into position by forces from the poles of the magnet. Scientifically speaking, there is a **magnetic field** around the magnet, and this exerts forces on magnetic materials in it.

Magnetic field patterns

Magnetic fields can be investigated using a small **compass**. The 'needle' is a tiny magnet which is free to turn on its spindle. When near a magnet, the needle is turned by forces between its poles and the poles of the magnet.
The needle comes to rest so that the turning effect is zero.

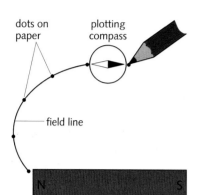

dots on paper
plotting compass
field line
N   S

The diagram on the left shows how a small compass can be used to plot the field around a bar magnet. Starting with the compass near one end of the magnet, the needle position is marked using two dots. Then the compass is moved so that the needle lines up with the previous dot... and so on. When the dots are joined up, the result is a magnetic **field line**. More lines can be drawn by starting with the compass in different positions.

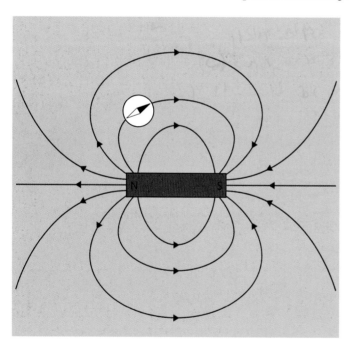

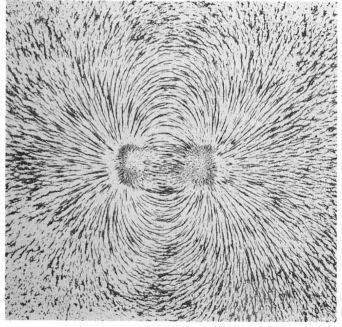

Magnet essentials

A magnet has a north-seeking (N) pole at one end and a south-seeking (S) pole at the other. When two magnets are brought together:

like poles repel, unlike poles attract.

In the diagram above, a selection of field lines has been used to show the magnetic field around a bar magnet:
- The field lines run from the N pole to the S pole of the magnet. The field direction, shown by an arrowhead, is defined as the direction in which the force on a N pole would act. It is the direction in which the N end of a compass needle would point.
- The magnetic field is strongest where the field lines are closest together.

If two magnets are placed near each other, their magnetic fields combine to produce a single field. Two examples are shown at the top of the next page. At the **neutral point**, the field from one magnet exactly cancels the field from the other, so the magnetic force on anything at this point is zero.

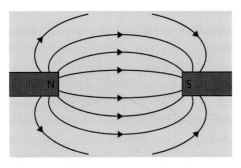

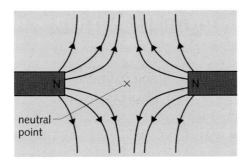

◀ Between magnets with unlike poles facing, the combined field is almost uniform (even) in strength. However, between like poles, there is a neutral point where the combined field strength is zero.

The Earth's magnetic field★

The Earth has a magnetic field. No one is sure of its cause, although it is thought to come from electric currents generated in the Earth's core. The field is rather like that around a large, but very weak, bar magnet.

With no other magnets near it, a compass needle lines up with the Earth's magnetic field. The N end of the needle points north. But an N pole is always attracted to an S pole. So it follows that the Earth's magnetic S pole must be in the north! It lies under a point in Canada called **magnetic north**.

Magnetic north is over 1200 km away from the Earth's geographic North Pole. This is because the Earth's magnetic axis is not quite in line with its north–south axis of rotation.

Magnetic screening

Some electronic equipment is easily upset by magnetic fields from nearby generators, motors, transformers, or the Earth. The equipment can be screened (shielded) by enclosing it in a layer of a soft magnetic material, such as iron or nickel. This redirects the field so that it does not pass through the equipment.

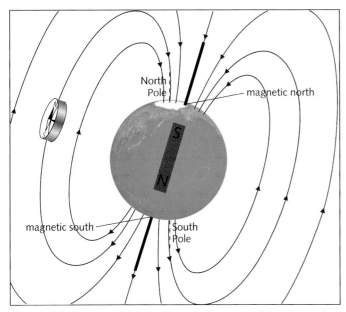

The Earth behaves as if it has a large but very weak bar magnet inside it.

A compass is of no use in polar regions because the Earth's magnetic field lines are vertical.

1 In the diagrams on the right, the same compass is being used in both cases.
   a) Copy diagram A. Label the N and S ends of the compass needle.
   b) Copy diagram B. Mark in the poles of the magnet to show which is N and which is S. Then draw an arrowhead on the field line to show its direction.
   c) In diagram B, at which position, X or Y, would you expect the magnetic field to be stronger?

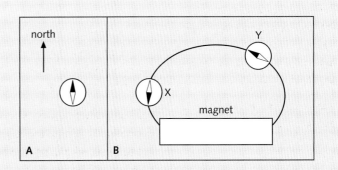

9.03 Magnetic effect of a current

Magnetic field around a wire

If an electric current is passed through a wire, as shown below left, a weak magnetic field is produced. The field has these features:

- the magnetic field lines are circles
- the field is strongest close to the wire
- increasing the current increases the strength of the field.

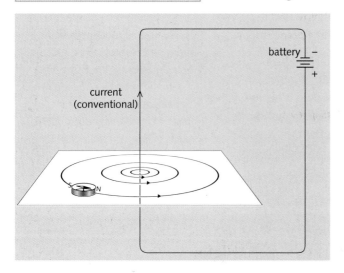

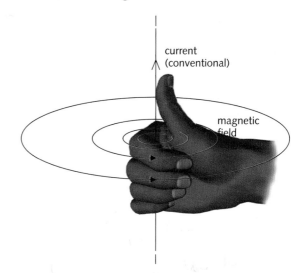

A rule for field direction The direction of the magnetic field produced by a current is given by the **right-hand grip rule** shown above right. Imagine gripping the wire with your right hand so that your thumb points in the conventional current direction. Your fingers then point in the same direction as the field lines.

Magnetic fields from coils

A current produces a stronger magnetic field if the wire it flows through is wound into a coil. The diagrams below show the magnetic field patterns produced by two current-carrying coils. One is just a single turn of wire. The other is a long coil with many turns. A long coil is called a **solenoid**.

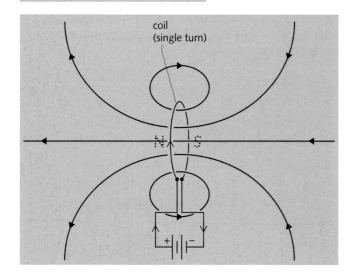

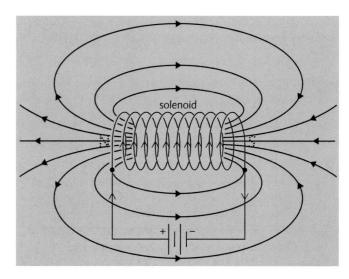

The magnetic field produced by a current-carrying coil has these features:
- the field is similar to that from a bar magnet, and there are magnetic poles at the ends of the coil
- increasing the current increases the strength of the field
- increasing the number of turns on the coil increases the strength of the field.

A rule for poles★ To work out which way round the poles are, you can use another **right-hand grip rule**, as shown on the right. Imagine gripping the coil with your right hand so that your fingers point in the conventional current direction. Your thumb then points towards the N pole of the coil.

Magnets are made – and demagnetized – using coils, as shown below. In audio and video cassette recorders, tiny coils are used to put magnetic patterns on tape. The patterns store sound and picture information.

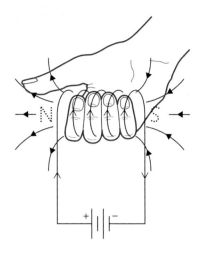

Right-hand grip rule for poles

Making a magnet

Above, a steel bar has been placed in a solenoid. When a current is passed through the solenoid, the steel becomes magnetized and makes the magnetic field much stronger than before. And when the current is switched off, the steel stays magnetized. Nearly all permanent magnets are made in this way.

Demagnetizing a magnet

Above, a magnet is slowly being pulled out of a solenoid through which an alternating current is passing. Alternating current (AC) flows backwards, forwards, backwards, forwards... and so on. It produces a magnetic field which changes direction very rapidly and throws the atoms in the magnet out of line.

---

1 The coil in diagram A is producing a magnetic field.
   a) Give *two* ways in which the strength of the field could be increased.
   b) How could the direction of the field be reversed?
   c) Copy the diagram. Show the conventional current direction and the N and S poles of the coil.
2 Redraw diagram B to show which way the compass needles point when a current flows through the wire. (Assume that the black end of each compass needle is a N pole, the conventional current direction is away from you, into the paper, and that the only magnetic field is that due to the current.)

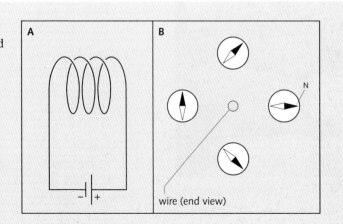

Related topics: current in a circuit 8.04; alternating current 8.12; magnetic poles 9.01; magnetic fields 9.02

9.04 Electromagnets

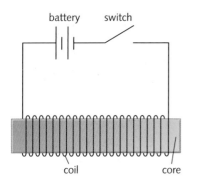

A simple electromagnet

Unlike an ordinary magnet, an **electromagnet** can be switched on and off. In a simple electromagnet, a **coil**, consisting of several hundred turns of insulated copper wire, is wound round a **core**, usually of iron or Mumetal. When a current flows through the coil, it produces a magnetic field. This magnetizes the core, creating a magnetic field about a thousand times stronger than the coil by itself. With an iron or Mumetal core, the magnetism is only temporary, and is lost as soon as the current through the coil is switched off. Steel would not be suitable as a core because it would become permanently magnetized.

The strength of the magnetic field is increased by:
- increasing the current
- increasing the number of turns in the coil.

Reversing the current reverses the direction of the magnetic field.

The following all make use of electromagnets.

The magnetic relay

A magnetic relay is a switch operated by an electromagnet. With a relay, a small switch with thin wires can be used to turn on the current in a much more powerful circuit – for example, one with a large electric motor in it:

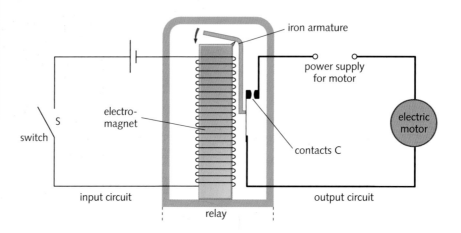

When the switch S in the input circuit is closed, a current flows through the electromagnet. This pulls the iron armature towards it, which closes the contacts C. As a result, a current flows through the motor.

The relay above is of the 'normally open' type: when the input switch is OFF, the output circuit is also OFF. A 'normally closed' relay works the opposite way: when the input switch is OFF, the output circuit is ON. In practice, most relays are made so that they can be connected either way.

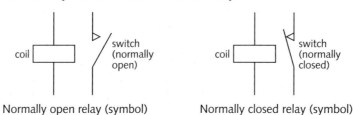

Normally open relay (symbol)   Normally closed relay (symbol)

> **Magnetic essentials**
>
> A **hard** magnetic material (for example, steel) is one which, when magnetized, does not readily lose its magnetism.
>
> A **soft** magnetic material (for example, iron) quickly loses its magnetism when the magnetizing field is removed.

With a relay, a small switch can be used to turn on a powerful starter motor.

The circuit breaker

A circuit breaker is an automatic switch which cuts off the current in a circuit if this rises above a specified value. It has the same effect as a fuse but, unlike a fuse, can be reset (turned ON again) after it has tripped (turned OFF).

In the type shown on the right, the current flows through two contacts and also through an electromagnet. If the current gets too high, the pull of the electromagnet becomes strong enough to release the iron catch, so the contacts open and stop the current. Pressing the reset button closes the contacts again.

Magnetic storage★

Although CDs and DVDs are popular, many people still use magnetic tape for recording sounds and TV pictures. The tape consists of a long, thin, plastic strip, coated with a layer of iron oxide or similar material. Magnetically, iron oxide is between soft and hard. Once magnetized it keeps its magnetism, but is relatively easy to demagnetize, ready for another recording.

The diagram below shows how sound is recorded on tape. The hard drive in a computer also stores data as a pattern of varying magnetism. In both examples, an electromagnet creates the varying magnetic field needed for the recording. Later, a playback head can read the pattern and produce a varying current.

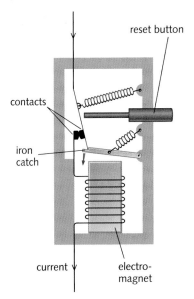

▲ Circuit breaker

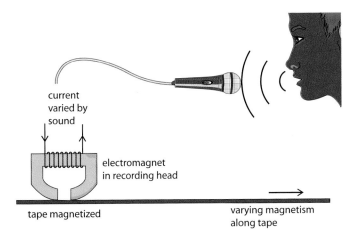

▲ **Recording on magnetic tape** The incoming sound waves are used to vary the current through a tiny electromagnet in the recording head. As the tape moves past the head, a track of varying magnetism is created along the tape.

▲ **Computer hard drive** The recording head is at the end of the arm. It contains a tiny electromagnet which is used to create tracks of varying magnetism on a spinning disc. The disc is made of aluminium or glass, and is coated with a layer of magnetic material similar to that on a tape.

 Q

1 An electromagnet has a core.
   a) What is the purpose of the core?
   b) Why is iron a better material for the core than steel?
   c) Write down *two* ways of increasing the strength of the magnetic field from an electromagnet.

2 In the diagram on the opposite page, an electric motor is controlled by a switch connected to a relay.
   a) What is the advantage of using a relay, rather than a switch in the motor circuit itself?
   b) Why does the motor start when switch S is closed?

3 The diagram at the top of the page shows a circuit breaker.
   a) What is the purpose of the circuit breaker?
   b) How do you think the performance of the circuit breaker would be affected if the coil of the electromagnet had more turns?

4 Sounds can be recorded on tape.
   a) Why is an electromagnet needed for this?
   b) Why must the coating on the tape be between soft and hard magnetically?

9.05 Magnetic force on a current

In the experiment shown below, a length of copper wire has been placed in a magnetic field. Copper is non-magnetic, so it is feels no force from the magnet. However, with a current passing through it, there is a force on the wire. The force arises because the current produces its own magnetic field which acts on the poles of the magnet. In this case, the force on the wire is upwards (see box below left). It would be downwards if either the magnetic field or the current were reversed. Whichever way the experiment is done, the wire moves *across* the field. It is *not* attracted to either pole.

The force is increased if:
- the current is increased
- a stronger magnet is used
- the length of wire in the field is increased.

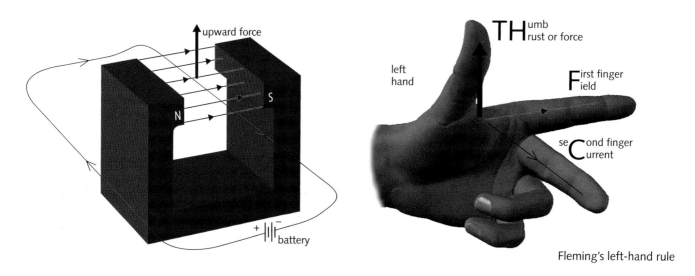

Fleming's left-hand rule

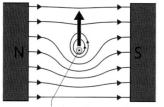

current direction out of paper

By itself, the current in a straight wire produces a circular magnetic field pattern. However, when the wire is between the poles of a magnet, the combined field is as above. In situations like this, the field lines tend to straighten. So, in this case, the wire gets pushed upwards.

Fleming's left-hand rule

In the above experiment, the direction of the force can be predicted using **Fleming's left-hand rule**, as illustrated above right. If you hold the thumb and first two fingers of your left hand at right angles, and point the fingers as shown, the thumb gives the direction of the force.

In applying the rule, it is important to remember how the field and current directions are defined:
- The field direction is from the N pole of a magnet to the S pole.
- The current direction is from the positive (+) terminal of a battery round to the negative (−). This is called the *conventional* current direction.

Fleming's left-hand rule only applies if the current and field directions are at right angles. If they are at some other angle, there is still a force, but its direction is more difficult to predict. If the current and field are in the *same* direction, there is *no* force.

Several devices use the fact that there is a force on a current-carrying conductor in a magnetic field. They include the loudspeaker and meter described on the next page and the electric motors on the next spread.

The moving-coil loudspeaker★

Most loudspeakers are of the moving-coil type shown on the right. The cylindrical magnet produces a strong radial ('spoke-like') magnetic field at right angles to the wire in the coil. The coil is free to move backwards and forwards and is attached to a stiff paper or plastic cone.

The loudspeaker is connected to an amplifier which gives out alternating current. This flows backwards, forwards, backwards... and so on, causing a force on the coil which is also backwards, forwards, backwards.... As a result, the cone vibrates and gives out sound waves. The sound you hear depends on how the amplifier makes the current alternate.

Turning effect on a coil

The coil below lies between the poles of a magnet. The current flows in opposite directions along the two sides of the coil. So, according to Fleming's left-hand rule, one side is pushed *up* and the other side is pushed *down*. In other words, there is a turning effect on the coil. With more turns on the coil, the turning effect is increased.

The meter in the photograph uses the above principle. Its pointer is attached to a coil in the field of a magnet. The higher the current through the meter, the further the coil turns against the springs holding it, and the further the pointer moves along the scale.

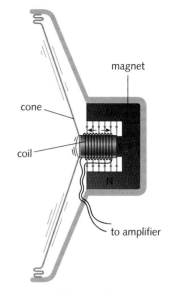

Moving-coil loudspeaker

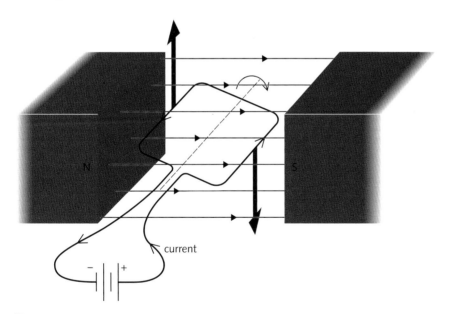

Moving-coil meter

Q

1 There is a force on the wire in the diagram on the right.
   a) Give *two* ways in which the force could be increased.
   b) Use Fleming's left-hand rule to work out the direction of the force.
   c) Give *two* ways in which the direction of the force could be reversed.
2 Explain why the cone of a loudspeaker vibrates when alternating current passes through its coil.
3 The diagram above shows a current-carrying coil in a magnetic field. What difference would it make if
   a) there were more turns of wire in the coil
   b) the direction of the current were reversed?

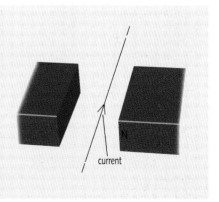

Related topics: sound waves **6.03**; current in a circuit **8.04**; magnetic fields **9.02**; field around a wire **9.03**; using a loudspeaker **10.01**

211

Electric motors

Turning effect on a coil

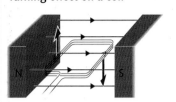

When a current flows through this coil, there is an upward force on one side and a downward force on the other. The direction of each force is given by Fleming's left-hand rule, explained on the previous spread.

If a coil is carrying a current in a magnetic field, as on the left, the forces on it produce a turning effect. Many electric motors use this principle.

A simple DC motor

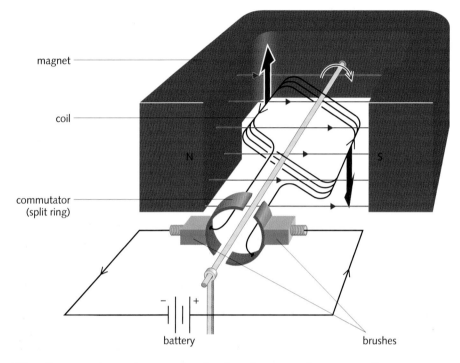

magnet

coil

commutator (split ring)

battery

brushes

The diagram above shows a simple electric motor. It runs on direct current (DC), the 'one-way' current that flows from a battery.

The coil is made of insulated copper wire. It is free to rotate between the poles of the magnet. The **commutator**, or split-ring, is fixed to the coil and rotates with it. Its action is explained below and in the diagrams on the left. The **brushes** are two contacts which rub against the commutator and keep the coil connected to the battery. They are usually made of carbon.

When the coil is horizontal, the forces are furthest apart and have their maximum turning effect (leverage) on the coil. With no change to the forces, the coil would eventually come to rest in the vertical position. However, as the coil overshoots the vertical, the commutator changes the direction of the current through it. So the forces change direction and push the coil further round until it is again vertical... and so on. In this way, the coil keeps rotating clockwise, half a turn at a time. If either the battery or the poles of the magnet were the other way round, the coil would rotate anticlockwise.

The action of the commutator

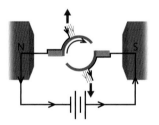

When the coil is nearly vertical, the forces cannot turn it much further...

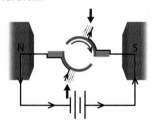

...but when the coil overshoots the vertical, the commutator changes the direction of the current through it, so the forces change direction and keep the coil turning.

The turning effect on the coil can be increased by:
- increasing the current
- using a stronger magnet
- increasing the number of turns on the coil
- increasing the area of the coil. (A longer coil means higher forces because there is a greater length of wire in the magnetic field; a wider coil gives the forces more leverage.)

Practical motors★

The simple motor on the opposite page produces a low turning effect and is jerky in action, especially at low speeds. Practical motors give a much better performance for these reasons:

- Several coils are used, each set at a different angle and each with its own pair of commutator segments (pieces), as shown on the right. The result is a greater turning effect and smoother running.
- The coils contain hundreds of turns of wire and are wound on a core called an **armature**, which contains iron. The armature becomes magnetized and increases the strength of the magnetic field.
- The pole pieces are curved to create a radial ('spoke-like') magnetic field. This keeps the turning effect at a maximum for most of the coil's rotation.

In some motors, the field is provided by an electromagnet rather than a permanent magnet. One advantage is that the motor can be run from an alternating current (AC) supply. As the current flows backwards and forwards in the coil, the field from the electromagnet changes direction to match it, so the turning effect is always the same way and the motor rotates normally. The mains motors in drills and food mixers work like this.

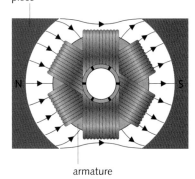

curved pole piece

armature

Practical motors have curved pole pieces, and several coils wound on an iron armature.

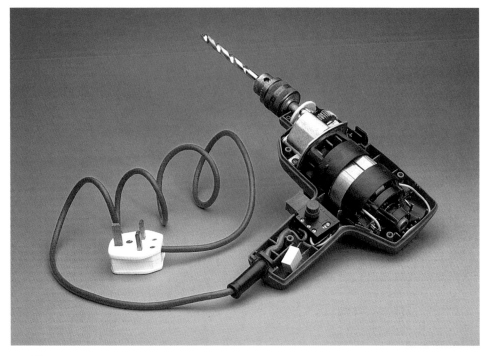

◄ In this electric drill, the motor is in the centre. Note the commutator segments at the right hand end, and the electromagnet.

1 Which part(s) of an electric motor
   a) connect the power supply to the split-ring and coil
   b) changes the current direction every half-turn?
2 On the right, there is an end view of the coil in a simple electric motor.
   a) Redraw the diagram to show the position of the coil when the turning effect on it is i) maximum ii) zero.
   b) Give *three* ways in which the maximum turning effect on the coil could be increased.
   c) Use Fleming's left-hand rule to work out which way the coil will turn.
3 What is the advantage of using an electromagnet in an electric motor, rather than a permanent magnet?

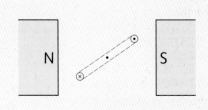

⊗ = current into paper
⊙ = current out of paper

Related topics: current **8.04**; AC and DC **8.12**; magnetic fields **9.02**; electromagnets **9.04**; Fleming's left-hand rule and turning effect **9.05**

Electromagnetic induction

A current produces a magnetic field. However, the reverse is also possible: a magnetic field can be used to produce a current.

Induced EMF and current in a moving wire

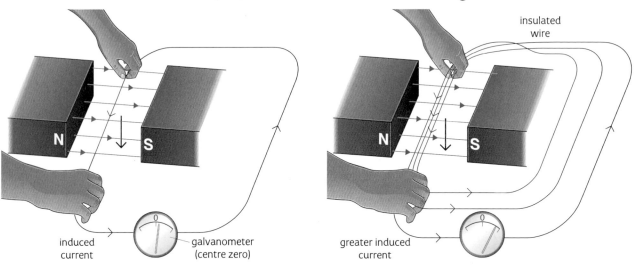

induced current — galvanometer (centre zero)

greater induced current

insulated wire

Circuit essentials

For a current to flow in a circuit, the circuit must be complete, with no breaks in it. Also, there must a source of EMF (voltage) to provide the energy. A battery is one such source. Others include a wire moving through a magnetic field, as explained on the right.

EMF stands for electromotive force. It is measured in volts.

When a wire is moved across a magnetic field, as shown above left, a small EMF (voltage) is generated in the wire. The effect is called **electromagnetic induction**. Scientifically speaking, an EMF is **induced** in the wire. If the wire forms part of a complete circuit, the EMF makes a current flow. This can be detected by a meter called a **galvanometer**, which is sensitive to very small currents. The one shown in the diagram is a centre-zero type. Its pointer moves to the left or right of the zero, depending on the current direction.

The induced EMF (and current) can be increased by:

● moving the wire faster
● using a stronger magnet
● increasing the length of wire in the magnetic field – for example, by looping the wire through the field several times, as shown above right.

The above results are summed up by **Faraday's law of electromagnetic induction**. In simplified form, this can be stated as follows:

> The EMF induced in a conductor is proportional to the rate at which magnetic field lines are cut by the conductor.

In applying this law, remember that field lines are used to represent the strength of a magnetic field as well as its direction. The closer together the lines, the stronger the field.

Either of the following will reverse the direction of the induced EMF and current:

● moving the wire in the opposite direction
● turning the magnet round so that the field direction is reversed.

If the wire is not moving, or is moving parallel to the field lines, there is no induced EMF or current.

Magnet essentials

The N and S poles of one magnet exert forces on those of another:

like poles repel, unlike poles attract.

The magnetic field around a magnet can be represented by field lines. These show the direction in which the force on an N pole would act.

Induced EMF and current in a coil

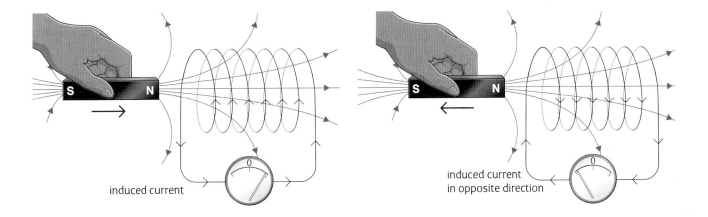

induced current

induced current
in opposite direction

If a bar magnet is pushed into a coil, as shown above left, an EMF is induced in the coil. In this case, it is the magnetic field that is moving rather than the wire, but the result is the same: field lines are being cut. As the coil is part of a complete circuit, the induced EMF makes a current flow.

The induced EMF (and current) can be increased by:
● moving the magnet faster
● using a stronger magnet
● increasing the number of turns on the coil (as this increases the length of wire cutting through the magnetic field).

Experiments with the magnet and coil also give the following results.

● If the magnet is pulled *out of* the coil, as shown above right, the direction of the induced EMF (and current) is reversed.
● If the S pole of the magnet, rather than the N pole, is pushed into the coil, this also reverses the current direction.
● If the magnet is held still, no field lines are cut, so there is no induced EMF or current.

The playback heads in audio and video cassette recorders contain tiny coils. A tiny, varying EMF is induced in the coil as the magnetized tape passes over it and field lines are cut by the coil. In this way, the magnetized patterns on the tape are changed into electrical signals which can be used to recreate the original sound or picture.

The pick-ups under the strings of this guitar are tiny coils with magnets inside them. The steel strings become magnetized. When they vibrate, current is induced in the coils, boosted by an amplifier, and used to produce sound.

1 The wire on the right forms part of a circuit. When the wire is moved downwards, a current is induced in it. What would be the effect of
a) moving the wire upwards through the magnetic field
b) holding the wire still in the magnetic field
c) moving the wire parallel to the magnetic field lines?
2 In the experiment at the top of the page, what would be the effect of
a) moving the magnet faster
b) turning the magnet round, so that the S pole is pushed into the coil
c) having more turns on the coil?

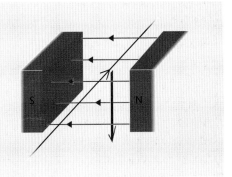

9.08 # More about induced currents

Magnetic essentials

Like magnetic poles repel; unlike ones attract. Magnetic field lines run from the N pole of a magnet to the S pole.

In diagrams, the conventional current direction is used. This runs from the + of the supply to the −.

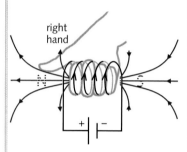

right hand

A current-carrying coil produces a magnetic field. The **right-hand grip rule** above tells you which end is the N pole. It is the end your thumb points at when your fingers point the same way as the current.

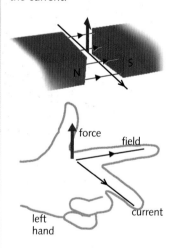

force
field
left hand
current

If a current-carrying wire is in a magnetic field as above, the direction of the force is given by **Fleming's left-hand rule**.

If a conductor is moving through a magnetic field, or in a changing field, an EMF (voltage) is induced in it.

Induced current direction: Lenz's law

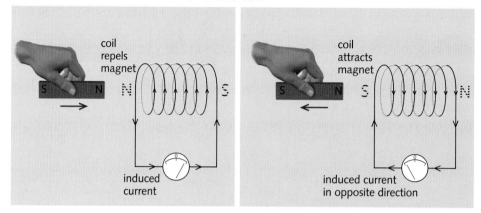

coil repels magnet — induced current

coil attracts magnet — induced current in opposite direction

If a magnet is moved in or out of a coil, a current is induced in the coil. The direction of this current can be predicted using **Lenz's law**:

> An induced current always flows in a direction such that it opposes the change which produced it.

Above, for example, the induced current turns the coil into a weak electromagnet whose N pole *opposes* the approaching N pole of the magnet. When the magnet is pulled *out of* the coil, the induced current alters direction and the poles of the coil are reversed. This time, the coil attracts the magnet as it is pulled away. So, once again, the change is opposed.

Lenz's law is an example of the law of conservation of energy. Energy is spent when a current flows round a circuit, so energy must be spent to induce the current in the first place. In the example above, you have to spend energy to move the magnet against the opposing force.

Induced current direction: Fleming's right-hand rule★

If a straight wire (in a complete circuit) is moving at right angles to a magnetic field, the direction of the induced current can be found using **Fleming's right-hand rule**, as shown below:

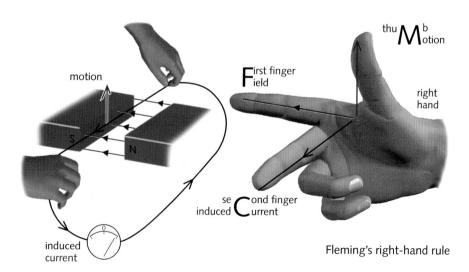

motion

induced current

thu**M**b otion

First finger ield

right hand

se**C**ond finger
induced urrent

Fleming's right-hand rule

On the opposite page, there is information about Fleming's right-hand and left-hand rules. The two rules apply to different situations:

● when a *current* causes *motion*, the *left*-hand rule applies
● when *motion* causes a *current*, the *right*-hand rule applies.

Fleming's right-hand rule follows from the left-hand rule and Lenz's law. The diagram on the right illustrates this. Here, the upward motion induces a current in the wire. The induced current is in the magnetic field, so there is a force on it whose direction is given by the *left*-hand rule. The force must be downwards to *oppose* the motion, so you can use this fact and the left-hand rule to work out which way the current must flow. However, the *right*-hand rule gives the same result – without you having to reason out all the steps!

motion

force on induced current opposes motion

Eddy currents★

spinning aluminium disc

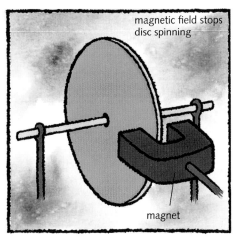

magnetic field stops disc spinning

magnet

If the aluminium disc above is set spinning, it may be many seconds before frictional force finally brings it to rest. However, if it spinning between the poles of a magnet, it stops almost immediately. This is because the disc is a good conductor and currents are induced in it as it moves through the magnetic field. These are called **eddy currents**. They produce a magnetic field which, by Lenz's law, opposes the motion of the disc. Eddy currents occur wherever pieces of metal are in a changing magnetic field – for example, in the core of a transformer.

Metal detectors rely on eddy currents. Typically, a pulse of current through a flat coil produces a changing magnetic field. This induces eddy currents in any metal object underneath. The eddy currents give off their own changing field which induces a second pulse in the coil. This is detected electronically.

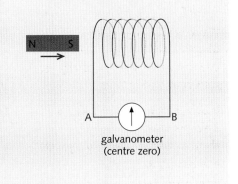

A metal detector creates eddy currents in metal objects and then detects the magnetic fields produced.

1 Look at the diagrams on the opposite page, illustrating Fleming's right-hand rule. If the directions of the magnetic field and the motion were both reversed, how would this affect the direction of the induced current?
2 On the right, a magnet is being moved towards a coil.
   a) As current is induced in the coil, what type of pole is formed at the left end of the coil? Give a reason for your answer.
   b) In which direction does the (conventional) current flow through the meter, AB or BA?
3 Aluminium is non-magnetic. Yet a freely spinning aluminium disc quickly stops moving if a magnet is brought close to it. Explain why.

N   S

A   B

galvanometer (centre zero)

Generators

Most of our electricity comes from huge **generators** in power stations. There are smaller generators in cars and on some bicycles. These generators, or dynamos, all use electromagnetic induction. When turned, they induce an EMF (voltage) which can make a current flow. Most generators give out alternating current (AC). AC generators are also called **alternators**.

A simple AC generator

The diagram below shows a simple AC generator. It is providing the current for a small bulb. The coil is made of insulated copper wire and is rotated by turning the shaft. The **slip rings** are fixed to the coil and rotate with it. The **brushes** are two contacts which rub against the slip rings and keep the coil connected to the outside part of the circuit. They are usually made of carbon.

When the coil is rotated, it cuts magnetic field lines, so an EMF is generated. This makes a current flow. As the coil rotates, each side travels upwards, downwards, upwards, downwards... and so on, through the magnetic field. So the current flows backwards, forwards... and so on. In other words, it is AC. The graph shows how the current varies through one cycle (rotation). It is a maximum when the coil is horizontal and cutting field lines at the fastest rate. It is zero when the coil is vertical and cutting no field lines.

The following all increase the maximum EMF (and the current):
- increasing the number of turns on the coil
- increasing the area of the coil
- using a stronger magnet
- rotating the coil faster.

Faster rotation also increases the frequency of the AC. Mains generators must keep a steady frequency – for example, 50 Hz (cycles per second) in the UK.

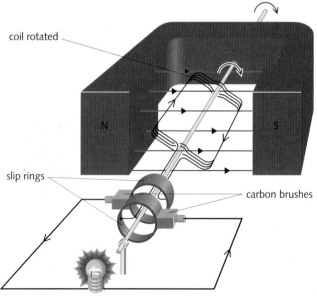

Simple AC generator, connected to a bulb

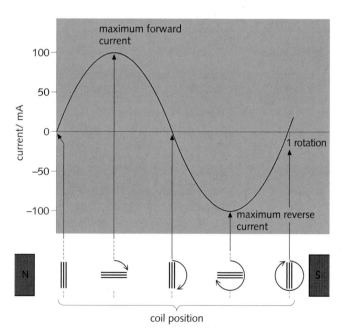

Graph showing the generator's AC output

Practical generators*

▲ Alternator from a car

◀ One of the alternators (AC generators) in a large power station. It is turned by a turbine, blown round by the force of high-pressure steam. It generates an EMF of over 30 000 volts, although consumers get their supply at a much lower voltage than this.

Unlike the simple generator on the opposite page, most AC generators have a fixed set of coils arranged around a rotating electromagnet. The various coils are made from many hundreds of turns of wire. To create the strongest possible magnetic field, they are wound on specially shaped cores containing iron. Slip rings and brushes are still used, but only to carry current to the spinning electromagnet. As the other coils are fixed, the current delivered by the generator does not have to flow through sliding contacts. (Sliding contacts can overheat if the current is very high.)

Direct current (DC) is 'one-way' current like that from a battery. DC generators are similar in construction to DC motors, with a fixed magnet, rotating coil, brushes, and a commutator to reverse the connections to the outside circuit every half-turn. When the coil is rotated, alternating current is generated. However, the action of the commutator means that the current in the outside circuit always flows the same way – in other words, it is DC.

Cars need DC for recharging the battery and running other circuits. To produce current, the engine turns a generator. However, an alternator is used, rather than a DC generator, because it can deliver more current. A device called a **rectifier** changes its AC output to DC.

Moving-coil microphone

Like generators, some microphones use the principle of electromagnetic induction.

In a moving-coil microphone, incoming sound waves strike a thin metal plate called a diaphragm and make it vibrate. The vibrating diaphragm moves a tiny coil backwards and forwards in a magnetic field. As a result, a small alternating current is induced in the coil. When amplified (made larger), the current can be used to drive a loudspeaker.

Q

1 The diagram on the right shows the end view of the coil in a simple generator. The coil is being rotated. It is connected through brushes and slip rings to an outside circuit.
a) What type of current is generated in the coil, AC or DC? Explain why it is this type of current being generated.
b) Give *three* ways in which the current could be increased.
c) The current varies as the coil rotates. What is the position of the coil when the current is a maximum? Why is the current a maximum in this position?
d) What is the position of the coil when the current is zero? Why is the current zero in this position?

2 Give *three* differences between the simple AC generator on the opposite page and most practical AC generators.

Related topics: EMF **8.05**; mains AC **8.12**; electromagnets **9.04**; DC motors **9.06**; electromagnetic induction **9.07**; rectifiers **10.02**

9.10 Coils and transformers (1)

Electromagnetic induction

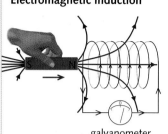

galvanometer (centre zero)

If a magnet is pushed in or out of a coil, the coil cuts through magnetic field lines, so an EMF (voltage) is induced in it. This is an example of electromagnetic induction. If the coil is in a complete circuit, the induced voltage makes a current flow.

A *moving* magnetic field can induce an EMF (voltage) in a conductor, as on the left. A *changing* magnetic field can have the same effect.

Mutual induction

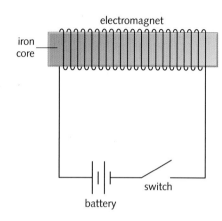
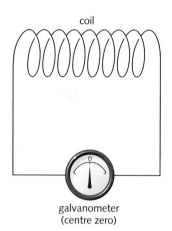

As the electromagnet above is switched on, an EMF is induced in the other coil, but only for a fraction of a second. The effect is equivalent to pushing a magnet towards the coil very fast. With a steady current through the electromagnet, no EMF is induced because the magnetic field is not changing. As the electromagnet is switched off, an EMF is induced in the opposite direction. The effect is equivalent to pulling a magnet away from the coil very fast.

The induced EMF at switch-on or switch-off is increased if:
● the core of the electromagnet goes right through the second coil
● the number of turns on the second coil is increased.

When coils are magnetically linked, as above, so that a changing current in one causes an induced EMF in the other, this is called **mutual induction**.

Using mutual induction, 40 000 volts (or more) for spark plugs is produced from a 12 volt supply. The high voltage is induced in a coil by switching an electromagnet on and off electronically.

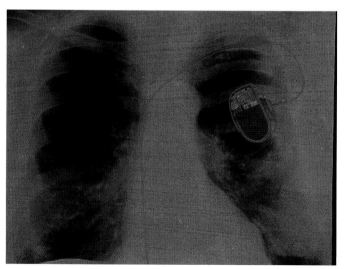

A heart pacemaker uses mutual induction. Pulses of current through a coil in the pacemaker unit induce pulses in a coil fitted in the patient's chest. These trigger heartbeats.

A simple transformer

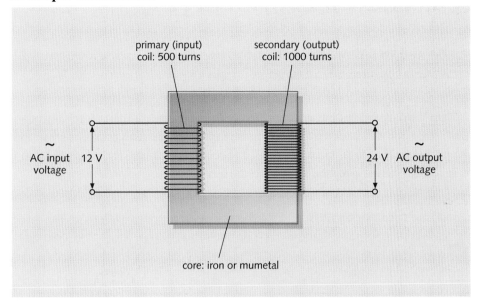

primary (input)
coil: 500 turns

secondary (output)
coil: 1000 turns

~
AC input   12 V
voltage

24 V   AC output
voltage

core: iron or mumetal

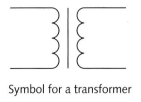

Symbol for a transformer

AC voltages can be increased or decreased using a **transformer**. A simple transformer is shown in the diagram above. It works by mutual induction.

When alternating current flows through the **primary** (input) coil, it sets up an alternating magnetic field in the core and, therefore, in the **secondary** (output) coil. This changing field induces an alternating voltage in the output coil. Provided all the field lines pass through both coils, and the coils waste no energy because of heating effects, the following equation applies:

$$\frac{\text{output voltage}}{\text{input voltage}} = \frac{\text{turns on output coil}}{\text{turns on input coil}}$$

In symbols:
$$\frac{V_2}{V_1} = \frac{n_2}{n_1}$$

For the transformer above, $n_2/n_1 = 1000/500 = 2$. The transformer has a **turns ratio** of 2. The same ratio links the voltages: $V_2/V_1 = 24/12 = 2$. Put in words, the output coil has twice the number of turns of the input coil, so the output voltage is twice the input voltage.

A transformer does not give you something for nothing. If it increases voltage, it reduces current. This is explained in the next spread.

DC and AC

Direct current (DC) flows one way only.

Alternating current (AC) flows alternately backwards and forwards.

PD, EMF, and voltage

PD (potential difference) is the scientific name for voltage. The PD produced within a battery or other source is called the EMF (electromotive force).

For convenience, engineers often use the word voltage rather than PD or EMF, especially when dealing with AC.

Voltages in AC circuits are commonly called AC voltages, although, strictly speaking, an 'alternating current voltage' doesn't make much sense!

Q

1 In the experiment on the right, what happens when
   a) the switch is closed (turned ON)
   b) the switch is left in the closed (ON) position
   c) the switch is then opened (turned OFF)?
2 In the experiment on the right, what would be the effect of
   a) extending the iron core so that it goes through both coils
   b) replacing the battery and switch by an AC supply?
3 A transformer has a turns ratio of 1/4 (quarter). Its input coil is connected to a 12 volt AC supply. Assuming there are no energy or field line losses:
   a) What is the output voltage?
   b) What turns ratio would be required for an output voltage of 36 volts?

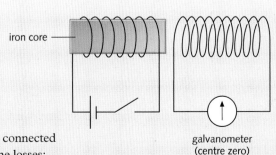

iron core

galvanometer
(centre zero)

Related topics: PD and EMF **8.05**; magnetic field lines **9.02**; electromagnets **9.04**; electromagnetic induction **9.07**; DC and AC **9.09**

221

Coils and transformers (2)

Step-up and step-down transformers

Depending on its turns ratio, a transformer can increase or decrease an AC voltage.

Step-up transformers have more turns on the output coil than on the input coil, so their output voltage is *more* than the input voltage. The transformer in the diagram below is a step-up transformer. Large step-up transformers are used in power stations to increase the voltage to the levels needed for overhead power lines. The next spread explains why.

Step-down transformers have fewer turns on the output coil than on the input coil, so the output voltage is *less* than the input voltage. In battery chargers, computers, and other electronic equipment, they reduce the voltage of the AC mains to the much lower levels needed for other circuits.

Both types of transformer work on AC, but not on DC. Unless there is a *changing* current in the input coil, no voltage is induced in the output coil. Connecting a transformer to a DC supply can damage it. A high current flows through the input coil, which can make it overheat.

Power through a transformer

If no energy is wasted in a transformer, the power (energy per second) delivered by the output coil will be the same as the power supplied to the input coil. So:

> input voltage × input current = output voltage × output current

In symbols:

$$V_1 I_1 = V_2 I_2$$

As voltage × current is the same on both sides of a transformer, it follows that a transformer which *increases* the voltage will *reduce* the current in the same proportion, and vice versa. The figures in the diagram below illustrate this.

A transformer connected to local power lines

Power essentials

Energy is measured in joules (J).

Power is measured in watts (W).

An appliance with a power output of 1000 W delivers energy at the rate of 1000 joules per second.

In circuits, power can be calculated using this equation:

power	=	voltage	×	current
(watts)		(volts)		(amperes)
(W)		(V)		(A)

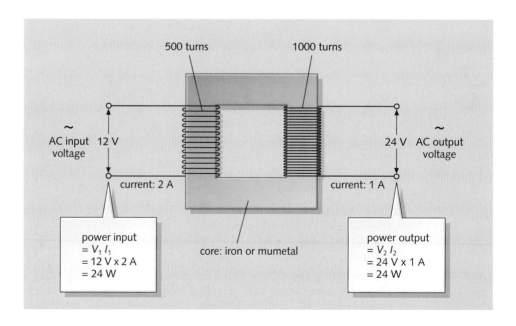

Practical transformers★

The diagram on the right shows two ways of arranging the coils and core in a practical transformer. Both methods are designed to trap the magnetic field in the core so that all the field lines from one coil pass through the other.

All transformers waste some energy because of heating effects in the core and coils. Here are two of the causes:

● The coils are not perfect electrical conductors and heat up because of their resistance. To keep the resistance low, thick copper wire is used where possible.

● The core is itself a conductor, so the changing magnetic field induces currents in it. These circulating **eddy currents** have a heating effect. To reduce them, the core is laminated (layered): it is made from thin, insulated sheets of iron or Mumetal, rather than a solid block.

Large, well-designed transformers can have efficiencies as high as 99%. In other words, their useful power output is 99% of their power input.

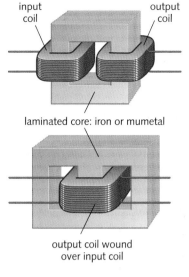

▲ Practical transformers

Solving problems

Example Assuming that the transformer on the right has an efficiency of 100%, calculate **a)** the supply voltage **b)** the current through the input coil.

a) This is solved using the transformer equation: $\dfrac{V_2}{V_1} = \dfrac{n_2}{n_1}$

where V_1 is the supply voltage to be calculated.

Substituting values: $\dfrac{10\,\text{V}}{\text{supply voltage}} = \dfrac{100}{2000}$

Rearranged, this gives: supply voltage $= 200\,\text{V}$

b) This is solved using the power equation: $V_1 I_1 = V_2 I_2$

where $V_2 I_2$ is already known to be 40 W.

Substituting values: $200\,\text{V} \times \text{input current} = 40\,\text{W}$

Rearranged, this gives: input current $= 0.2\,\text{A}$

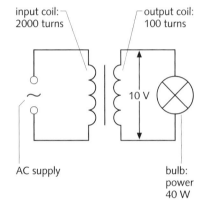

Q

1 How does a step-up transformer differ from a step-down transformer?
2 Explain each of the following:
   a) a transformer will not work on DC
   b) the core of a transformer needs to be laminated
   c) if a transformer increases voltage, it reduces current.
3 In the circuit on the right, a transformer connected to the 230 V AC mains is providing power for a low-voltage heater. Using the information in the diagram, and assuming that the efficiency is 100%, calculate
   a) the voltage across the heater
   b) the power supplied by the mains
   c) the power delivered to the heater
   d) the current through the heater.

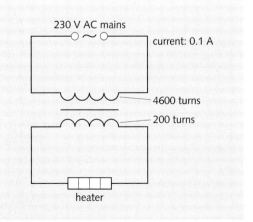

Related topics: resistance **8.06**; power calculations **8.11**; eddy currents **9.08**; DC and AC **9.09**; power transmission **9.12**

223

9.12 Power across the country

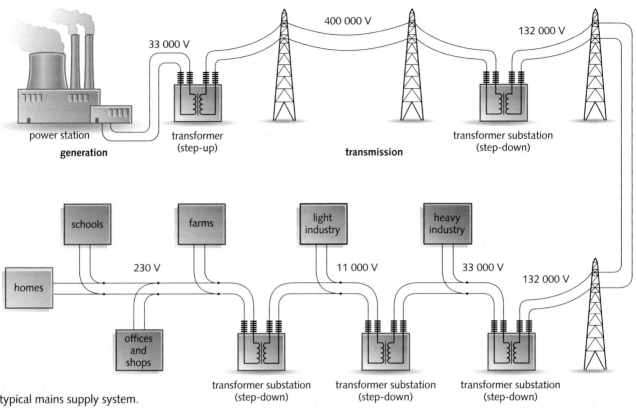

400 000 V

33 000 V

132 000 V

power station

generation

transformer
(step-up)

transmission

transformer substation
(step-down)

schools

farms

light
industry

heavy
industry

230 V

11 000 V

33 000 V

homes

132 000 V

offices
and
shops

transformer substation
(step-down)

transformer substation
(step-down)

transformer substation
(step-down)

distribution

▲ A typical mains supply system.
Actual voltages may differ,
depending on the country.

Power essentials

An appliance with a power output
of 1000 watts (W) delivers energy
at the rate of 1000 joules per
second.

In circuits

power	= voltage	× current
(watts)	(volts)	(amperes)
(W)	(V)	(A)

Transformer essentials

Transformers are used to increase
or decrease AC voltages. If a
transformer is 100% efficient, its
power output and input are equal.
So if it increases voltage, it
reduces current in the same
proportion so that 'voltage ×
current' stays the same.

Power for the AC mains is *generated* in power stations, *transmitted* (sent) through
long-distance cables, and then *distributed* to consumers.

Typically, a large power station might contain four generators, each producing a
current of 20 000 amperes at a voltage of 33 000 volts. The current from each
generator is fed to a huge step-up transformer which transfers power to overhead
cables at a greatly increased voltage (275 000 V or 400 000 V in the UK). The
reason for doing this is explained on the next page. The cables feed power to a
nationwide supply network called a **grid**. Using the grid, power stations in areas
where the demand is low can be used to supply areas where the demand is high.
Also power stations can be sited away from heavily populated areas.

Power from the grid is distributed by a series of **substations**. These contain
step-down transformers which reduce the voltage in stages to the level needed by
consumers. Depending on the country, this might be between 110 V and 240 V
for home consumers, although industry normally uses a higher voltage.

Transmission issues

AC or DC? Alternating current (AC) is used for the mains. On a large scale, it
can be generated more efficiently than 'one-way' direct current (DC). However,
the main advantage of AC is that voltages can be stepped up or down using
transformers. Transformers will not work with DC.

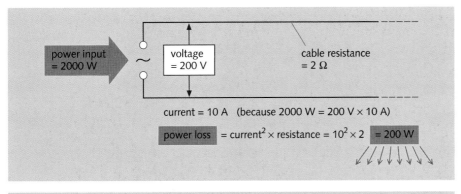

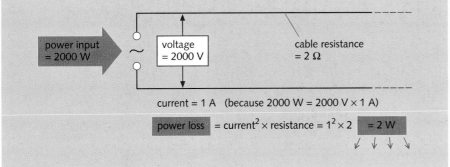

Resistance and power loss

The resistance of a conductor is measured in ohms (Ω).

When a current flows through a resistance, it has a heating effect, so power is wasted. The power loss can be calculated like this:

power loss = current^2 $\times$ resistance

◄ These calculations show the power losses in a cable when the same amount of power is sent at two different voltages (for simplicity, some units have been omitted).

High or low voltage? Transmission cables are good conductors, but they still have significant resistance – especially when they are hundreds of kilometres long. This means that energy is wasted because of the heating effect of the current. The calculations above demonstrate why less power is lost from a cable if power is transmitted through it at high voltage. By using a transformer to increase the voltage, the current is reduced, so thinner, lighter, and cheaper cables can be used.

Overhead or underground?★ There are two ways of running high-voltage transmission cables across country. They can be suspended overhead from tall towers called pylons, or they can be put underground.

In countries where power has to be transmitted very long distances, overhead cables are more common because they are cheaper. They are easier to insulate because, over most of their length, the air acts as an insulator. Also, costly digging operations are avoided. However, pylons and overhead cables spoil the environment. They are often not allowed in densely populated areas or in areas of outstanding natural beauty. So underground cables (called land lines) are used instead.

Pylons and overhead cables are not usually permitted in areas like this.

1 What is the main advantage of a grid network?
2 What are substations for?
3 Explain each of the following.
  a) AC rather than DC is used for transmitting mains power.
  b) The voltage is stepped up before power from a generator is fed to overhead transmission cables.
4 Give an example of where underground transmission cables might be used instead of overhead ones, despite the extra cost.

5 The second paragraph on the opposite page describes the output of the four generators in a typical, large power station. Calculate the power station's total power output in MW. (1 MW = 1 000 000 W)
6 The diagram at the top of this page compares power losses from a cable at two different voltages. Calculate the power loss if the same power is sent at 20 000 V.
7 4 kW of power is fed to a transmission cable of resistance 5 Ω. Calculate the power loss in the cable if the power is transmitted at **a)** 200 V **b)** 200 000 V.

1 An electromagnet is made by winding wire around an iron core.

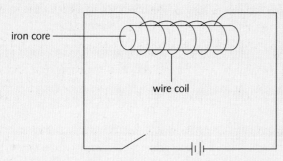

iron core

wire coil

The diagram shows an electromagnet connected to a circuit.

a) State **two** ways of making the strength of the electromagnet weaker.   [2]

b) Explain why the core is made of iron instead of steel.   [1]

WJEC

2 **A**, **B**, **C** and **D** are small blocks of different materials. The table below shows what happens when two of the blocks are placed near one another.

Arrangement of blocks		Effect
A	B	attraction
B	C	attraction
A	C	no effect
B	D	no effect

a magnet	a magnetic material	a non-magnetic material

Use one of the phrases in the above boxes to describe the magnetic property of each block. Each phrase may be used once, more than once or not at all.

a) Block **A** is _____

b) Block **B** is _____

c) Block **C** is _____

d) Block **D** is _____   [4]

WJEC

3 The figure below shows a circuit, which includes an electrical relay, used to switch on a motor M.

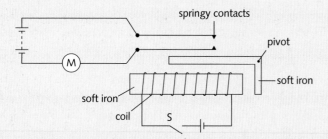

springy contacts

pivot

soft iron

M

soft iron

coil   S

Explain, in detail, how closing switch S causes the motor M to start.   [4]

UCLES

4 The diagram shows a long wire placed between the poles of a magnet. When current **I** flows through the wire, a force acts on the wire causing it to move.

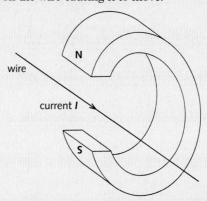

N

wire

current **I**

S

a) Use Fleming's left-hand rule to find the direction of the force on the wire. Copy the diagram and show the direction of the force on your copy with an arrow labelled **F**.   [1]

b) State what happens to the force on the wire when
  (i)   the size of the current through the wire is increased,   [1]
  (ii) a weaker magnet is used,   [1]
  (iii) the direction of the current is reversed.   [1]

c) Name **one** practical device which uses this effect.   [1]

WJEC

5 The diagram below shows a permanent magnet being moved towards a coil whose ends are connected to a sensitive ammeter. As the magnet approaches, the ammeter needle gives a **small** deflection to the **left**.

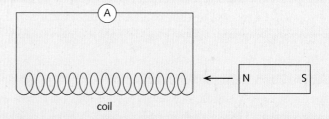

A

N   S

coil

a) State what you would expect the ammeter to show if, in turn,
  (i)   the magnet was pulled away from the coil
  (ii) the magnet was reversed so that the S pole was moved towards the coil
  (iii) the magnet was now pulled away from the coil, at a much higher speed.   [4]

b) Give the name of the process which is illustrated by these experiments.   [1]

UCLES

6 a) The chemical energy stored in a fossil fuel produces heat energy when the fuel is burned. Describe how this heat energy is then used to produce electrical energy at a power station.   [2]

b) Power stations use transformers to increase the voltage to very high values before transmitting it to all parts of the country. Explain why electricity is transmitted at very high voltages. [1]

c) A power station produces electricity at 25 000 V which is increased by a transformer to 400 000 V. The transformer has 2000 turns on its primary coil. Use the formula

$$\frac{\text{voltage across primary coil}}{\text{voltage across secondary coil}} = \frac{\text{number of turns on primary coil}}{\text{number of turns on secondary coil}}$$

to calculate the number of turns on its secondary coil. [2]

WJEC

7 The diagram shows a simple transformer.

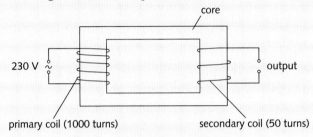

The transformer is a **step-down** transformer.

a) What is a **step-down** transformer? [1]

b) How can you tell from the diagram that this is a **step-down** transformer? [1]

c) Calculate the output voltage of this transformer. [3]

d) Explain why transformers are used when power needs to be transmitted over long distances. [3]

e) What is the core of a transformer usually made of? [2]

MEG

8 The diagram shows the main parts of one type of ammeter. There are two short iron bars inside a coil of insulated wire. One bar is fixed and cannot move and the other is on the end of a pivoted pointer. The diagram shows the ammeter in use and measuring a current of 1.5 amperes (A).

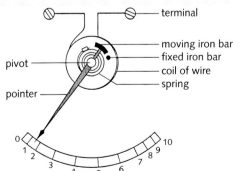

a) How much electrical charge will pass through this ammeter in one minute? Include in your answer the equation you are going to use. Show clearly how you get to your final answer and give the unit. [3]

b) (i) Apart from heat, what will be produced by the coil of wire when the electricity passes through it? [1]

(ii) What effect will this have on the two iron bars? What causes the effect? Draw one or more diagrams if this will help you to explain. [4]

To answer part a), you will need information from Section 8

SEG

9 a) When a coil rotates in a magnetic field, an alternating voltage is produced. Explain how the voltage is produced [2]

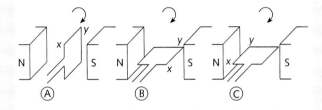

b) The diagrams A B and C show three positions of a coil as it rotates clockwise in a magnetic field produced by two poles.

The graph below shows how the voltage produced changes as the coil rotates.

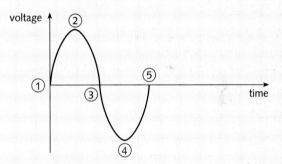

When the coil is in the position shown by diagram A, the output voltage is zero and is marked as 1 on the voltage–time graph. State which point on the voltage–time graph corresponds to the coil position shown by

(i) diagram B, [1]

(ii) diagram C. [1]

c) State **one** way of increasing the size of the voltage produced by **this** coil rotating in a magnetic field. [1]

WJEC

227

Photocopy the list of topics below and tick the boxes of the ones that are included in your examination syllabus. (Your teacher should be able to tell you which they are.) Use your list when you revise. The spread number in brackets tells you where to find more information.

❏ **1** The two types of magnetic pole. (9.01)

❏ **2** The properties of magnets. (9.01)

❏ **3** The direction of the force between magnetic poles. (9.01)

❏ **4** Induced magnetism: permanent and temporary magnets. (9.01)

❏ **5** Methods of making a magnet. (9.01 and 9.03)

❏ **6** Magnetic and non-magnetic materials. (9.01)

❏ **7** Hard and soft magnetic materials. (9.01)

❏ **8** Demagnetizing a magnet. (9.01 and 9.03)

❏ **9** The magnetic field around a magnet. (9.02)

❏ **10** Finding field patterns with a plotting compass. (9.02)

❏ **11** The Earth's magnetic field. (9.02)

❏ **12** Magnetic screening. (9.02)

❏ **13** The magnetic field around a current-carrying wire. (9.03)

❏ **14** The magnetic field around a current-carrying coil. (9.03)

❏ **15** Factors affecting the magnetic field from a coil. (9.03)

❏ **16** The right-hand grip rule for poles. (9.03)

❏ **17** The electromagnet. (9.04)

❏ **18** Factors affecting an electromagnet's field strength. (9.04)

❏ **19** How a magnetic relay works. (9.04)

❏ **20** How a circuit breaker works. (9.04)

❏ **21** Magnetic storage on tape and disc. (9.04)

❏ **22** The force on a current in a magnetic field, and the factors affecting it. (9.05)

❏ **23** Fleming's left-hand rule. (9.05)

❏ **24** How a moving-coil loudspeaker works. (9.05)

❏ **25** The turning effect on a current-carrying coil in a magnetic field. (9.05)

❏ **26** How a simple DC motor works. (9.06)

❏ **27** Features of practical electric motors. (9.06)

❏ **28** Electromagnetic induction and the factors affecting it. (9.07)

❏ **29** Demonstrating an induced EMF (voltage) in a coil. (9.07)

❏ **30** Faraday's law of electromagnetic induction. (9.07)

❏ **31** Lenz's law. (9.08)

❏ **32** Fleming's right-hand rule. (9.08)

❏ **33** Eddy currents and their effects. (9.08)

❏ **34** How a simple AC generator works. (9.09)

❏ **35** Features of practical generators. (9.09)

❏ **36** Demonstrating mutual induction. (9.10)

❏ **37** The action of a transformer. (9.10)

❏ **38** The equation linking a transformer's input and output voltages. (9.10)

❏ **39** The difference between a step-up and step-down transformer. (9.11)

❏ **40** The equation linking a transformer's input and output powers. (9.11)

❏ **41** Practical transformer design. (9.11)

❏ **42** The transmission and distribution of mains power across country. (9.12)

❏ **43** Why AC is used for power transmission. (9.12)

❏ **44** Why power is transmitted at high voltage. (9.12)

10 Electrons and Electronics

- ELECTRONIC COMPONENTS
- TRANSDUCERS
- DIODES AND RECTIFICATION
- POTENTIAL DIVIDER
- TRANSISTOR SWITCHES
- LOGIC GATES
- ELECTRON BEAMS
- OSCILLOSCOPE

A grain weevil emerging from a grain of wheat. This picture, which is 50 times actual size, was produced by a scanning electron microscope. The instrument uses a narrow beam of electrons, rather than light, and the image is constructed by a computer. Unlike a light microscope, an electron microscope gives a three-dimensional image. However, the colours are false, and added by the computer.

10.01 Electronic essentials

Essential ideas

Before working through this section, you need to understand the basic principles of circuits, covered in spreads 8.04–8.10.

Circuits with microchips and other semiconductor devices are called **electronic** circuits. They include the circuits in TV sets, computers, CD players, and amplifiers. Most handle very low currents, although they can control much more powerful circuits.

An electronic control system

The sound amplification system below is electronic. When you speak into the microphone, the sound waves cause tiny changes in the current through it. These changes are called **signals**. They are amplified (magnified) by the amplifier so that the loudspeaker gives out a louder version of the original sound. The extra power needed comes from the power supply.

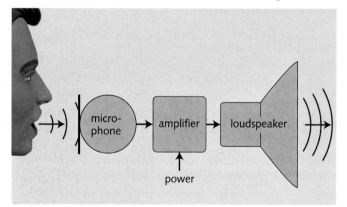

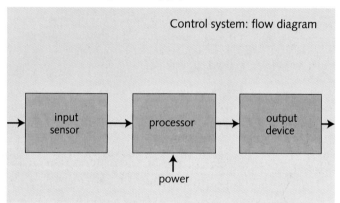

Control system: flow diagram

This is an example of a **control system**, as shown in the flow diagram above right. An **input sensor** (the microphone) sends signals to a **processor** (the amplifier) which uses them to control the flow of power to an **output device** (the loudspeaker). There are some more examples of input sensors and output devices on the opposite page.

▲ Integrated circuit (IC) package with connecting pins. Components like this are used in many electronic circuits.

Analogue and digital signals

In the system above, the current varies continuously, just like the incoming sound waves. Continuous variations like this are called **analogue** signals. But many systems use signals of a different type. For example, in the clock below, electronic circuits create the number display by switching strips on or off so that they light up in different combinations. Here, the signals represent only two states – on and off. They are **digital** signals.

For more about analogue and digital signals and their uses, see Spread 7.13.

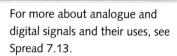

Components

Here are some of the components (parts) used in electronic circuits:

Resistors keep currents and voltages at the levels needed for other components to work properly.

Capacitors store small amounts of electric charge. They are used in smoothing circuits and time-delay circuits, in the tuning circuits in radios and TVs, and for passing on signals from one circuit to another.

If two capacitors are connected across the same battery, the one with the higher **capacitance**, in microfarads (µF), will store most charge.

Diodes let current flow through in one direction only. Most are made from specially treated crystals of silicon, a semiconductor.

Light-emitting diodes (LEDs) glow when a small current passes through them. They are used as indicator (on/off) lights and in some alphanumeric (letter and number) displays like those on digital clocks.

Transistors are used for amplifying signals and for switching. Most are made from specially treated crystals of silicon.

Integrated circuits (ICs), or 'microchips', contain many complete circuits, with resistors, transistors, other components, and connections all formed on a tiny chip of silicon only a few millimetres square.

Relays are electromagnetic switches. With a relay, a high-power circuit can be switched on (or off) by a tiny current from an electronic circuit.

symbol

Capacitor

symbol

Light-emitting diode (LED)

> When choosing a component, it must have a suitable **power rating**. This is the maximum power it can take without being damaged by overheating.

Input sensors examples	**Output devices** examples	
pressure switch (switch operated by pressing it)	LED (light-emitting diode)	relay
reed switch (switch operated by a magnet)	bulb	electric heater
variable resistor	buzzer	electric bell
thermistor (temperature-dependent resistor)	loudspeaker	electric motor
LDR (light-dependent resistor)		
microphone		

These are all **transducers** – devices that convert electrical signals into some other form, or vice versa.

1 *relay   IC   resistor   LED   capacitor*
Which of the above components
a) stores small amounts of electric charge
b) emits light when a small current flow through it
c) contains many circuits, with components formed on a tiny chip of silicon
d) uses the small current from an electronic circuit to switch a more powerful circuit?
2 Give *three* examples of transducers.

3 When a door bell is rung, what is being used as
a) the input sensor **b)** the output device?
4 Electronic systems handle either analogue signals or digital ones. Which type is used by each of these?
a) A simple sound amplification system, where you speak into a microphone and a louder version of your voice comes out of a speaker.
b) A system which automatically opens a shop door when someone approaches

Related topics: analogue and digital **7.13**; resistors **8.06**; resistor colour code **page 325**; diodes **8.07** and **10.02**; relay **9.04** and **10.02**; capacitors **10.02** and **10.04**; transistors **10.03–10.04**

10.02 More on components

Diode

Diodes

Diodes allow current to flow through them in one direction only. The circuits below show what happens when a diode is connected into a circuit one way round and then the other:

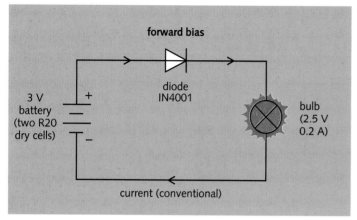

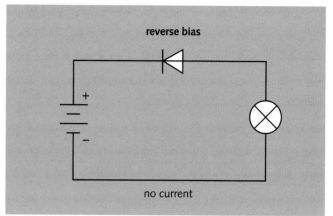

When the diode is **forward biased**, it has an extremely *low* resistance, so a current flows through it and the bulb lights up. In this case, the arrowhead in the symbol points the same way as the conventional (plus-to-minus) current direction.

When the diode is **reverse biased**. It has an extremely *high* resistance and the bulb does not light. In effect, the diode blocks the current.

Circuit essentials

AC (alternating current) flows alternately backwards and forwards. DC (direct current) flows one way only.

When resistors are in series, each has the same current through it. The resistor with the highest resistance has the greatest PD (voltage) across it.

Diodes can be used to change AC to DC. This process is called **rectification**. The diodes that do it are known as **rectifiers**. A simple rectifier circuit is shown below. The diode lets the forward parts of the alternating current through, but blocks the backward parts. So the current through the resistor flows one way only. It has become a rather jerky form of DC.

A CRO (oscilloscope) can be used to show how the circuit changes the AC input. The bottom half of the output waveform is missing. The current is flowing in surges, with short periods of no current between.

The pulsing current from a rectifier can be smoothed by connecting a capacitor across the output. The capacitor collects charge during the surges and releases it when the current from the rectifier falls. More complicated circuits with extra diodes give an even smoother current, more like the steady DC from a battery.

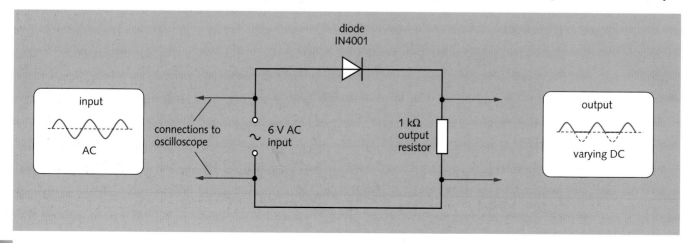

Potential divider

A **potential divider** is an arrangement that delivers only a proportion of the voltage from a battery (or other source). Circuit A shows the principle:

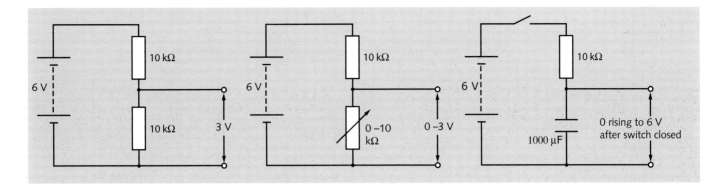

A In this potential divider, the lower resistor has half the total resistance of the two resistors, so its share of the battery's voltage is also a half.

B If one of the resistors is replaced by a variable resistor, the output voltage can be varied. Here, it can range from 0 to 3 V, depending on the setting on the variable resistor.

C Here, one of the resistors has been replaced by a capacitor. If you close the switch, the output voltage takes around half a minute to rise from 0 to 6 V as the capacitor gradually charges up.

Circuit B could be used as a volume control in a radio.

Some electronic circuits are designed to switch on when a voltage reaches a set value. If the variable resistor in circuit B were replaced by an **LDR** (light-dependent resistor), then the circuit controlling a lamp could be switched on when it got dark. Similarly, a fire alarm could be switched on by a potential divider containing a **thermistor** (temperature-dependent resistor).

Circuit C, with the capacitor, could be used in a time-delay circuit: for example, a circuit that switches on a light or a buzzer after a set time.

Reed switch

A **reed switch** is operated by a magnetic field. In the example on the right, the contacts close if a magnet is brought near, then open again if it is moved away. Burglar alarm circuits often contain reed switches. The magnets are attached to the moving parts of windows and doors.

With a coil round it, a reed switch becomes a **reed relay**. The current in one circuit (through the coil) switches on another circuit (through the contacts).

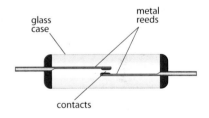

A reed switch. When the magnet is moved near, the reeds become magnetized and attract each other.

1 What is the purpose of a rectifier?
2 Look at circuits X and Y on the right. In which one
   a) does the bulb light up
   b) does the diode have a high resistance?
3 Give *two* uses of a capacitor, described in this spread.
4 If, in circuit A above, the lower resistor were replaced with one of 5 kΩ, how would this affect the output voltage of the potential divider?
5 How would you close the contacts in a reed switch?

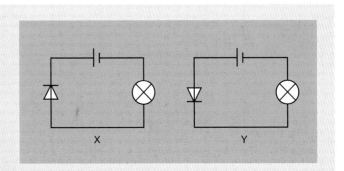

Related topics: current direction **8.04**; resistance **8.06**; relay **9.04** and **10.04**; AC and DC **9.09**; capacitors **10.01**; time-delay and temperature-sensitive switches **10.04**; oscilloscope (CRO) **10.08**

10.03 Transistors (1)

Transistors are used in TVs, CD players, and just about every other piece of electronic equipment. They can join circuits so that a small current in one controls a larger current in the next. Most form part of the integrated circuits in microchips. However, they are also manufactured as separate components.

Transistors

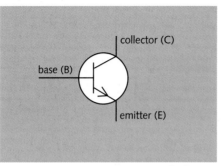

npn junction transistor: symbol

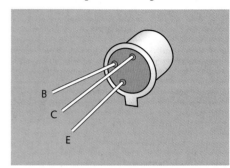

Type BC108 transistor

The transistor above is an **npn transistor**. It has three terminals, called the **emitter**, the **base**, and the **collector**. The circuits on this spread and the next show how it can be used as a switch.

Switching a transistor on

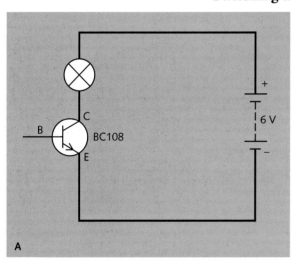

A

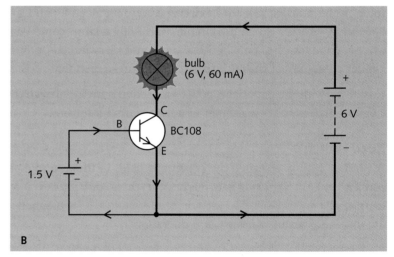

B

PD, voltage, and current

Potential difference (PD) is the scientific name for voltage. It is measured in volts (V). However, for convenience, engineers dealing with electronic circuits sometimes use the term voltage rather than PD.

Current is measured in amperes (A).

1000 mA = 1 A

In diagram A, the transistor has a blocking effect. In other words, it is switched off. No current can flow through it, so the bulb does not light.

Diagram B shows what happens when a small battery is connected between the base and the emitter. A tiny current now flows through the base and this alters the way the transistor behaves. The blocking effect vanishes, the transistor conducts, and the bulb lights up. However, the transistor will only switch on like this if the + and − connections to the batteries are the right way round. In the base and collector circuits, the conventional current direction must match the arrow direction in the transistor symbol.

You do not need an extra battery to switch a transistor on. The job can be done using a proportion of the main battery's voltage instead. This is supplied by a potential divider (see previous spread).

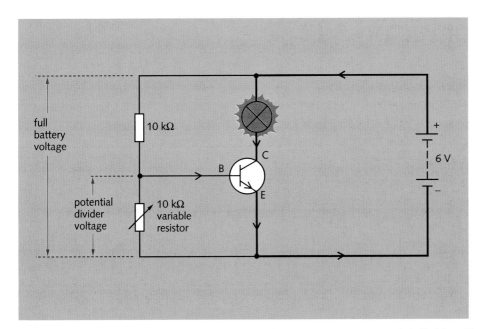

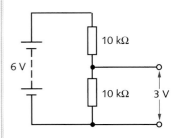
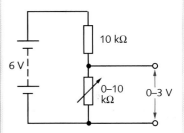
Above, a resistor and a variable resistor are being used as a potential divider. If the variable resistor has no resistance, the voltage across it is zero and the transistor stays switched off. However, if the resistance of the variable resistor is increased, its share of the battery's voltage also increases. When it passes about 0.6 V, the transistor switches on and the bulb lights. The above principle is used in the transistor switching circuits on the next spread.

Currents and current gain★

Typical values for the emitter current I_E, base current I_B, and collector current I_C through a transistor are shown below. As the current flowing out of the transistor must equal the total current flowing in: $I_E = I_B + I_C$.

Once a transistor is conducting, a small change in base current causes a much larger change in the collector current. This means that a transistor can be used to amplify signals. The **current gain** is defined as shown below right.

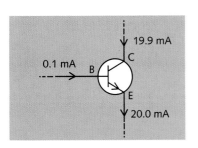

$$I_E = I_B + I_C$$

$$20.0\,\text{mA} = 0.1\,\text{mA} + 19.9\,\text{mA}$$

$$\text{current gain} = \frac{I_C}{I_B}$$

$$= \frac{19.9\,\text{mA}}{0.1\,\text{mA}} = 199$$

1 In the circuit on the right:
   a) What do the letters E, B, and C stand for?
   b) What do the two resistors do?
   c) Is the bulb ON or OFF?
   d) What would be the effect of replacing the lower resistor with a short piece of connecting wire?
2 If, in the circuit on the right, the base current is 0.05 mA and the collector current is 4.95 mA, what is:
   a) the emitter current **b)** the current gain?

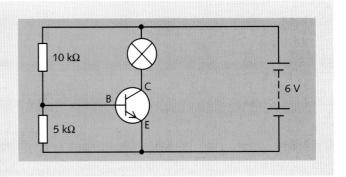

Related topics: current **8.04**; voltage (PD) **8.05**; resistors **8.06**; circuit rules **8.09**; resistors in series **8.10**; potential divider **10.02**

Transistors (2)

A light-sensitive switch

The circuit below contains a **light-dependent resistor** (**LDR**), a special type of resistor whose resistance falls when light shines on it. When the LDR is put in the dark, the bulb lights up. The principle is used in lamps which come on automatically at night.

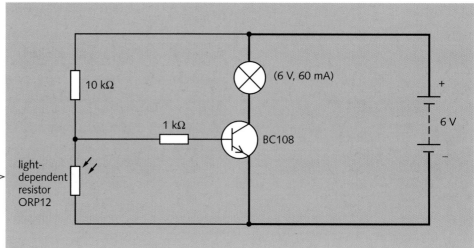

The LDR is part of a potential divider. In daylight, the LDR has a low resistance, and a low share of the battery voltage – too low to switch the transistor on. In darkness, the resistance of the LDR rises considerably, and so does its share of the battery voltage. Now, the voltage across the LDR is high enough to switch the transistor on, so the bulb lights up.

> In the circuit diagrams on this spread, there is an extra resistor next to the base connection on the transistor. This is to prevent too large a base current flowing through the transistor and damaging it.

A time-delay switch

When you close the main switch in the circuit below, there is a time delay before the bulb lights up. It is caused by the capacitor in the potential divider. The capacitor charges up slowly, and it is several seconds before the voltage across it is high enough to switch the transistor on. The delay can be increased by increasing the capacitance, or increasing the resistance of the upper resistor.

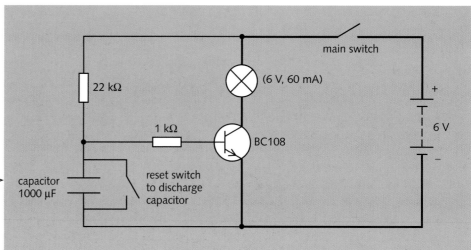

A temperature-sensitive switch

The circuit below contains a **thermistor**, a special type of resistor whose resistance falls considerably when its temperature rises. When the thermistor is heated, the bell rings. The principle is used in automatic fire alarms.

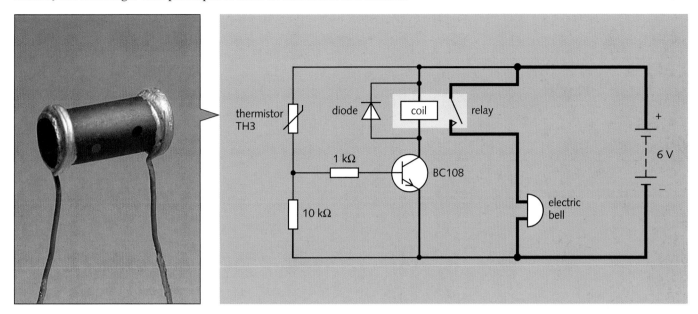

The thermistor is part of a potential divider. At room temperature, the thermistor has a high resistance and the major share of the battery voltage. As a result, the voltage across the lower resistor is not enough to switch the transistor on. When the thermistor is heated, its resistance falls, and the lower resistor gets a much larger share of the battery voltage. So the transistor is switched on and the bell starts to ring.

The circuit has two extra features:
- The transistor does not switch on the bell directly. Instead it switches on a relay, which switches on the bell. As the current in the bell circuit does not have to flow through the transistor, a more powerful bell can be used – or even a mains-operated bell in a completely separate circuit.
- The relay has a diode across it. This protects the transistor from the voltages generated (by electromagnetic induction) when the relay coil is switched on and off.

A relay could also be used in either of the circuits on the opposite page.

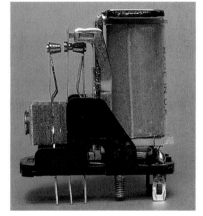

Relay

1 Give one practical use of
   a) a light-sensitive switch circuit **b)** a temperature-sensitive switch circuit.
2 Why is a relay often used with a transistor switch?
3 In the light-sensitive switch circuit on the opposite page, what would be the effect of interchanging the LDR and the 10 kΩ resistor so that the LDR is at the bottom?
4 In the temperature-sensitive switch circuit above:
   a) What would be the effect of replacing the 10 kΩ resistor with one of lower value?
   b) What change(s) would you make to the circuit so that you could vary the temperature level at which the bell sounds?

10.05 Logic gates (1)

▲ You are using logic gates when you press the buttons on a video recorder.

Video recorders, security lamps, alarm systems, and washing machines are just some of the things controlled by electronic switches called **logic gates**. The diagram below shows a simple form of gate – although this one is not electronic. It is just two ordinary switches, A and B, in a box:

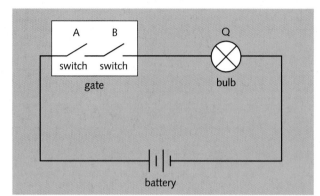

Truth table

inputs		output
A	B	Q
0	0	0
0	1	0
1	0	0
1	1	1

There has to be an unbroken circuit for the bulb to light up. So, if A and B are both ON (closed), the bulb is ON. But if either A or B is OFF (open), the bulb is OFF. The **truth table** gives the results of all the possible switch settings. It uses two **logic numbers**: 0 for OFF and 1 for ON.

In practice, logic gates work electronically, using tiny transistors as switches. They are manufactured as integrated circuits (ICs), with each chip holding several gates. The chip also needs a DC power supply. Typically, this is a 5 volt supply, with one terminal marked +5 V and the other 0 V.

AND, OR, and NOT gates

There are different types of logic gate. Three of them are shown in symbol form on the opposite page. To help you remember which is which, the words AND, OR, and NOT have been written inside the symbols. However, these words are not really part of the symbols. For simplicity, the input and output connections are shown as single wires rather than complete circuits. Also, connections to the power supply have been omitted.

▲ A logic IC package. Twelve of the 'pins' make connections to the gates on the chip. The other two are for the power supply.

- Each input is made either HIGH (for example, +5 V) or LOW (0 V). As a result, the output is either HIGH or LOW, depending on the input state(s).
- In the truth tables, the output and input states are represented by the logic numbers 1 (HIGH) and 0 (LOW).

Gates can be combined so that the output of one becomes the input of another. The diagram below shows one example:

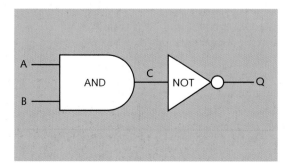

inputs			output
A	B	C	Q
0	0	0	1
0	1	0	1
1	0	0	1
1	1	1	0

AND gate

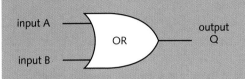

This has two inputs and one output.

For the output to be HIGH,
both inputs must be HIGH.
In other words...

Output Q is HIGH, if inputs A **AND** B are HIGH.

inputs		output
A	B	Q
0	0	0
0	1	0
1	0	0
1	1	1

OR gate

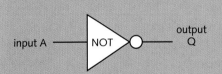

This has two inputs and one output.

For the output to be HIGH,
at least one of the inputs must be HIGH.
In other words...

Output Q is HIGH if input A **OR** B (or both)
is HIGH.

inputs		output
A	B	Q
0	0	0
0	1	1
1	0	1
1	1	1

NOT gate   (also called an **inverter**)

This has one input and one output.

The output is HIGH if
the input is LOW, and vice versa.
In other words...

Output Q is HIGH if input A is **NOT** HIGH.

input	output
A	Q
0	1
1	0

Using a gate

The diagram on the right shows one use for a logic gate.
The recorder will only start recording if the 'record' and
'play' buttons are pressed together.

For most practical applications, combinations of gates are
needed (see the next spread). Often, each input sensor
forms part of a potential divider, as in a transistor switch,
and the small output current switches on lamps, motors,
and other devices via a relay.

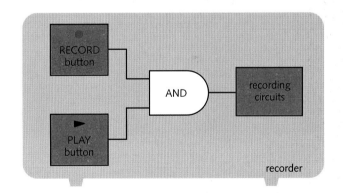

1 Look at the simple two-switches-in-a-box gate on the
   opposite page. Decide what type of gate it is.
2 The upper diagram on the right shows another two-
   switches-in-a-box gate. Write a truth table for this gate
   and decide what type of gate it is.
3 The lower diagram shows a combination of gates.
   a) Write a truth table for this combination, showing all
   the possible states of A, B, C, and Q.
   b) What must the states of inputs A and B be for the
   output to be HIGH?

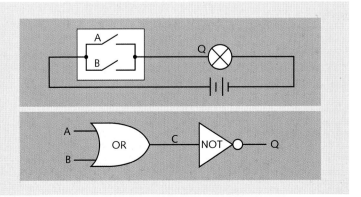

10.06 Logic gates (2)

Using a combination of gates

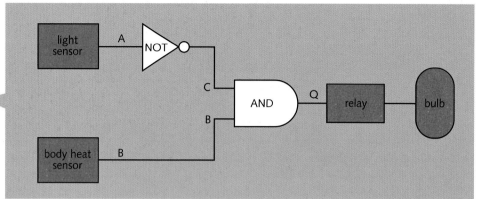

The diagram above shows how sensors and logic gates can be used to control a security lamp. The sensors and gates have been connected so that, if it is dark and someone approaches, the lamp comes on automatically. The last gate cannot provide enough power for the lamp, so it switches on a relay instead. This switches on a separate circuit with the lamp in it.

To check that the combination behaves as intended, you can write a truth table for it (see question 1). In this case, the light sensor's output is LOW (0) in the dark; the body heat sensor's output is HIGH (1) if someone approaches; the final output Q must be HIGH for the lamp to come on.

NAND and NOR gates

Two more gates are shown below. These are especially useful because all the other gates can be made by connecting or combining NAND gates (or alternatively NOR gates) in different ways. Examples are given in question 2.

logic 0 = LOW (OFF)   logic 1 = HIGH (ON)

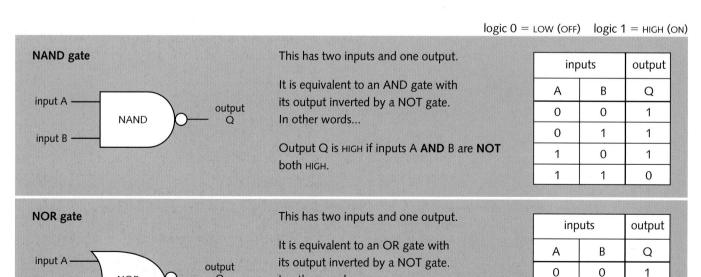

NAND gate

This has two inputs and one output.

It is equivalent to an AND gate with its output inverted by a NOT gate. In other words...

Output Q is HIGH if inputs A **AND** B are **NOT** both HIGH.

inputs		output
A	B	Q
0	0	1
0	1	1
1	0	1
1	1	0

NOR gate

This has two inputs and one output.

It is equivalent to an OR gate with its output inverted by a NOT gate. In other words...

Output Q is HIGH if neither input A **NOR** input B is HIGH.

inputs		output
A	B	Q
0	0	1
0	1	0
1	0	0
1	1	0

The bistable*

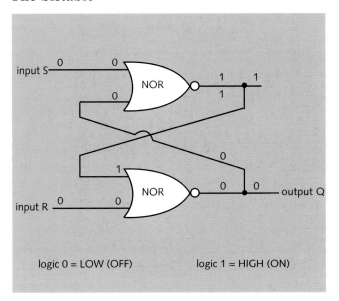

logic 0 = LOW (OFF)   logic 1 = HIGH (ON)

The astable, or multivibrator*

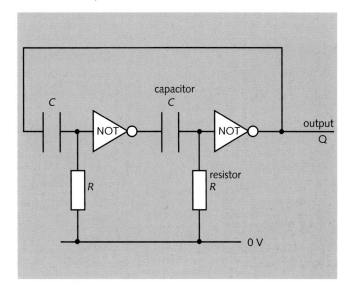

The arrangement above is called a **bistable**. Two NOR gates have been cross-coupled so that the output of each is fed back to one input on the other. This means that, with both inputs LOW (0), there are *two* possible output states: the output can be LOW (0) as shown or HIGH (1).

The bistable can be flipped from one state to the other. For example, if the top input is made HIGH, the bistable flips to the other output state. One pulse is enough to make the change: the bistable 'remembers' its new output state. This feature makes it very useful in computers, where data is stored in binary code as a series of 0's and 1's.

The two inputs are called S and R because one is used to *set* the system (change the output from LOW to HIGH), while the other is used to *reset* it.

The **astable** above delivers pulses at a regular rate: its output keeps switching HIGH, LOW, HIGH, LOW… and so on. Without the resistors and capacitors, the circuit could be in either of two stable states, giving either a HIGH or a LOW output. With the resistors and capacitors, the circuit oscillates between the two. Each time the output from one gate goes HIGH, the voltage applied to the input of the next gate changes as the capacitor between them charges up. After a delay, the voltage is enough to make the second gate change state. After another delay, this switches the first gate… and so on. Increasing R or C increases the time between the pulses – in other words, it reduces their frequency. Low-frequency pulses could be used to make lights flash on and off. Very high frequency pulses are used in the 'clock' that controls operations in a computer.

1 Look at the security lamp system on the opposite page.
   a) Write a truth table for the system, showing all the possible states of A, B, C, and Q.
   b) From your table, work out the state of Q when it is dark and someone is approaching the sensors.
2 NAND gates can be used to make other types of gate. There are two examples on the right.
   a) Write a truth table for each arrangement, showing all the possible states of A, (B, C), and Q.
   b) Decide what type of gate each is equivalent to.
3 Redraw the diagram of the bistable (top left on this page) to show the other state of Q when both inputs are 0.

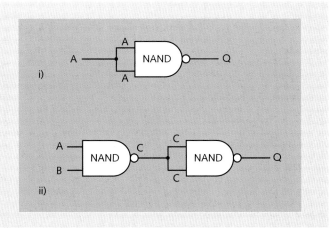

Electron beams (1)

Given enough energy, electrons can escape from a conductor and move through a vacuum (empty space). Beams of electrons like this are used in oscilloscopes, TV picture tubes, computer monitors, and X-ray tubes.

Thermionic emission

If a tungsten filament is heated to about 2000 °C, some of the electrons in the white hot metal gain enough energy to escape from its surface. The effect is called **thermionic emission** and it occurs in other metals and metal oxides as well. The diagram below shows an experiment to demonstrate the effect.

In the vacuum tube below, there are two electrodes, called the **anode** (+) and the **cathode** (−). The cathode in this case is a tungsten filament. Normally, electrons cannot cross the gap to the anode, so the meter reads zero. However, when the filament is switched on, a current starts to flow as electrons (−) escape from the hot surface and are attracted across to the anode (+). (With air in the tube, rather than a vacuum, the electrons would collide with gas molecules. Also, the white hot filament would burn up.)

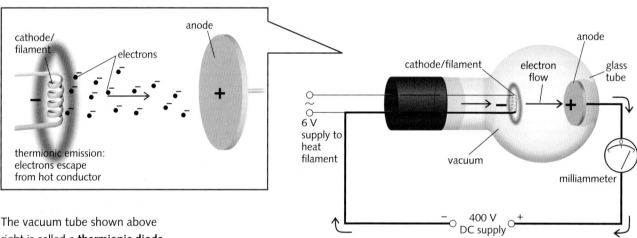

thermionic emission: electrons escape from hot conductor

The vacuum tube shown above right is called a **thermionic diode**. Electrons can pass from the hot cathode to the anode, but not the other way, so the tube acts as a 'one-way valve' for current.

Current and electronic charge★

The electron has a negative charge of 1.6×10^{-19} C (coulombs). This is called the **electronic charge**. It is represented by the symbol e.

A flow of electrons is a flow of charge, so it is a current. The following example illustrates the link between electron flow and current:

Example If the current between a cathode and an anode is 40 mA (0.04 A), how many electrons leave the cathode every second?

In 1 s: total charge leaving cathode = current × time = 0.04 A × 1 s = 0.04 C
But: total charge = charge on one electron × number of electrons
So: 0.04 C = 1.6×10^{-19} C × number of electrons

Rearranged this gives: number of electrons = $\dfrac{0.04}{1.6 \times 10^{-19}} = 2.5 \times 10^{17}$

So 2.5×10^{17} electrons leave the cathode every second.

The unit of charge is the coulomb (C). This is the quantity of charge delivered when a current of 1 ampere (A) flows for 1 second (s):

charge = current × time
(C)   (A)   (s)

Electron energy★

If an electron accelerates freely from a cathode to an anode, it loses potential energy, all of which is changed into kinetic energy (energy of motion). The energy transformed can be found using the equation in the panel on the right.

For an electron (charge e) moving through a voltage V:
energy transformed = charge × voltage = eV

So: gain in kinetic energy = eV

Deflection tube

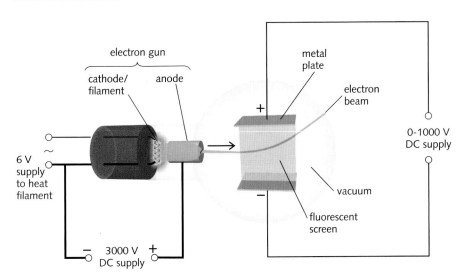

electron gun

cathode/filament   anode

metal plate

electron beam

6 V supply to heat filament

− 3000 V + DC supply

0-1000 V DC supply

vacuum

fluorescent screen

The properties of a beam of electrons can be investigated using the **deflection tube** above. The **electron gun** contains a heated filament/cathode and an anode with a hole in it. It produces a narrow beam of electrons. The screen is coated with a **fluorescent** material which glows when electrons strike it. It shows the path of the beam. Above and below the beam, there are two metal plates. When a voltage is applied across these, the beam is deflected (bent) towards the positive (+) plate.

The beam can also be deflected by a magnetic field, produced by passing a current through a pair of coils as on the right. The direction of the force is given by Fleming's left-hand rule (remembering that the conventional current direction is *opposite* to that of the electron flow). If the field direction is reversed, the force direction is also reversed.

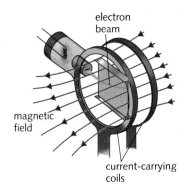

electron beam

magnetic field

current-carrying coils

Magnetic deflection

1 Look at the experiment with the vacuum tube (the thermionic diode) on the opposite page.
a) What is *thermionic emission*, and where is it taking place?
b) Why has all the air been removed from the tube?
c) Why are electrons attracted to the anode?
d) If the connections to the 400 V DC supply were reversed, would there still be a current in the main circuit? Give a reason for your answer.

2 In an experiment similar to the one on the opposite page, the meter reads 100 mA and the voltage between the cathode and anode is 500 V. The electronic charge is 1.6×10^{-19} C. Calculate
a) the kinetic energy of an electron as it reaches the anode
b) the charge (in C) reaching the anode every second
c) the number of electrons reaching the anode every second.

Related topics: kinetic energy **4.01**; charge and current **8.04**; potential energy and voltage **8.05**; Fleming's left-hand rule **9.05**

10.08 Electron beams (2)

Cathode rays

Beams of fast-moving electrons are known as cathode rays. The name dates from the late 19th century when 'invisible rays' were found coming from the cathode. Experiments showed that the rays were negatively (−) charged, and led to the discovery of the electron.

The oscilloscope

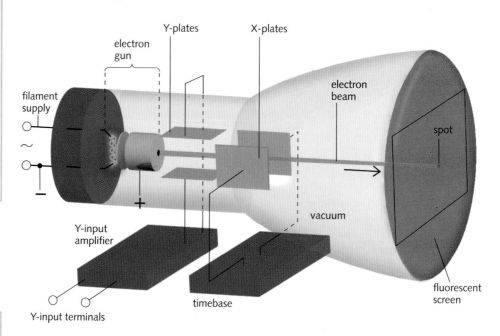

AC and DC essentials

Alternating current (AC) flows backwards and forwards. Direct current (DC) flows in one direction only.

Frequency essentials

Frequency is measured in hertz (Hz). If an AC supply has a frequency of 50 Hz, then the current flows backwards and forwards 50 times per second, causing 50 complete cycles of the waveform to be traced out on the screen of a CRO.

The **cathode ray oscilloscope** (**CRO**) above uses a narrow beam of electrons to trace out waveforms and other signals on a fluorescent screen. There is a bright spot on the screen where the beam strikes it. By deflecting the beam, the spot can be moved about. If it moves fast enough, it appears as a line. To deflect the beam, there are two sets of **deflection plates**:

The **Y-plates** move the beam vertically. This happens when an external source of voltage – for example, an AC supply – is connected across the Y-input terminals. The amount of vertical movement can be amplified (magnified) by turning up the **gain control**.

The **X-plates** move the beam horizontally. Normally, the movement is produced by a circuit called the **timebase** inside the oscilloscope. The timebase automatically applies a changing voltage across the plates so that the spot moves from left to right across the screen at a steady speed, flicks back to the start, moves across again... and so on, over and over again.

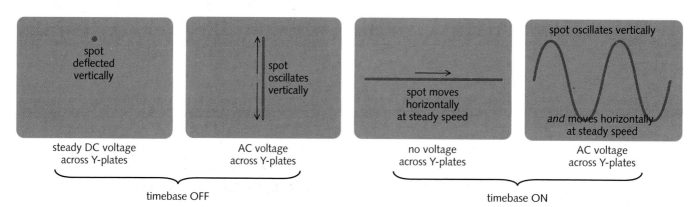

Measuring with an oscilloscope

The Y-input of the oscilloscope on the right is connected to an AC supply. From the waveform on the screen and the settings on the oscilloscope's controls, the voltage and frequency of the supply can be found as follows.

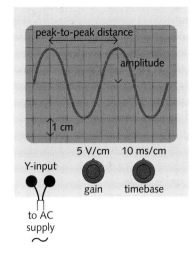

Measuring peak voltage The gain control is set at 5 V/cm. This means that the spot is deflected 1 cm vertically for every 5 volts across the Y-input terminals. The *peak* voltage is represented by the distance marked 'amplitude' on the waveform. As the amplitude is 2.0 cm:

peak voltage = 2.0 cm × 5 V/cm = 10.0 V

Measuring time and frequency The timebase control is set at 10 ms/cm. This means that the spot takes 10 milliseconds (0.01 s) to move 1 cm horizontally. As the horizontal peak-to-peak distance is 4.0 cm:

peak-to-peak time = 4.0 cm × 10 ms/cm = 40 ms = 0.04 s

This is the time taken for the spot to trace out one complete cycle of the waveform. It is known as the **period**.
So: number traced out per second = 1/0.04 = 25
So: frequency = 25 Hz

Television*

At one time, all TVs used a cathode ray tube (CRT). Many still do. In a colour tube, magnetic deflection coils pull three electron beams in a zig-zag path over the screen. The screen is coated with fluorescent strips that glow red, green, and blue when the beams strike them. Together, these make up a full colour picture. Modern flat panel sets do not have tubes. Instead, their screens contain over half a million tiny bubbles or crystals that light up in different combinations.

X-ray tube*

When very fast electrons are suddenly stopped, most of their kinetic energy is changed into thermal energy (heat). However, X-rays are also produced. This principle is used in the X-ray tube shown on the right, where X-rays are emitted from a tungsten target when electrons strike it. (Tungsten is used because of its high melting point.) The higher the accelerating voltage between cathode and anode, the shorter the wavelength of the X-rays and the more penetrating they are.

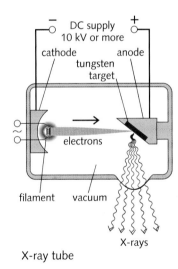

X-ray tube

1 Name *two* devices that use electron beams.
2 Diagrams A, B, and C on the right represent the screen and deflection plates of an oscilloscope.
   a) Copy diagrams A and B. Label the X-plates and the Y-plates. Then indicate which plates are being used to move the spot by marking them + and −.
   b) Explain how the line in diagram C is produced.
3 The oscilloscope in the diagram on the right has an AC supply connected to its Y-input. Find
   a) the amplitude of the waveform
   b) the peak voltage of the supply
   c) the time between two peaks on the waveform
   d) the frequency of the AC supply.

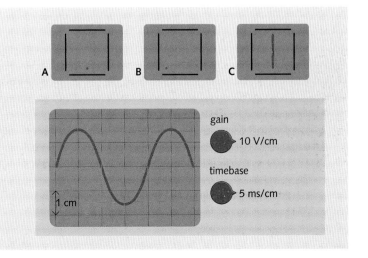

Related topics: amplitude and frequency **6.01**; sounds on an oscilloscope **6.05**; X-rays **7.11–7.12**; AC voltage **9.09**

245

1 The circuit shown below allows a computer to turn an electric heater on and off. Some of the connections have been missed out.

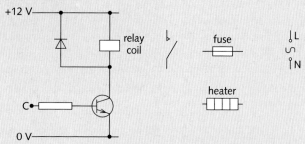

a) (i) Copy and complete the diagram to show how the heater and fuse should be connected to the relay contacts and the mains electricity supply. [2]

(ii) The outside casing of the heater has been earthed. Explain how this increases the safety of the system. [2]

b) The heater must come on when the voltage at C goes above 3.0 V. The relay contacts close when the current in the relay coil is 95 mA. The current gain of the transistor is 100 and its base-emitter voltage drop is 0.7 V when it is switched on.

(i) Calculate the minimum base current required for the heater to come on. [2]

(ii) State the formula linking voltage, current and resistance. [1]

(iii) Calculate a suitable value for the base resistor. [3]

MEG

2

thermistor	AND gate	pressure pad	microphone	relay switch
loudspeaker	light-emitting diode	light-dependent resistor	OR gate	

Select **from the above list**:

a) an input device which detects changes in light; [1]

b) an output device which produces a sound; [1]

c) a processing device which only gives an output when both inputs are high. [1]

WJEC

3 a) All electronic systems have *input sensors*, *processors* and an *output device*. Explain the function of

(i) input sensors, [1]

(ii) processors. [1]

b) The block diagram below shows an electronic system that can be used as a burglar alarm.

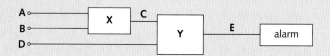

A, B and **D** are the inputs. The processor contains logic gates **X** and **Y**. The alarm is the output device. The truth table for the circuit is shown at the top of the next column.

A	B	C	D	E
0	0	0	0	0
0	1	1	0	0
1	0	1	0	0
1	1	1	0	0
0	0	0	1	0
0	1	1	1	1
1	0	1	1	1
1	1	1	1	1

(i) Use a truth table to identify the logic gates. [2]

(ii) State an input, **A**, **B** or **D** which could be connected to a sensor in order to detect a burglar. [1]

(iii) Name a suitable device which could be used as an input processor. [1]

WJEC

4 A car is fitted with a security device. To start the car, two keys are needed. The first key operates a 'hidden' switch, the second key operates the ignition switch.

If the 'hidden' switch and the ignition switch are both on, the engine will start. Turning the ignition switch without operating the 'hidden' switch will not allow the engine to start and will activate an alarm.

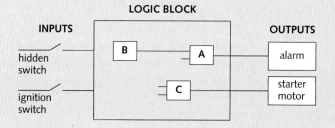

A, B and **C** are the three gates needed for the design of the logic block.

a) Identify gate

(i) A [1]

(ii) B [1]

(iii) C [1]

b) Copy the diagram and show how the **inputs** and **logic gates** are connected together. [2]

c) The logic gates used in the logic block are called digital devices. Explain what this means. [1]

WJEC

5 The block diagram below shows a burglar alarm.

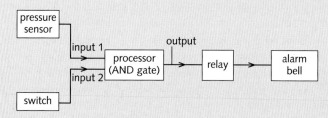

The pressure sensor is under a doormat. When a person is standing on the doormat, **input** 1 to the processor is **HIGH** (**ON**).

A burglar steps on the doormat and the alarm bell rings. State what happens when the burglar steps off the doormat. Explain your answer. [3]

NEAB

6 The devices below are all used in electronic circuits.

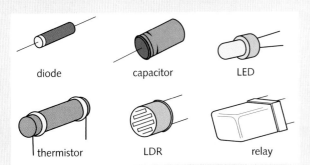

Which of the above is best described by each of the following statements?
a) Glows when a small current flows through it. [1]
b) Links two circuits so that a small current in one can switch on or off a larger current in the other. [1]
c) Has a lower resistance when it is heated. [1]
d) Has a lower resistance when light shines on it. [1]
e) Lets current through in one direction only. [1]

7 The diagram shows a circuit for a temperature sensor.

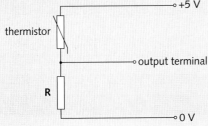

The temperature of the thermistor rises.
a) What happens to the resistance of the thermistor? [1]
b) What happens to the voltage across resistor **R**? [1]

MEG

8 A simple burglar alarm has two sensors:
• a heat source, which gives a high output when someone is nearby.
• a light sensor, which gives a high output when light shines on it.
A thermistor is used in the heat sensor.
a) What happens to the thermistor to cause a change in the sensor output? [1]
b) Suggest a suitable component for the **light sensor**. [1]

MEG

9 The diagram shows a simple electron deflection tube. Electrons are produced by the filament and accelerated through the anode down the tube.

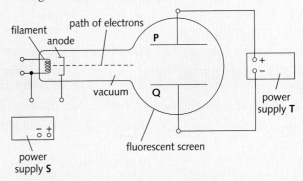

a) Explain how the filament is made to produce electrons as shown. [2]
b) What charge do electrons carry, positive, negative or zero? [1]
c) Electrons are made to accelerate towards the anode by use of the power supply **S**. Copy the diagram and show suitable connections on it. [2]
P and **Q** are two deflection plates. **P** and **Q** are connected to the terminals of a power supply **T**.
d) On your copy of the diagram, complete the path of the electrons between the plates. [2]
e) Why is a vacuum needed in the deflection tube? [1]

MEG

10 The diagram below shows a cathode ray oscilloscope (CRO). Electrons leaving the cathode are accelerated towards the anode. They pass between two sets of deflector plates before striking the screen.

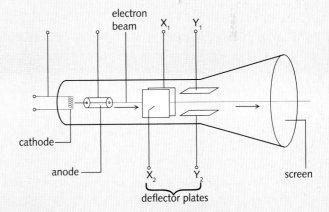

a) What makes the cathode give off electrons? What is the effect called? [2]
b) What makes the electrons accelerate from the cathode towards the anode? [2]
c) How can the beam be made to bend upwards so that it strikes the screen nearer the top? [2]

247

Photocopy the list of topics below and tick the boxes of the ones that are included in your examination syllabus. (Your teacher should be able to tell you which they are.) Use your list when you revise. The spread number in brackets tells you where to find more information.

❏ **1** The parts of an electronic control system. (10.01)

❏ **2** Analogue and digital signals. (10.01)

❏ **3** Input sensors and output devices. (10.01)

❏ **4** The function of components such as resistors, capacitors, diodes, light-emitting diodes (LEDs), transistors, integrated circuits (ICs), and relays. (10.01)

❏ **5** Transducers. (10.01)

❏ **6** The power rating of a component. (10.01)

❏ **7** Changing AC to DC: rectification. (10.02)

❏ **8** The use of diodes as rectifiers. (10.02)

❏ **9** The potential divider. (10.02)

❏ **10** The reed switch and reed relay. (10.02)

❏ **11** The npn junction transistor. (10.03)

❏ **12** The action of a transistor. (10.03)

❏ **13** Using a potential divider to switch a transistor on. (10.03)

❏ **14** The meaning of current gain. (10.03)

❏ **15** Using a transistor and LED in a light-sensitive switch. (10.04)

❏ **16** Using a transistor and capacitor in a time-delay switch. (10.04)

❏ **17** Using a transistor and thermistor in a temperature-sensitive switch. (10.04)

❏ **18** Logic gates and their symbols. (10.04)

❏ **19** AND, OR, and NOT gates and their truth tables. (10.05)

❏ **20** Working out truth tables for combinations of gates. (10.05–10.06)

❏ **21** NAND and NOR gates and their truth tables. (10.06)

❏ **22** The bistable. (10.06)

❏ **23** The astable (multivibrator). (10.06)

❏ **24** Thermionic emission. (10.09)

❏ **25** The charge on the electron. (10.07)

❏ **26** Calculating the energy gained by an electron. (10.07)

❏ **27** The deflection tube. (10.07)

❏ **28** The effects of electric and magnetic fields on a beam of electrons. (10.07)

❏ **29** The cathode ray oscilloscope and its parts. (10.08)

❏ **30** Measuring voltage, time, and frequency with an oscilloscope. (10.08)

❏ **31** How an oscilloscope displays a waveform. (10.08)

❏ **32** How an X-ray tube works. (10.08)

Atoms and Radioactivity

The *aurora borealis* ('northern lights') in the night sky over Alaska, USA. The shimmering curtain of light is produced when atomic particles streaming from the Sun strike atoms and molecules high in the Earth's atmosphere. The Earth's magnetic field concentrates the incoming atomic particles above the north and south polar regions, so that is where aurorae are normally seen.

11.01 Inside atoms

Charge essentials

There are two types of electric charge: positive (+) and negative (−). Like charges repel; unlike charges attract.

A simple model of the atom

Everything is made of atoms. Atoms are far too small to be seen with any ordinary microscope – there are more than a billion billion of them on the surface of this full stop. However, by shooting tiny atomic particles through atoms, scientists have been able to develop **models** (descriptions) of their structure. In advanced work, scientists use a mathematical model of the atom. However, the simple model below is often used to explain the basic ideas.

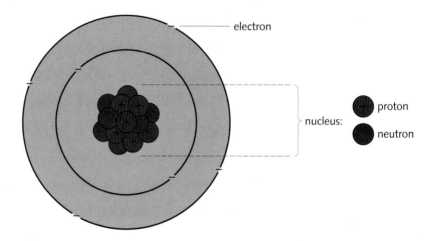

▶ A simple model of the atom. In reality, the nucleus is far too small to be shown to its correct scale. If the atom were the size of a concert hall, its nucleus would be smaller than a pill!

An atom is made up of smaller particles:

● There is a central **nucleus** made up of **protons** and **neutrons**. Around this, **electrons** orbit at high speed. The numbers of particles depends on the type of atom.

● Protons have a positive (+) charge. Electrons have an equal negative (−) charge. Normally, an atom has the same number of electrons as protons, so its total charge is zero.

● Protons and neutrons are called **nucleons**. Each is about 1800 times more massive than an electron, so virtually all of an atom's mass is in its nucleus.

● Electrons are held in orbit by the force of attraction between opposite charges. Protons and neutrons are bound tightly together in the nucleus by a different kind of force, called the **strong nuclear force**.

Elements and atomic number

All materials are made from about 100 basic substances called **elements**. An atom is the smallest 'piece' of an element you can have. Each element has a different number of protons in its atoms: it has a different **atomic number** (sometimes called the **proton number**). There are some examples on the left. The atomic number also tells you the number of electrons in the atom.

Isotopes and mass number

The atoms of any one element are not all exactly alike. Some may have more neutrons than others. These different versions of the element are called **isotopes**. They have identical chemical properties, although their atoms have different masses. Most elements are a mixture of two or more isotopes. You can see some examples in the chart on the opposite page.

element	chemical symbol	atomic number (proton number)
hydrogen	H	1
helium	He	2
lithium	Li	3
beryllium	Be	4
boron	B	5
carbon	C	6
nitrogen	N	7
oxygen	O	8
radium	Ra	88
thorium	Th	90
uranium	U	92
plutonium	Pu	94

The total number of protons and neutrons in the nucleus is called the **mass number** (or **nucleon number**). Isotopes have the *same* atomic number but *different* mass numbers. For example, the metal lithium (atomic number 3) is a mixture of two isotopes with mass numbers 6 and 7. Lithium-7 is the more common: over 93% of lithium atoms are of this type. On the right, you can see how to represent an atom of lithium-7 using a symbol and numbers. Each different type of atom, lithium-7 for example, is called a **nuclide**.

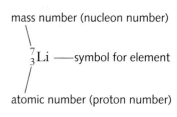

mass number (nucleon number)

$^{7}_{3}\text{Li}$ ——symbol for element

atomic number (proton number)

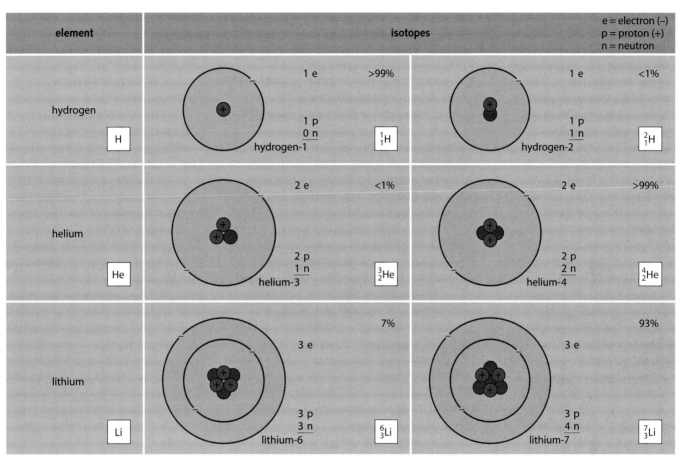

element	isotopes	e = electron (–) p = proton (+) n = neutron
hydrogen H	1 e 1 p 0 n hydrogen-1 $^{1}_{1}\text{H}$ >99%	1 e 1 p 1 n hydrogen-2 $^{2}_{1}\text{H}$ <1%
helium He	2 e 2 p 1 n helium-3 $^{3}_{2}\text{He}$ <1%	2 e 2 p 2 n helium-4 $^{4}_{2}\text{He}$ >99%
lithium Li	3 e 3 p 3 n lithium-6 $^{6}_{3}\text{Li}$ 7%	3 e 3 p 4 n lithium-7 $^{7}_{3}\text{Li}$ 93%

Electron shells

Electrons orbit the nucleus at certain fixed levels only, called **shells**. There is a limit to how many electrons each shell can hold – for example, no more than 2 in the first shell and 8 in the second. It is an atom's outermost electrons which form the chemical bonds with other atoms, so elements with similar electron arrangements have similar chemical properties.

> The **periodic table** is a chart of all the elements. Elements in the same group have similar electron arrangements and similar chemical properties.

For questions 4 and 5, you will need data from the table of elements on the opposite page.

1 An atoms contains *electrons*, *protons*, and *neutrons*. Which of these particles
  a) are outside the nucleus b) are uncharged
  c) have a negative charge d) are nucleons
  e) are much lighter than the others?
2 An aluminium atom has an atomic number of 13 and a mass number of 27. How many
  a) protons b) electrons c) neutrons does it have?

3 Chlorine is a mixture of two isotopes, with mass numbers 35 and 37. What is the difference between the two types of atom?
4 In symbol form, nitrogen-14 can be written $^{14}_{7}\text{N}$ How can each of the following be written?
  a) carbon-12 b) oxygen-16 c) radium-226
5 Atom X has 6 electrons and a mass number of 12. Atom Y has 6 electrons and a mass number of 14. Atom Z has 7 neutrons and a mass number of 14. Identify the elements X, Y, and Z.

Related topics: electric charge 8.01–8.02; experimental evidence for nucleus 11.09

Nuclear radiation (1)

11.02

Isotope essentials

Different versions of the same element are called isotopes. Their atoms have different numbers of neutrons in the nucleus.

For example, lithium is a mixture of two isotopes: lithium-6 (with 3 protons and 3 neutrons in the nucleus) and lithium-7 (with 3 protons and 4 neutrons).

Some materials contain atoms with unstable nuclei. In time, each unstable nucleus disintegrates (breaks up). As it does so, it shoots out a tiny particle and, in some cases, a burst of wave energy as well. The particles and waves 'radiate' from the nucleus, so they are somtimes called **nuclear radiation**. Materials which emit nuclear radiation are known as **radioactive** materials. The disintegration of a nucleus is called **radioactive decay**.

Some of the materials in nuclear power stations are highly radioactive. But nuclear radiation comes from natural sources as well. Although it is convenient to talk about 'radioactive materials', it is really particular isotopes of an element that are radioactive. Here are some examples:

isotopes		
stable nuclei	unstable nuclei, radioactive	found in
carbon-12 carbon-13	carbon-14	air, plants, animals
potassium-39	potassium-40	rocks, plants, sea water
	uranium-234 uranium-235 uranium-238	rocks

Ionizing radiation

Ions are charged atoms (or groups of atoms). Atoms become ions when they lose (or gain) electrons. Nuclear radiation can remove electrons from atoms in its path, so it has an **ionizing** effect. Other forms of ionizing radiation include ultraviolet and X-rays.

If a gas becomes ionized, it will conduct an electric current. In living things, ionization can damage or destroy cells (see the next spread).

Alpha, beta, and gamma radiation

There are three main types of nuclear radiation: **alpha particles**, **beta particles**, and **gamma rays**. Gamma rays are the most penetrating and alpha particles the least, as shown below:

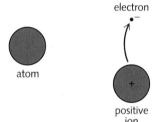

If an atom loses (or gains) an electron, it becomes an ion.

Discovering radiaoctivity

Henri Becquerel discovered radioactivity, by accident, in 1896. When he left some uranium salts next to a wrapped photographic plate, he found that the plate had become 'fogged', and realized that some invisible, penetrating radiation must be coming from the uranium.

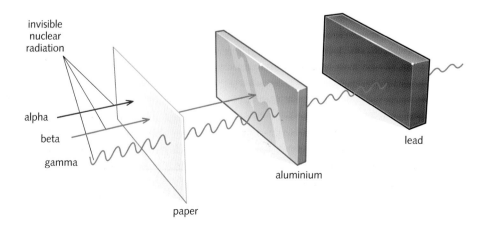

type of radiation	alpha particles (α)	beta particles (β)	gamma rays (γ)
	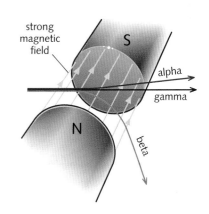 each particle is 2 protons + 2 neutrons (it is identical to a nucleus of helium-4)	each particle is an electron (created when the nucleus decays)	electromagnetic waves similar to X-rays
relative charge compared with charge on proton	+2	-1	0
mass	high, compared with betas	low	–
speed	up to 0.1 × speed of light	up to 0.9 x speed of light	speed of light
ionizing effect	strong	weak	very weak
penetrating effect	not very penetrating: stopped by a thick sheet of paper, or by skin, or by a few centimetres of air	penetrating, but stopped by a few millimetres of aluminium or other metal	very penetrating: never completely stopped, though lead and thick concrete will reduce intensity
effects of fields	deflected by magnetic and electric fields	deflected by magnetic and electric fields	not deflected by magnetic or electric fields

The nature and main properties of the three types of radiation are given in the chart above. The diagram on the right shows how the different types are affected by a magnetic field. The alpha beam is a flow of positively (+) charged particles, so it is equivalent to an electric current. It is deflected in a direction given by Fleming's left-hand rule – the rule used for working out the direction of the force on a current-carrying wire in a magnetic field. The beta particles are much lighter than the alpha particles and have a negative (–) charge, so they are deflected more, and in the opposite direction. Being uncharged, the gamma rays are not deflected by the field.

Alpha and beta particles are also affected by an electric field – in other words, there is a force on them if they pass between oppositely charged plates.

Q

1 Name a radioactive isotope which occurs naturally in living things.
2 *alpha   beta   gamma*
   Which of these three types of radiation
   a) is a form of electromagnetic radiation
   b) carries positive charge
   c) is made up of electrons
   d) travels at the speed of light
   e) is the most ionizing

   f) can penetrate a thick sheet of lead
   g) is stopped by skin or thick paper
   h) has the same properties as X-rays
   i) is not deflected by an electric or magnetic field?
3 What is the difference between the atoms of an isotope that is radioactive and the atoms of an isotope that is not?
4 How is an ionized material different from one that is not ionized?

Related topics: electromagnetic waves **7.11–7.12**; X-rays **7.12** and **10.08**; Fleming's left-hand rule **9.05**; isotopes **11.01**

11.03 Nuclear radiation (2)

Radiation dangers

Nuclear radiation can damage or destroy living cells, and stop organs in the body working properly. It can also upset the chemical instructions in cells so that these grow abnormally and cause cancer. The greater the intensity of the radiation, and the longer the exposure time, the greater the risk.

Radioactive gas and dust are especially dangerous because they can be taken into the body with air, food, or drink. Once absorbed, they are difficult to remove, and their radiation can cause damage in cells deep in the body. Alpha radiation is the most harmful because it is the most highly ionizing.

Normally, there is much less risk from radioactive sources *outside* the body. Sources in nuclear power stations and laboratories are well shielded, and the intensity of the radiation decreases as you move away from the source. Beta and gamma rays are potentially the most harmful because they can penetrate to internal organs. Alpha particles are stopped by the skin.

Background radiation

There is a small amount of radiation around us all the time because of radioactive materials in the environment. This is called **background radiation**. It mainly comes from natural sources such as soil, rocks, air, building materials, food and drink – and even space.

In some areas, over a half of the background radiation comes from radioactive radon gas (radon-222) seeping out of rocks – especially some types of granite. In high risk areas, houses may need extra underfloor ventilation to stop the gas collecting or, ideally, a sealed floor to stop it entering in the first place.

Geiger-Müller (GM) tube

This can be used to detect alpha, beta, and gamma radiation. Its structure is shown below. The 'window' at the end is thin enough for alpha particles to pass through. If an alpha particle enters the tube, it ionizes the gas inside. This sets off a high-voltage spark across the gas and a pulse of current in the circuit. A beta particle or burst of gamma radiation has the same effect.

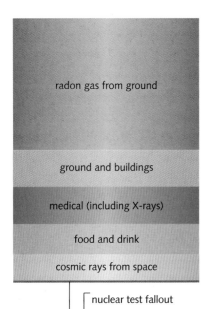

radon gas from ground

ground and buildings

medical (including X-rays)

food and drink

cosmic rays from space

nuclear test fallout
nuclear power stations
nuclear waste
other

▲ Where background radiation comes from (average proportions).

▲ This nuclear laboratory worker is about to use a GM tube and ratemeter to check for any traces of radioactive dust on her clothing.

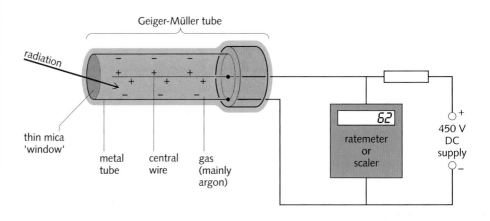

Geiger-Müller tube

radiation

thin mica 'window'

metal tube

central wire

gas (mainly argon)

62

ratemeter or scaler

450 V DC supply

The GM tube can be connected to the following:

- **A ratemeter** This gives a reading in counts per second. For example, if 50 alpha particles were detected by the GM tube every second, the ratemeter would read 50 counts per second.
- **A scaler** This counts the *total* number of particles (or bursts of gamma radiation) detected by the tube.
- **An amplifier and loudspeaker** The loudspeaker makes a 'click' when each particle or burst of gamma radiation is detected.

When the radiation from a radioactive source is measured, the reading always *includes* any background radiation present. So an average reading for the background radiation alone must also be found and subtracted from the total.

Cloud chamber

This is useful for studying alpha particles because it makes their tracks visible. The chamber has cold alcohol vapour in the air inside it. The alpha particles make the vapour condense, so you see a trail of tiny droplets where each particle passes through. At one time, cloud chambers were widely used in nuclear research, but they have since been replaced by other devices.

Safety in the laboratory

Experiments with weak radioactive sources are sometimes carried out in school and college laboratories. Such sources are normally sealed so that no radioactive fragments or dust can escape. For safety, a source should be

- stored in a lead container, in a locked cabinet.
- picked up with tongs, not by hand.
- kept well away from the body, and not pointed at other people.
- left out of its container for as short a time as possible.

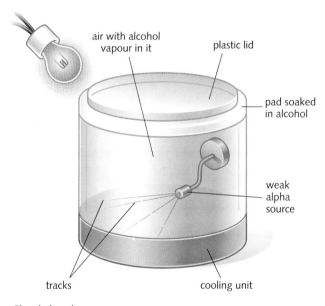

Cloud chamber

labels: air with alcohol vapour in it; plastic lid; pad soaked in alcohol; weak alpha source; tracks; cooling unit

Tracks of alpha particles in a cloud chamber. The colours are false and have been added to the picture. The green and yellow lines are the alpha tracks. The red line is the track of a nitrogen nucleus that has been hit by an alpha particle.

Q

1 What, on average, is the biggest single source of background radiation?
2 Radon gas seeps out of rocks underground. Why is it important to stop radon collecting in houses?
3 Which is the most dangerous type of radiation
   a) from radioactive sources outside the body
   b) from radioactive materials absorbed by the body?
4 In the experiment on the right:
   a) What is the count rate due to background radiation?
   b) What is the count rate due to the source alone?
   c) If the source emits one type of radiation only, what type is it? Give a reason for your answer.

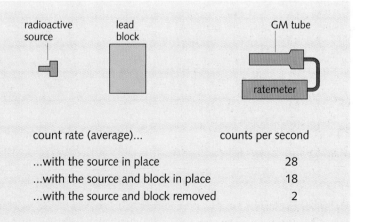

count rate (average)...	counts per second
...with the source in place	28
...with the source and block in place	18
...with the source and block removed	2

labels: radioactive source; lead block; GM tube; ratemeter

11.04 Radioactive decay (1)

The symbol system used for representing atoms can also be used for nuclei and other particles

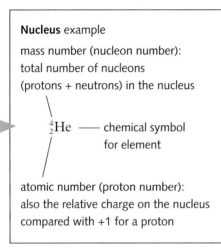

Nucleus example

mass number (nucleon number):
total number of nucleons
(protons + neutrons) in the nucleus

$^{4}_{2}\text{He}$ —— chemical symbol for element

atomic number (proton number):
also the relative charge on the nucleus
compared with +1 for a proton

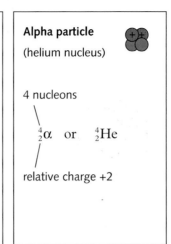

Alpha particle
(helium nucleus)

4 nucleons

$^{4}_{2}\alpha$ or $^{4}_{2}\text{He}$

relative charge +2

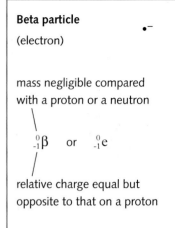

Beta particle
(electron)

mass negligible compared
with a proton or a neutron

$^{0}_{-1}\beta$ or $^{0}_{-1}\text{e}$

relative charge equal but
opposite to that on a proton

If an isotope is radioactive, it has an unstable arrangement of neutrons and protons in its nuclei. The emission of an alpha or beta particle makes the nucleus more stable, but alters the numbers of protons and neutrons in it. So it becomes the nucleus of a different element. The original nucleus is called the **parent** nucleus. The nucleus formed is the **daughter** nucleus. The daughter nucleus and any emitted particles are the **decay products**.

Alpha decay

Radium-226 (atomic number 88) decays by alpha emission. The loss of the alpha particle leaves the nucleus with 2 protons and 2 neutrons less than before. So the mass number drops to 222 and the atomic number to 86. Radon has an atomic number of 86, so radon is the new element formed:

Nuclear essentials

Atoms of any one element all have the same number of protons in their nucleus.

Elements exist in different versions, called isotopes. For example, lithium is a mixture of two isotopes: lithium-6 (with 3 protons and 3 neutrons in the nucleus) and lithium-7 (with 3 protons and 4 neutrons).

Any one particular type of atom, for example lithium-7, is called a **nuclide**. However the word 'isotope' is commonly used instead of nuclide.

Radioactive isotopes have unstable nuclei. In time each nucleus decays (breaks up) by emitting an alpha or beta particle and, in some cases, a burst of gamma radiation as well.

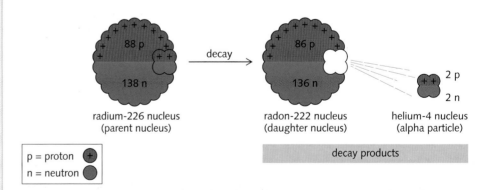

radium-226 nucleus
(parent nucleus)

radon-222 nucleus
(daughter nucleus)

helium-4 nucleus
(alpha particle)

p = proton
n = neutron

decay products

The decay process can be written as a nuclear equation:

$$^{226}_{88}\text{Ra} \rightarrow {}^{222}_{86}\text{Rn} + {}^{4}_{2}\alpha$$

During alpha decay:

● the top numbers balance on both sides of the equation (226 = 222 + 4), so the nucleon number is conserved (unchanged)
● the bottom numbers balance on both sides of the equation (88 = 86 + 2), so charge is conserved
● a new element is formed, with an atomic number 2 less than before. The mass number is 4 less than before.

256

Beta decay

Iodine-131 (atomic number 53) decays by beta emission. When this happens, a neutron changes into a proton, an electron, and an uncharged, almost massless relative of the electron called an antineutrino. The electron and antineutrino leave the nucleus at high speed. As a proton has replaced a neutron in the nucleus, the atomic number rises to 54. This means that a nucleus of xenon-131 has been formed:

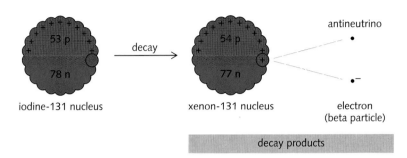

iodine-131 nucleus   xenon-131 nucleus   electron (beta particle)

decay products

The decay process can be written as a nuclear equation:

$$^{131}_{53}\text{I} \rightarrow {}^{131}_{54}\text{Xe} + {}^{0}_{-1}\beta + {}^{0}_{0}\bar{\nu} \qquad (\bar{\nu} = \text{antineutrino})$$

During this type of beta decay:

- the top numbers balance on both sides of the equation ($131 = 131 + 0 + 0$), so the nucleon number is conserved
- the bottom numbers balance on both sides of the equation ($53 = 54 - 1 + 0$), so charge is conserved
- a new element is formed, with an atomic number 1 more than before. The mass number is unchanged.

Gamma emission

With some isotopes, the emission of an alpha or beta particle from a nucleus leaves the protons and neutrons in an 'excited' arrangement. As the protons and neutrons rearrange to become more stable, they lose energy. This is emitted as a burst of gamma radiation.

- Gamma emission by itself causes no change in mass number or atomic number.

Beta⁻ and beta⁺

There is a less common form of beta decay, in which the emitted beta particle is a **positron**. This is the **antiparticle** of the electron, with the same mass, but opposite charge ($+1$). During this type of decay, a proton changes into a neutron, a positron, and a neutrino. The element formed has an atomic number one less than before.

To distinguish the two types of beta decay, they are sometimes called beta⁻ decay (electron emitted) and beta⁺ decay (positron emitted).

1 The following equation represents the radioactive decay of thorium-232. A, Z, and X are unknown.

$$^{232}_{90}\text{Th} \rightarrow {}^{A}_{Z}X + {}^{4}_{2}\alpha$$

a) What type of radiation is being emitted?
b) What are the values of A and Z?
c) Use the table on page 250 to decide what new element is formed by the decay process.
d) Rewrite the above equation, replacing A, Z, and X with the numbers and symbols you have found.
e) What are the decay products?

2 When radioactive sodium-24 decays, magnesium-24 is formed. The following equation represents the decay process, but the equation is incomplete:

$$^{24}_{11}\text{Na} \rightarrow {}^{24}_{12}\text{Mg} + \underline{\quad\quad}$$

Assuming that only one charged particle is emitted:

a) What is the mass number of this particle?
b) What is the relative charge of this particle?
c) What type of particle is it?

Related topics: nuclei and isotopes **11.01**; alpha, beta, and gamma radiation **11.02**; more on beta decay **11.10**

11.05 Radioactive decay (2)

Radioactive decay happens spontaneously (all by itself) and at random. There is no way of predicting when a particular nucleus will disintegrate, or in which direction a particle will be emitted. Also, the process is unaffected by pressure, temperature, or chemical change. However, some types of nucleus are more unstable than others and decay at a faster rate.

Rate of decay and half-life

Iodine-131 is a radioactive isotope of iodine. The chart below illustrates the decay of a sample of iodine-131. On average, 1 nucleus disintegrates every second for every 1 000 000 nuclei present.

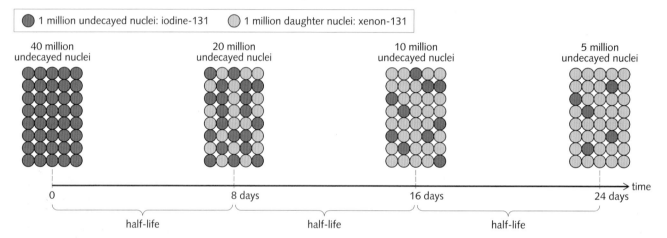

To begin with, there are 40 million undecayed nuclei. 8 days later, half of these have disintegrated. With the number of undecayed nuclei now halved, the number of distintegrations over the next 8 days is also halved. It halves again over the next 8 days... and so on. Iodine-131 has a **half-life** of 8 days.

radioactive isotope	half-life
boron-12	0.02 seconds
radon-220	52 seconds
iodine-128	25 minutes
radon-222	3.8 days
strontium-90	28 years
radium-226	1602 years
carbon-14	5730 years
plutonium-239	24 400 years
uranium-235	7.1×10^8 years
uranium-238	4.5×10^9 years

> The half-life of a radioactive isotope is the time taken for half the nuclei present in any given sample to decay.

The half-lives of some other radioactive isotopes are given on the left. It might seem strange that there should be any short-lived isotopes still remaining. However, some are radioactive daughters of long-lived parents, while others are produced artificially in nuclear reactors.

Activity and half-life

In a radioactive sample, the average number of disintegrations per second is called the **activity**. The SI unit of activity is the **becquerel** (**Bq**). An activity of, say, 100 Bq means that 100 nuclei are disintegrating per second.

The graph at the top of the next page shows how, on average, the activity of a sample of iodine-131 varies with time. As the activity is always proportional to the number of undecayed nuclei, it too halves every 8 days. So 'half-life' has another meaning as well:

> The half-life of a radioactive isotope is the time taken for the activity of any given sample to fall to half its original value.

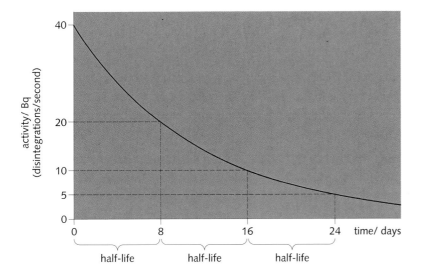

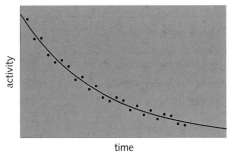

◄ Radioactive decay of iodine-131. Iodine-131 has a half-life of 8 days. From any point on the curve, it always takes 8 (days) along the time axis for the activity to halve.

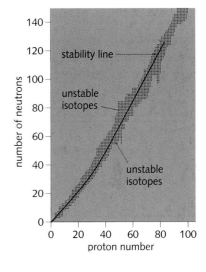

▲ Radioactive decay is a random process. So, in practice, the curve is a 'best fit' of points which vary irregularly like this.

To obtain a graph like the one above, a GM tube is used to detect the particles emitted by the sample. The number of counts per second recorded by the ratemeter is proportional to the activity – though not equal to it, because not all of the emitted particles are detected.

Stability of the nucleus★

In a nucleus, some proportions of neutrons to protons are more stable than others. If the number of neutrons is plotted against the number of protons for all the different isotopes of all the elements, the general form of the graph is as shown on the right. It has these features:

- Stable isotopes lie along the stability line.
- Isotopes above the stability line have too many neutrons to be stable. They decay by beta$^-$ (electron) emission because this reduces the number of neutrons.
- Isotopes below the stability line have too few neutrons to be stable. They decay by beta$^+$ (positron) emission because this increases the number of neutrons.
- The heaviest isotopes (proton numbers > 83) decay by alpha emission.

Q

To answer questions 1 and 2, you will need information from the table of half-lives on the opposite page.

1 If samples of strontium-90 and radium-226 both had the same activity today, which would have the lower activity in 10 years' time?
2 If the activity of a sample of iodine-128 is 800 Bq, what would you expect the activity to be after
   a) 25 minutes **b)** 50 minutes **c)** 100 minutes?
3 The graph on the right shows how the activity of a small radioactive sample varied with time.
   a) Why are the points not on a smooth curve?
   b) Estimate the half-life of the sample.

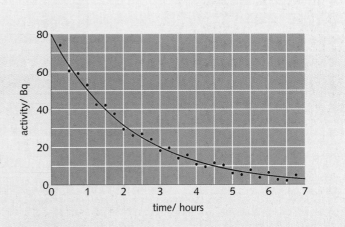

Related topics: nuclei and isotopes **11.01**; GM tube **11.03**

Nuclear energy★

> **Nuclear essentials**
>
> Atoms of any one element all have the same number of protons in the nucleus. If this number is altered in some way, an atom of a completely different element is formed.
>
> Elements exist in different versions, called isotopes, with different numbers of neutrons in the nucleus. Radioactive isotopes have unstable nuclei. In time, these decay (break up) by emitting one or more particles and, in some cases, gamma radiation as well.

▶ A chain reaction. A neutron causes a uranium-235 nucleus to split, producing more neutrons, which cause more nuclei to split... and so on.

> **Nuclear safety**
>
> Nuclear power stations have safety procedures to
>
> ● shield people from direct nuclear radiation
>
> ● keep people's time of exposure to radiation as short as possible
>
> ● prevent radioactive materials from getting into the body.
>
> Concrete, steel, and lead shielding reduce radiation, and radioactive materials are kept in sealed containers to prevent gas, dust, or liquid escaping.

When alpha or beta particles are emitted by a radioactive isotope, they collide with surrounding atoms and make them move faster. In other words, the temperature rises as nuclear energy (potential energy stored in the nucleus) is transformed into thermal energy (heat).

In radioactive decay, the energy released per atom is around a million times greater than that from a chemical change such as burning. However, the rate of decay is usually very slow. Much faster decay can happen if nuclei are made more unstable by bombarding them with neutrons. Whenever a particle penetrates and changes a nucleus, this is called a **nuclear reaction**.

Fission

Natural uranium is a dense radioactive metal consisting mainly of two isotopes: uranium-238 (over 99%) and uranium-235 (less than 1%). The diagram below shows what can happen if a neutron strikes and penetrates a nucleus of uranium-235. The nucleus becomes highly unstable and splits into two lighter nuclei, shooting out two or three neutrons as it does so. The splitting process is called **fission**, and the fragments are thrown apart as energy is released. If the emitted neutrons go on to split other nuclei... and so on, the result is a **chain reaction**, and a huge and rapid release of energy.

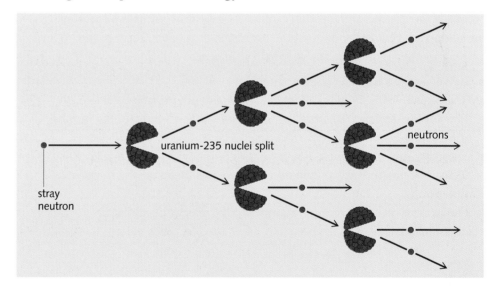

stray neutron

uranium-235 nuclei split

neutrons

For a chain reaction to be maintained, the uranium-235 has to be above a certain **critical mass**, otherwise too many neutrons escape. In the first atomic bombs, an uncontrolled chain reaction was started by bringing two lumps of pure uranium-235 together so that the critical mass was exceeded. In present-day nuclear weapons, plutonium-239 is used for fission.

Fission in a nuclear reactor

In a **nuclear reactor** in a nuclear power station, a controlled chain reaction takes place and thermal energy (heat) is released at a steady rate. The energy is used to make steam for the turbines, as in a conventional power station. In many reactors, the nuclear fuel is uranium dioxide, the natural uranium being enriched with extra uranium-235. The fuel is in sealed cans (or tubes).

To maintain the chain reaction in a reactor, the neutrons have to be slowed down, otherwise many of them get absorbed by the uranium-238. To slow them, a material called a **moderator** is needed. Graphite is used in some reactors, water in others. The rate of the reaction is controlled by raising or lowering **control rods**. These contain boron or cadmium, materials which absorb neutrons.

Nuclear waste

After a fuel can has been in a reactor for three of four years, it must be removed and replaced. The amount of uranium-235 in it has fallen and the fission products are building up. Many of these products are themselves radioactive, and far too dangerous to be released into the environment. They include the following isotopes, none of which occur naturally.

- Strontium-90 and iodine-131, which are easily absorbed by the body. Strontium becomes concentrated in the bones; iodine in the thyroid gland.
- Plutonium-239, which is produced when uranium-238 is bombarded by neutrons. It is itself a nuclear fuel and is used in nuclear weapons. It is also highly toxic. Breathed in as dust, the smallest amount can kill.

Spent fuel cans are taken to a reprocessing plant where unused fuel and plutonium are removed. The remaining waste, now a liquid, is sealed off and stored with thick shielding around it. Some of the isotopes have long half-lives, so safe storage will be needed for thousands of years. The problem of finding acceptable sites for long-term storage has still not been solved.

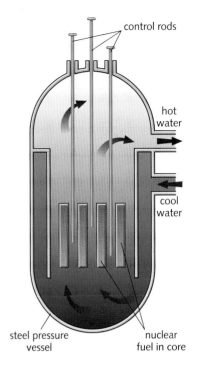

A pressurized water reactor (PWR). For safety, the reactor is housed inside a sealed containment building made of steel and concrete.

Energy and mass

According to Albert Einstein (1905), energy itself has mass. If an object gains energy, its mass increases; if it loses energy, its mass decreases. The mass change m (kg) is linked to the energy change E (joules) by this equation:

$$E = mc^2$$

(where c is the speed of light, 3×10^8 m/s)

The value of c^2 is so high that energy gained or lost by everyday objects has a negligible effect on their mass. However, in nuclear reactions, the energy changes per atom are much larger, and produce detectable mass changes. For example, when the fission products of uranium-235 are slowed down in a nuclear reactor, their total mass is found to be reduced by about 0.1%.

The steel flasks on this train contain waste from a nuclear reactor.

1 The high temperatures deep underground are caused by the decay of radioactive isotopes in the rocks. Why does radioactive decay cause a rise in temperature?
2 What is meant by **a)** fission **b)** a chain reaction?
3 Give one example of
   a) a controlled chain reaction
   b) an uncontrolled chain reaction.
4 **a)** Where does plutonium-239 come from?
   b) Why is plutonium-239 so dangerous?
5 In a typical fission process, uranium-235 absorbs a neutron, creating a nucleus which splits to form barium-141, krypton-92, and three neutrons.

	mass/kg
neutron	1.674×10^{-27}
uranium-235 nucleus	390.250×10^{-27}
barium-141 nucleus	233.964×10^{-27}
krypton-92 nucleus	152.628×10^{-27}

a) Using the data above, calculate the total mass of the uranium nucleus and the neutron.
b) Calculate the total mass of the barium and krypton nuclei and the three neutrons.
c) Use $E = mc^2$ to calculate the energy released per decay by the fission process.

Related topics: energy **4.01**; power stations **4.05–4.06**; radiation dangers **11.02–11.03**; radioactive decay **11.04–11.05**; half-life **11.05**

261

Fusion future★

Nuclear essentials

The nucleus of an atom is made up of protons and (in most cases) neutrons.

Each element has a different number of protons in the nucleus of its atom. The lightest element, hydrogen has just one.

Elements exist in different versions, called isotopes. These have different numbers of neutrons in the nucleus.

Hydrogen

Hydrogen is the most plentiful element, in the Universe. The Sun is 75% hydrogen. There is also lots of hydrogen on Earth, though most has combined with oxygen to form water (H_2O).

In the nucleus of an atom, the protons and neutrons are held tightly together by a force called, simply, a **strong nuclear force**. However, in some nuclei, they are more tightly held than in others. To get release energy, the trick is to make the protons and neutrons regroup into more tightly held arrangements than before. Protons and neutrons in 'middleweight' nuclei tend to be the most tightly held, so splitting very heavy nuclei releases energy: that happens in nuclear fission. However, energy can also be released by fusing (joining) very light nuclei together to make heavier ones. This is called **nuclear fusion**. It is the process that powers the stars. One day, it may drive power stations on Earth.

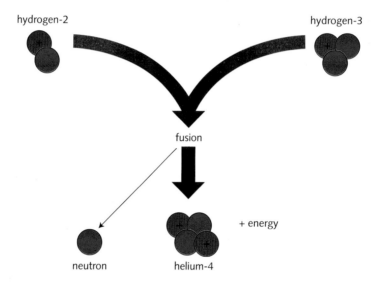

The diagram above shows the fusion of two hydrogen nuclei to form helium. Fusion is difficult to achieve because the nuclei are charged, and repel each other. To beat the repulsion and join up, they must travel very fast – which means that the gas must be much hotter than any temperatures normally achieved on Earth.

Building a fusion reactor

Scientists and engineers are trying to design fusion reactors for use as an energy source on Earth. But there are huge problems to overcome. Hydrogen must be heated to at least 40 million degrees Celsius, and kept hot and compressed, otherwise fusion stops. No ordinary container can hold a superhot gas like this, so scientists are developing reactors that trap the nuclei in a magnetic field.

Fusion reactors will have huge advantages over today's fission reactors. They will produce more energy per kilogram of fuel. Their hydrogen fuel can be extracted from sea water. Their main waste product, helium, is not radioactive. And they have built-in safety: if the system fails, fusion stops.

Fusion in a star

The Sun is a star. Like most other stars, it gets its energy from the fusion of hydrogen into helium. Deep in its core, the heat output and huge gravitational pull keep the hydrogen hot and compressed enough to maintain fusion. It has enough hydrogen left to keep it shining for another 6 billion years.

This magnetic containment vessel, called a **tokamak**, is being used to investigate fusion.

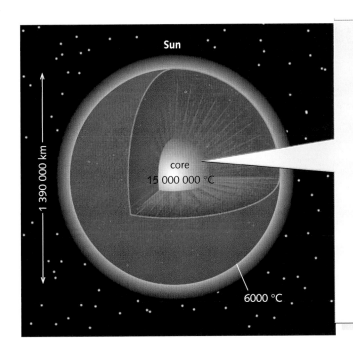

Sun

1 390 000 km

core
15 000 000 °C

6000 °C

Fusion in the Sun's core

Energy is released as hydrogen is converted into helium.

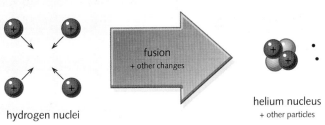

hydrogen nuclei

fusion
+ other changes

helium nucleus
+ other particles

Four hydrogen nuclei fuse together for each helium nucleus formed. This is a multi-stage process which also involves the creation of two neutrons from two protons.

In the Sun, fusion happens at 'only' 15 million degrees Celsius. But the Sun uses different fusion reactions from those being tried on Earth. If the Sun were scaled down to the size of a nuclear reactor, its power output would be too low to be useful.

Scientists think that the Sun formed about 4500 million years ago in a huge cloud of gas (mainly hydrogen) and dust called a **nebula**. There, gravity slowly pulled the material into blobs. In the centre, one blob grew bigger than all the rest. Around it, smaller blobs formed. These would become planets and moons.

As more and more material was pulled into the central blob, gravitational potential energy was changed into thermal energy, so the blob became hotter and hotter. Eventually, its core became so hot and compressed that fusion started and it 'lit up' to become a star – the Sun. Other stars formed – and are being formed – in the same way.

The Great Nebula in the constellation of Orion. Stars form in huge clouds of gas and dust like this. The gas is mainly hydrogen.

 Q

1 *Splitting very heavy nuclei to form lighter ones*
   Joining very light nuclei to form heavier ones
   a) Which of the above statements describes what happens during nuclear fusion?
   b) What process does the other statement describe?
2 What advantages will power stations with fusion reactors have over today's nuclear power stations?
3 Why have fusion reactors have been so difficult to develop?

4 Nuclear reactions are taking place in the Sun's core.
   a) What substance does the Sun use as its nuclear fuel?
   b) What is the name of the process that supplies the Sun with its energy?
   c) What substance is made by this process?
5 A nebula is a huge cloud of gas and dust in space.
   a) Why does material in a nebula collect in blobs?
   b) Why, if a blob is large enough, will it eventually start to shine as a star?

Related Topics: gravity **2.09**; power stations **4.05**; energy resources **4.07–4.08**; atoms **11.01**; nuclear energy **11.06**

11.08 Using radioactivity

Radioactive isotopes are called **radioisotopes** (or **radionuclides**). Some are produced artificially in a nuclear reactor when nuclei absorb neutrons or gamma radiation. For example, all natural cobalt is cobalt-59, which is stable. If cobalt-59 absorbs a neutron, it becomes cobalt-60, which is radioactive.

Here are some of the practical uses of radioisotopes.

Tracers

Radioisotopes can be detected in very small (and safe) quantities, so they can be used as **tracers** – their movements can be tracked. Examples include:

- Checking the function of body organs. For example, to check thyroid function, a patient drinks a liquid containing iodine-123, a gamma emitter. Over the next 24 hours, a detector measures the activity of the tracer to find out how quickly it becomes concentrated in the thyroid gland.
- Tracking a plant's uptake of fertilizer from roots to leaves by adding a tracer to the soil water.
- Detecting leaks in underground pipes by adding a tracer to the fluid in the pipe.

For tests like those above, artificial radioisotopes with short half-lives are used so that there is no detectable radiation after a few days.

Radiotherapy

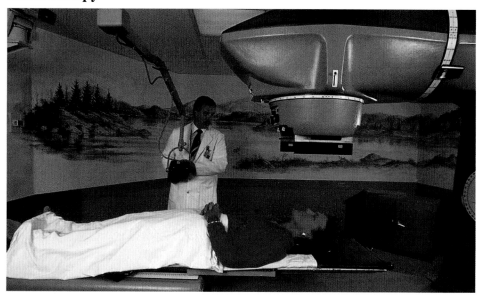

▶ A concentrated beam of gamma rays from a cobalt-60 source will be directed at one small area of this patient's body to kill the cancer cells in a tumour.

Cobalt-60 is a strong gamma emitter. Gamma rays can penetrate deep into the body and kill living cells. So a highly concentrated beam from a cobalt-60 source can be used to kill cancer cells. Treatment like this is called **radiotherapy**.

Testing for cracks

Gamma rays have the same properties as short-wavelength X-rays, so they can be used to photograph metals to reveal cracks. A cobalt-60 gamma source is compact and does not need electrical power like an X-ray tube.

Thickness monitoring

In some production processes a steady thickness of material has to be maintained. The diagram below shows one way of doing this.

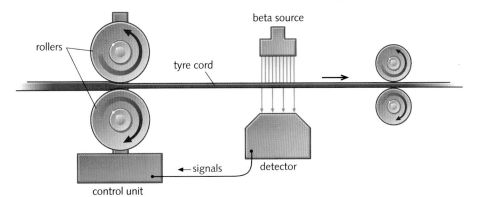

◀ The moving band of tyre cord has a beta source on one side and a detector on the other. If the cord from the rollers becomes too thin, more beta radiation reaches the detector. This sends signals to the control unit, which adjusts the gap between the rollers.

Carbon dating

There is carbon in the atmosphere (in carbon dioxide) and in the bodies of animals and plants. A small proportion is radioactive carbon-14 (half-life 5730 years). Although carbon-14 decays, the amount in the atmosphere changes very little because more is continually being formed as nitrogen in the upper atmosphere is bombarded by cosmic radiation from space. While plants and animals are living, feeding, and breathing, they absorb and give out carbon, so the proportion of carbon-14 in their bodies stays constant. But when they die, no more carbon is taken in and the proportion of carbon-14 is gradually reduced by radioactive decay. By measuring the activity of a sample, the age of the remains can be estimated. This is called **carbon dating**. It can be used to find the age of organic materials such as wood and cloth. However, it assumes that the proportion of carbon-14 in the atmosphere was the same hundreds or thousands of years ago as it is today.

Using carbon dating, scientists have discovered that these remains of a mammoth are 15 000 years old.

Dating rocks

When rocks are formed, some radioisotopes become trapped in them. For example, potassium-40 is trapped when molten material cools to form igneous rock. As the potassium-40 decays, more and more of its stable decay product, argon-40, is created. Provided none of this argon gas has escaped, the age of the rock (which may be hundreds of millions of years) can be estimated from the proportions of potassium-40 to argon-40. Igneous rock can also be dated by the proportion of uranium to lead isotopes – lead being the final, stable product of a series of decays that starts with uranium.

1 a) What are radioisotopes?
   b) How are artificial radioisotopes produced?
   c) Give *two* medical uses of radioisotopes.
2 Give *two* uses of gamma radiation.
3 In the thickness monitoring system shown above:
   a) Why is a beta source used, rather than an alpha or gamma source?
   b) What is the effect on the detector if the thickness of the tyre cord increases?

4 a) Give *two* uses of radioactive tracers.
   b) Why is it important to use radioactive tracers with short half-lives?
5 Carbon-14 is a radioactive isotope of carbon.
   a) What happens to the proportion of carbon-14 in the body of a plant or animal while it is alive?
   b) Why does the proportion of carbon-14 in the remains of dead plants and animals give clues about their age?

Related topics: alpha, beta, and gamma radiation **11.02–11.03**; radioactive decay **11.04–11.05**; half-life **11.05**

265

Atoms and particles (1)

11.09

Atoms are made up of even smaller particles. From experimental evidence collected over the past hundred years, scientists have been able to develop and improve their models (descriptions) of atoms and the particles in them.

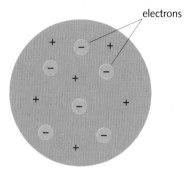

electrons

Thomson's 'plum pudding' model of the atom

Thomson's 'plum pudding' model

The **electron** was the first atomic particle to be discovered. It was identified by J. J. Thomson in 1897. The electron has a negative (−) electric charge, so an atom with electrons in it must also contain positive (+) charge to make it electrically neutral. Thomson suggested that an atom might be a sphere of positive charge with electrons dotted about inside it rather like raisins in a pudding. This became knows as the 'plum pudding' model.

Rutherford's nuclear model

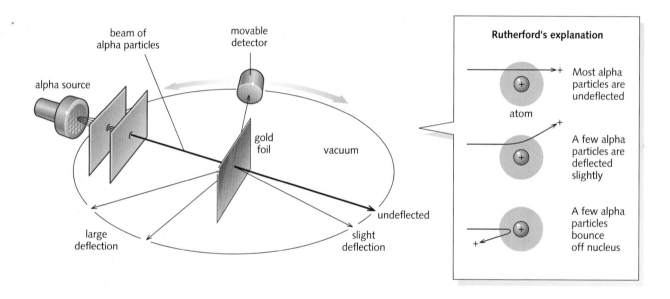

The above experiment was carried out in 1911 by Geiger and Marsden under the supervision of Ernest Rutherford. It produced results which could not be explained by the plum pudding model. Thin gold foil was bombarded with alpha particles, which are positively charged. Most passed straight through the gold atoms, but a few were repelled so strongly that they bounced back or were deflected through large angles. Rutherford concluded that the atom must be largely empty space, with its positive charge and most of its mass concentrated in a tiny **nucleus** at the centre. In his model, the much lighter electrons orbited the nucleus rather like the planets around the Sun.

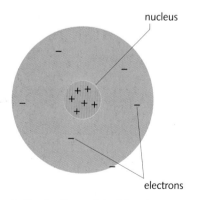

nucleus

electrons

Rutherford's model of the atom: electrons orbit a central nucleus. (If the nucleus were correctly drawn to scale, it would be too small to see.)

Discovering particles in the nucleus

Rutherford's model said nothing about what was inside the nucleus. However, in 1919, Rutherford bombarded nitrogen gas with fast alpha particles and found that positively charged particles were being knocked out. These were **protons**. In 1932, James Chadwick discovered that the nucleus also contained uncharged particles with a similar mass to protons. He called these **neutrons**.

The problem of spectral lines★

Light comes from atoms. The spectrum of white light is a continuous range of colour from red (the longest wavelength) to violet (the shortest). However, not all spectra are like this. For example, if there is an electric discharge through hydrogen, the glowing gas emits particular wavelengths only, so the spectrum is made up of lines, as shown below. As it stood, Rutherford's model could not explain why spectra like this occurred. To solve this problem, the model had to be modified.

◄ Part of the line spectrum of hydrogen. Each line represents light of a particular wavelength.

shorter...wavelength..longer

The Rutherford-Bohr model★

In 1913, Neils Bohr modified Rutherford's model by applying the **quantum theory** devised by Max Planck in 1900. According to this theory, energy cannot be divided into ever smaller amounts. It is only emitted (or absorbed) in tiny 'packets', each called a **quantum**. Bohr reasoned that electrons in higher orbits have more energy than those in lower ones. So, if only quantum energy changes are possible, only certain electron orbits are allowed. This modified model is known as the **Rutherford-Bohr model**.

Using the model, Bohr was able to explain why atoms emit light of particular wavelengths only (see the next spread). He even predicted the positions of the lines in the spectrum of hydrogen. However, his calculations did not work for substances with a more complicated electron structure. To deal with this problem, scientists have developed a **wave mechanics** model in which allowed orbits are replaced by allowed **energy levels**. However, this is an entirely mathematical approach, and the Rutherford-Bohr model is still used as a way of representing atoms in pictures.

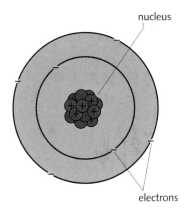

The Rutherford-Bohr model of the atom (with nuclear particles included). In this model, only certain electron orbits are allowed.

Q

1 What is the difference between Rutherford's model of the atom and Thomson's 'plum pudding' model?
2 What is the difference between the Rutherford-Bohr model of the atom and Rutherford's model?
3 On the right, a beam of alpha particles is being directed at a thin piece of gold foil. How does the Rutherford model of the atom explain why
   a) most of the alpha particles go straight through the foil
   b) some alpha particles are deflected at large angles?
4 Why do the results of the experiment on the right suggest that the nucleus has a positive charge?

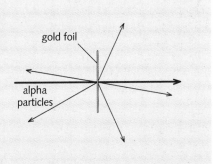

Related topics: light waves **7.01** and **7.11**; spectrum **7.04**; electric charge **8.01–8.02**; particles in the atom **11.01**

Atoms and particles (2)★

If an electron gains energy...

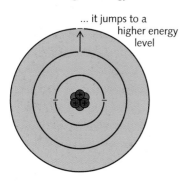

... it jumps to a higher energy level

When the electron drops back to a lower level...

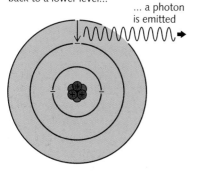

... a photon is emitted

▲ How an atom gives off light

How an atom gives off light

Bohr's explanation of how an atom gives off light was like this.

If an electron gains energy in some way – for example, because its atom collides with another one – it may jump to a higher energy level. But the atom does not stay in this **excited state** for long. Soon, the electron loses energy by dropping back to a lower level. According to the quantum theory, the energy is radiated as a pulse of light called a **photon**. The greater the energy change, the shorter the wavelength of the light.

As a line spectrum contains particular wavelengths only, it provides evidence that only certain energy changes are occurring within the atom – and therefore that only certain energy levels are allowed.

Fundamental particles

A **fundamental particle** is one which is *not* made up of other particles. An atom is not fundamental because it is made up of electrons, protons, and neutrons. But are these fundamental? To answer this and other questions, scientists carry out experiments with **particle accelerators**. They shoot beams of high-energy particles (such as protons) at nuclei, or at other beams, and detect the particles emerging from the collisions. In collider experiments, new particles are created as energy is converted into mass. However, most of these particles do not exist in the atoms of ordinary matter.

▶ Part of the proton-antiproton collider at the Fermilab research centre, Illinois, USA. Electromagnets are used to accelerate the particles round a circular path over 8 km long.

The present theory of particles is called the **standard model**. According to this model, electrons are fundamental particles. However, neutrons and protons are made up of other particles called **quarks**, as shown in the chart on the next page. In ordinary matter, there are two types of quark, called the **up quark** and the **down quark** for convenience. Each proton or neutron is made up of three quarks. The quarks have fractional charges compared with the charge on an electron.

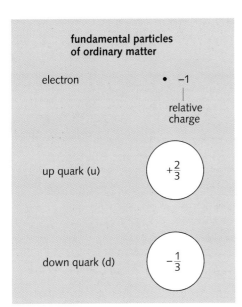

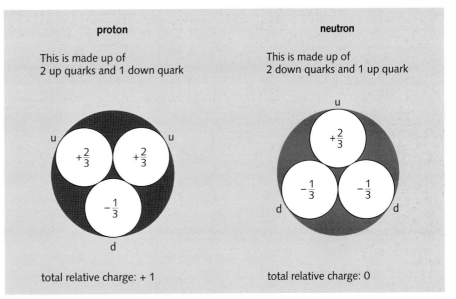

Individual quarks have never been detected. The existence of quarks has only been deduced from the patterns seen in the properties of other particles – for example, how high-energy particles are scattered.

Quark changes in beta decay

In the most common form of beta decay, a neutron decays to form a proton, an electron (the beta particle), and an antineutrino:

> neutron → proton + electron + antineutrino

If this is rewritten to show the quarks:

> up quark   up quark
> down quark → up quark + electron + antineutrino
> down quark   down quark

From the above, you can see that this type of beta decay occurs when a down quark changes into an up quark, as follows:

> down quark → up quark + electron + antineutrino
> $(-^1/_3)$ $(+^2/_3)$ (-1) (0)

The relative charges underneath the equation show that there is no change in total charge. In other words, charge is conserved.

In the less common form of beta decay, a proton decays to form a neutron, a positron (the beta particle), and a neutrino. This happens when an up quark in the proton changes into a down quark.

Decay essentials

The break-up of an unstable nucleus is called radioactive decay. During beta decay, a beta particle is shot out. In most cases, this particle is an electron (−). However, more rarely, it a positron (+), an antiparticle with the same mass as an electron, but opposite charge.

1 When an electron drops back to a lower energy in an atom, it loses energy.
a) What happens to this energy?
b) If the difference between the two energy levels was greater, how would this affect the wavelength of the light emitted?
c) Why do atoms emit certain wavelengths only?
2 What is meant by a fundamental particle?

3 Which of the following are thought to be fundamental particles?
electrons   protons   neutrons   quarks
4 Quarks have a fractional charge. Explain why, if a neutron is made up of three quarks, it is uncharged.
5 In one form of beta decay, an up quark changes into a down quark. Explain why, in this case, the beta particle emitted must be a positron and not an electron.

Related topics: light waves **7.01** and **7.11**; charge on electron **10.07**; particles in the atom **11.01**; beta decay **11.04–11.05**

269

1

electrons	nuclei	protons	waves

a) Copy and complete the following sentences using words from the above list. Each word may be used once, more than once or not at all.

   (i) Radioactive substances have atoms with unstable

      _____ .   [1]

   (ii) Beta particles are _____ .   [1]

   (iii) Gamma rays are _____ .   [1]

b) Name another type of radioactive particle not mentioned in part **a)**.   [1]

WJEC

2 The symbol $^{35}_{17}Cl$ represents one atom of chlorine.

a) State the names and numbers of the different types of particle found in one of these chlorine atoms.   [3]

b) State where these particles are to be found in the atom.   [2]

UCLES

3

proton number	26
mass number	59
radiation emitted	beta and gamma

The table above shows information about a radioisotope of iron called iron-59.

a) Calculate:

   (i) the number of neutrons in the nucleus;   [1]

   (ii) the total number of charged particles in a single atom of iron-59.   [1]

b) Iron-59 and iron-56 are both isotopes of iron. What are isotopes?   [1]

c) Iron-59 emits two types of radiation. Briefly explain how the gamma radiation could be separated from the beta radiation emitted.   [1]

WJEC

4 Phosphorus-32 is a radioactive isotope. It can be used to prove that plants absorb phosphorus from the soil around them.

a) (i) The stable isotope of phosphorus has a mass number of 31. State the structural difference between atoms of phosphorus-31 and phosphorus-32.   [2]

   (ii) Explain why both isotopes of phosphorus have identical chemical properties.   [1]

b) Phosphorus-32 is a **beta-emitter** with a **half-life** of 14 days.

   (i) What is a beta particle?   [1]

   (ii) The atomic number of phosphorus-32 atom is 15. State the new values of the atomic and mass numbers of the atom just after it has emitted a beta particle.   [2]

   (iii) Explain what is meant by the term **half-life**.   [1]

c) A solution of the isotope is watered onto the soil around the plant. Each day for the next week, a leaf is removed from the plan and tested for radioactivity.

   (i) State **three** safety precautions which should be adopted when doing experiments with phosphorus-32.   [3]

   (ii) Describe **two** methods which could be used to measure the activity of a leaf.   [2]

MEG

5 Phyl is in hospital. She is injected with the radioisotope technetium-99m.

This isotope is absorbed by the thyroid gland in her throat. A radiation detector placed outside her body and above her throat detects the radiation.

Technetium-99m has a half-life of 6 hours. It emits gamma radiation.

a) Why is an emitter of alpha radiation unsuitable?   [1]

b) (i) How long will it take for the activity of the technetium-99m to fall to a quarter of its original value?   [2]

   (ii) After 24 hours, how will the activity of the technetium-99m compare with its original value?   [2]

c) Eventually the level of radiation from the technetium-99m will fall to less than the level of the background radiation. State **two** naturally occurring sources of background radiation.   [2]

MEG

6 The radioactive isotopes in the Chernobyl fallout which caused most concern were iodine-131 and caesium-137. Both are beta and gamma emitters. Iodine-131, in rainfall, found its way into milk but caesium-137, with a half-life of 30 years, may cause more long term problems.

a) From which part of the atom do the beta and gamma rays come?   [1]

b) Explain what the number 131 tells you about the iodine atom. [2]

c) After the Chernobyl accident, a milk sample containing iodine-131 was found to have an activity of 1600 units per litre. The activity of the sample was measured every 7 days and the results are shown in the table below.

time (days)	0	7	14	21	28	35
activity (units per litre)	1600	875	470	260	140	77

   (i) Draw a graph of activity against time, using the grid on the next page as a guide.   [2]

   (ii) Estimate the half-life of iodine-131 and show on the graph how you arrived at your answer.   [2]

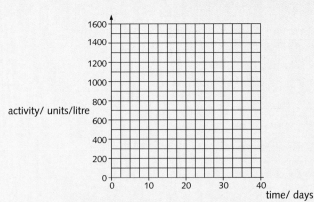

d) Give a reason why caesium-137 could cause longer-term problems than iodine-131. [2]

WJEC

7 a) (i) Explain why some substances are radioactive and some are not. [2]
(ii) State the cause of background radiation. [1]
(iii) Explain what you understand by the meaning of the **half-life** of a radioactive element. [2]
b) Technetium-99m is a radioactive material with a half-life of 6 hours. It is used to study blood flow around the body. A sample of technetium-99m has an activity of 96 counts per minute when injected into a patient's blood stream. Estimate
(i) its activity after 12 hours, [1]
(ii) how long it will take for the radioactivity from the injection to become undetectable. [1]
c) Technetium-99m is a gamma (γ) emitter and does not produce alpha (α) or beta (β) radiations. Explain why it is safe to inject technetium-99m into the body. [2]
d) Radioactive salt (sodium chloride) is also used in medicine. The radioactive sodium (Na) in the salt decays, according to the equation shown below, to form magnesium (Mg).

$$_{11}^{24}\text{Na} \longrightarrow _{12}^{24}\text{Mg} + \textbf{X} + \gamma \text{ radiation}$$

(i) Name the particle **X**. [1]
(ii) Use the information given in the equation above to find the
I. total number of charged particles in each sodium atom, [1]
II. number of neutrons in the nucleus of a sodium 24 atom. [1]

WJEC

8 This question is about information given in a leaflet about radon.
a) Extract 1 'Radon is a naturally occurring radioactive gas. It comes from uranium which occurs in rocks and soils.'
(i) Explain the meaning of the word **radioactive**.
(ii) Explain, in detail, the danger of breathing radon gas into the lungs. [4]

Extract 2 is a diagram showing how radon decays

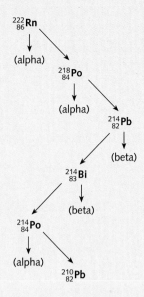

Two of the nuclei shown in the diagram are isotopes of polonium.
b) Explain the meaning of the word **isotope**. [1]
c) In the diagram, radon is shown as $_{86}^{222}\text{Rn}$. In a neutral radon atom, what is the number of
(i) protons,
(ii) electrons,
(iii) neutrons? [3]

NEAB

9 A radioactive isotope of gold has the symbol $_{79}^{196}\text{Au}$. If this isotope is injected into the bloodstream of a patient, it can be used by doctors as a tracer to monitor the way the patient's heart works. The isotope emits gamma radiation that is detected outside the patient's body.
a) Why would an isotope that emits alpha radiation be unsuitable as a tracer to monitor the working of the heart? [1]
b) Give one non-medical use for a radioactive tracer. [1]

SEG

10 Isotopes of the radioactive element uranium occur naturally in small proportions in some rocks. The table gives information about **one** uranium isotope.

nucleon (mass) number	238
proton (atomic) number	92
radiation emitted	alpha particle

a) How many neutrons are there in an atom of this uranium isotope? [1]
b) From which part of the uranium atom does the alpha particle come? [1]

SEG

Photocopy the list of topics below and tick the boxes of the ones that are included in your examination syllabus. (Your teacher should be able to tell you which they are.) Use your list when you revise. The spread number in brackets tells you where to find more information.

❏ 1 The particles in an atom. (11.01)

❏ 2 The particles in the nucleus of an atom. (11.01)

❏ 3 The forces binding the particles in an atom together. (11.01)

❏ 4 The meanings of atomic number (proton number) and mass number (nucleon number). (11.01)

❏ 5 Isotopes and nuclides (11.01)

❏ 6 Electron shells. (11.01)

❏ 7 Radioactive materials and the radiation they produce. (11.02)

❏ 8 Ionizing radiation. (11.02)

❏ 9 Alpha particles and their properties. (11.02)

❏ 10 Beta particles and their properties. (11.02)

❏ 11 Gamma rays and their properties. (11.02)

❏ 12 The ionizing effects and penetrating powers of alpha, beta, and gamma radiation. (11.02)

❏ 13 The dangers of nuclear radiation. (11.03)

❏ 14 The main sources of background radiation. (11.03)

❏ 15 Detecting radiation with a Geiger-Müller tube. (11.03)

❏ 16 Allowing for background radiation in experiments. (11.03)

❏ 17 Handling and storing radioactive materials safely. (11.03 and 11.06)

❏ 18 Using a cloud chamber to show the tracks of alpha particles. (11.03)

❏ 19 The changes in the nucleus that occur during alpha and beta decay. (11.04)

❏ 20 Writing nuclear equations. (11.04)

❏ 21 The random nature of radioactive decay. (11.05)

❏ 22 How the rate of radioactive decay changes with time. (11.05)

❏ 23 The meaning of half-life. (11.05)

❏ 24 The SI unit of activity: the becquerel.

❏ 25 Measuring the activity of a radioactive sample. (11.05)

❏ 26 Working out a half-life from a radioactive decay curve. (11.05)

❏ 27 Why some isotopes have more stable nuclei than others. (11.05)

❏ 28 The release of energy by radioactive decay. (11.06)

❏ 29 What happens in a nuclear reaction. (11.06)

❏ 30 What happens during fission. (11.06)

❏ 31 What a chain reaction is. (11.06)

❏ 32 How a controlled chain reaction is used in a nuclear reactor. (11.06)

❏ 33 The problems of nuclear waste. (11.06)

❏ 34 The link between energy and mass. (11.06)

❏ 35 Fusion reactors (11.07)

❏ 36 Fusion in stars. (11.07)

❏ 37 What radioisotopes are. (11.08)

❏ 38 Using radioisotopes
   – as tracers
   – in radiotherapy
   – to find cracks in metals
   – for thickness monitoring
   – for dating old materials. (11.08)

❏ 39 Evidence for a nucleus in an atom. (11.09)

❏ 40 Models of the atom. (11.09)

❏ 41 The quantum theory. (11.09)

❏ 42 Photons from atoms. (11.10)

❏ 43 Quarks. (11.10)

12

History of Key Ideas

- CHANGING IDEAS ON FORCE, MOTION, ENERGY, AND HEAT
- CHANGING IDEAS ON LIGHT, RADIATION, AND ATOMS
- CHANGING IDEAS ON MAGNETISM AND ELECTRICITY

This ancient stone circle at Stonehenge in Wiltshire, England, was built in about 1500 BC. Its builders left no written records to explain its purpose, although archaelogists think that it was probably used to observe the movements of the Sun and Moon. On Midsummer's Day, the Sun rises over the Heel stone (seen in the foreground), so Stonehenge may have been a gigantic calendar.

12.01 Force, motion, and energy*

Forces and motion

On Earth, unless there is a force to overcome friction, moving things eventually come to rest. Over 2300 years ago, this led Aristotle and other Greek philosophers to believe that a force was always needed for motion. The more speed something had, the more force it needed. But in the heavens, the Sun, Moon, and stars obeyed different rules. They moved in circles for ever and ever. These ideas were generally accepted until the early 1600s, when Galileo Galilei started to come up with new ideas about motion. From his observations, Galileo deduced that, without friction, sliding objects would keep their speed. Also, all falling objects, light or heavy, would gain speed at the same steady rate.

Our present-day ideas about forces and motion mainly come from Isaac Newton, who put forward his three **laws of motion** in 1687. Our definition of force is based on his second law: force = mass × acceleration. Newton also realized that 'heavenly bodies' did not obey different rules from everything else. The motion of the Moon around the Earth was controlled by the same force – gravity – that made objects fall downwards on Earth. Newton is supposed to have had this idea while watching an apple fall from a tree, although his mathematical treatment of gravity was much more complicated than this simple experience suggests.

In 1905, Albert Einstein put forward his **special theory of relativity**. From this, we now know that, near the speed of light, Newton's second law is no longer valid. However, at the speeds we normally measure on Earth, the law is quite accurate enough.

Energy and heat

The modern, scientific meaning of energy arose in the early 1800s when scientists and engineers were developing ways of measuring the performance of steam engines. Steam engines used forces to move things. They did **work**. To do this, they had to spend **energy**. So it made sense to measure energy and work in the same units (we now use the joule).

▲ It is said that Galileo investigated the laws of motion by dropping cannon balls from the top of Pisa's famous tower. There is no evidence to support this story. However, Galileo was born in Pisa, Italy (in 1564), and studied and lectured there.

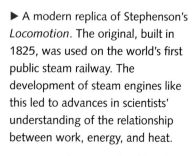

▶ A modern replica of Stephenson's *Locomotion*. The original, built in 1825, was used on the world's first public steam railway. The development of steam engines like this led to advances in scientists' understanding of the relationship between work, energy, and heat.

With the idea of energy established, people soon realized that energy could exist in different forms – electrical, potential, kinetic, and so on. However, the law of conservation of energy was not developed until 1847.

Today, we link heat with energy. However, scientists once thought that heat was an invisible, weightless fluid called 'caloric' which flowed out of hot things and was squeezed from solids when they were rubbed. In the 1790s, Count Rumford did some experiments which suggested that the caloric theory was wrong. While boring cannon barrels, he found that he could get an endless supply of heat by keeping the borer turning. If heat was a fluid, then the supply should run out. Instead, the amount of heat seemed to be directly linked with the amount of work being done. The link between work and heat was firmly established in 1849 by James Joule. He found that it always took 4.2 joules of work to produce 1 calorie of heat (an old unit, equivalent to the heat required to increase the temperature of 1 gram of water by 1 °C). However, Joule's work did not explain what heat *was*.

We now know that materials are made up of particles (atoms or molecules) which are in a state of random motion, and that heat is associated with that motion. In a solid or liquid, the particles vibrate. In a gas they move about freely at high speed. The higher the temperature, the faster the particles move. The random motion of the particles is called **thermal activity**, and the energy which an object has because of it is called **internal energy** (the sum of the kinetic and potential energies of all the particles.)

▲ Boring out cannon barrels made them hot. Count Rumford discovered that the amount of heat produced was related to the amount of work done during the boring process.

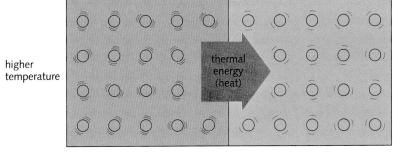

higher temperature

thermal energy (heat)

lower temperature

If a hot object is put in contact with a colder one, as above, energy is transferred from one to the other because of the temperature difference. This energy is called **heat**. So internal energy is the *total* amount of energy due to thermal activity, while heat represents an amount of energy *transferred*. However, for simplicity, both can be called **thermal energy**.

◄ In engines, releasing thermal energy by burning fuel is one stage in the process of producing motion.

Rays, waves, and particles★

Light and radiation

Over 2300 years ago, the Ancient Greeks knew that light travelled in straight lines. The Romans used water-filled glass spheres to magnify things. But it was not until the 1200s that glass lenses were first made, for spectacles. The telescope was invented in the early 1600s. In the same century, Snell discovered the law of refraction, Huyghens suggested that light was a form of wave motion, and Newton demonstrated that white light was a mixture of colours. Newton also tried to explain the nature of light. He thought that light was made up of millions of tiny 'corpuscles' (particles).

▶ One of Newton's experiments. Newton passed white sunlight into a glass prism, and produced a spectrum. When he recombined the colours with a second prism, he obtained white light again. He concluded that the colours must be from the white light and not produced by the glass.

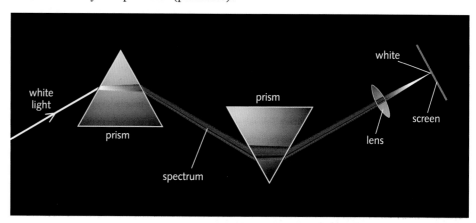

In the early 1800s, Thomas Young investigated the interference and diffraction of light and successfully used the wave theory to explain these effects. From his results, he was also able to calculate a value for the wavelength of light. Young's work seemed to put an end to Newton's 'corpuscles', but these were to appear later in another form. Some materials give off electrons when they absorb light. This is called the **photoelectric effect**. In 1905, Einstein was able to explain it by assuming that light consisted of particle-like bursts of wave energy, called **photons**. Light, it seemed, could behave like waves *and* particles.

James Clerk Maxwell was the first to put forward the idea that light was a type of **electromagnetic radiation**. He did this in 1864. From his theoretical work on electric and magnetic fields, he predicted the existence of electromagnetic waves, calculated what their speed should be, and found that it matched the speed of light. His equations also predicted the existence of radio waves, although 'real' radio waves were not detected until the 1880s. X-rays were discovered by Wilhelm Röntgen in 1895, but their electromagnetic nature was not established until 1912.

In 1896, Henri Becquerel detected a penetrating radiation coming from uranium salts. He had discovered **radioactivity**. Later, Marie Curie showed that the radiation came from within the atom and was not due to reactions with other materials. In 1899, Ernest Rutherford investigated radioactivity and identified two types of radiation, which he called alpha and beta. The following year, he discovered gamma rays. Today, we know that waves, such as gamma radiation, can behave like particles, and that particles can also behave like waves. Scientists call this **wave-particle duality**.

Marie Curie in her laboratory

Atoms and electrons

The word 'atom' comes from the Greek *atomos*, meaning indivisible. The first modern use of the word was by John Dalton, who put forward his atomic theory in 1803, in order to explain the rules governing the proportions in which different elements combined chemically. Dalton suggested that all matter consisted of tiny particles called atoms. Each element had its own type of atom, and atoms of the same element were identical. Atoms could not be created or destroyed, nor could they be broken into smaller bits.

But what were atoms made of? The first clues came in the 1890s, when scientists were studying the conduction of electricity through gases. They found that atoms could give out invisible, negatively charged rays. J. J. Thomson investigated the rays and deduced that they were particles much lighter than atoms. This was in 1897. Thomson had discovered the **electron**.

If an atom contained electrons, it must also contain positive charge to make it electrically neutral. But where was this charge? In 1911, a team led by Ernest Rutherford directed alpha particles at thin gold foil. They found that most passed straight through but a few were deflected at huge angles. To explain this, Rutherford suggested that each atom must be largely empty space, with its positive charge and most of its mass concentrated in a tiny **nucleus**.

Ernest Rutherford (right) and his assistant Hans Geiger, in 1912. They are standing next to the apparatus which they used for detecting alpha particles.

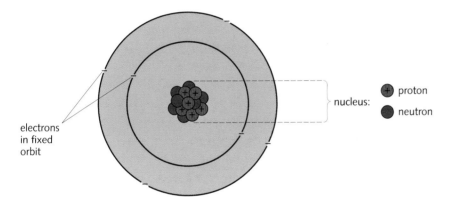

electrons in fixed orbit

nucleus:

⊕ proton

● neutron

◀ The Rutherford-Bohr model of the atom (with nuclear particles included). The picture is not to scale. With an atom of the size shown, the nucleus would be far too small to see.

In Rutherford's model (picture) of the atom, electrons orbited the nucleus like planets around the Sun. Unfortunately, the model had a serious flaw: according to classical theory, an orbiting electron ought to radiate energy continuously and spiral into the nucleus, so its orbit could not be stable. In 1913, Neils Bohr used the **quantum theory** to solve this problem. According to Bohr, electrons were in fixed orbits and could not radiate continuously. They could only lose energy by jumping to a lower orbit and emitting a quantum ('packet') of electromagnetic energy – in other words, a photon. Using this model, Bohr was able to predict the positions of the lines in the hydrogen spectrum. However, his calculations did not work for elements with a more complicated electron structure. To deal with this problem, scientists later developed a mathematical, **wave-mechanics** model of the atom.

The Rutherford-Bohr model of the atom said nothing about what was inside the nucleus. However, in 1919, Rutherford used alpha particles to knock positively charged particles out of the nucleus. These were **protons**. In 1932, James Chadwick discovered that the nucleus also contained **neutrons**. In recent years, experiments with particle accelerators have suggested that protons and neutrons are made from particles called **quarks**.

Richard Feynmann giving a lecture at the California Institute of Technology. Advances made by Professor Feynmann in quantum electrodynamics won him the 1965 Nobel Prize for Physics.

12.03 Magnetism and electricity★

Magnetism

Around 2600 years ago, the Ancient Greeks knew that a certain type of iron ore, now known as magnetite or lodestone, could attract small pieces of iron. They found the ore in a place called Magnesia, which is how magnetism got its name. The Chinese had also come across the mysterious ore and, by 200 BCE, knew that a piece of lodestone, if free to turn, would always point in the same direction.

By around 800 CE, the Chinese had discovered how to make magnetic needles by stroking small pieces or iron with lodestone. The first compasses probably consisted of a magnetized needle supported by a straw floating in a bowl of water. However, they were not really suitable for use on ships. Compasses with pivoted needles did not appear until the 1200s.

No signs of a mountain at the North Pole. At one time, sailors thought that a huge magnetic mountain here might be the source of the Earth's magnetism.

At this time, no one really understood why a compass needle points north. Some sailors believed that there was a huge mountain of lodestone at the North Pole, whose force was so strong that it would pull the iron nails out of a ship's hull. Then, in 1600, William Gilbert published the results of his experiments with magnets. He introduced the term **magnetic pole** and suggested that the Earth itself might behave like a bar magnet. But what caused magnetism? The answer to that would come from an understanding of electricity.

Electricity

The Ancient Greeks also knew of the strange properties of a solidified resin called amber. When rubbed, it attracted dust and other small things. The Greek word for amber is *elektron*, from which the word electricity comes.

Our modern knowledge of electricity really began in the 1600s, when experimenters started to investigate amber and other rubbed materials more closely. They found that it was possible to produce repulsion as well as attraction, and that there were two different kinds of **electric charge**. In 1752, Benjamin Franklin carried out a famous – and extremely dangerous – experiment in which he flew a kite in a thunderstorm and got sparks to jump from a key attached to the line. The sparks were just like those produced by rubbing amber. Here was evidence that lightning and electricity were the same thing.

Amber

At this time, electrical experiments were with 'static electricity' – charges on insulators that could be transferred in sudden jumps. However, in 1800, Alessandro Volta discovered that two metals with salt water between them could cause a continuous flow of charge – in other words, an **electric current**. He had made the first battery. Within 50 years, the electric motor, generator, and lamp had all been invented. However, no one had any evidence to explain what electricity really was until J. J. Thomson's discovery of the **electron** in 1897. From this, we now know that the current in a circuit is a flow of electrons.

Electromagnetism...

Until the early 1800s, electricity and magnetism were regarded as two different phenomena. Then in 1820, in Denmark, Hans Oersted demonstrated that a compass needle could be deflected by an electric current. The following year, Michael Faraday succeeded in using the force from a magnet on the current in a wire to produce rotation. He had made a very simple form of **electric motor**. Later, in the 1830s, he discovered **electromagnetic induction**, the effect in which a voltage is generated in a conductor by moving or varying a magnetic field around it. Today's generators and transformers make use of this idea.

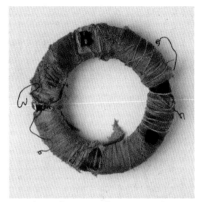

Faraday's first transformer, made in 1831

In the 1860s, James Clerk Maxwell linked electricity and magnetism mathematically. Later, following the discovery of the electron, the cause of magnetism became clear. As an electron orbits in an atom, it produces a magnetic field, rather as the current in a coil produces a field. In most materials, the various fields are in random directions and cancel each other out, but in a magnetized material, some of the fields line up and reinforce each other.

An aurora, another example of the link between magnetism and electricity. Charged particles (mainly electrons) streaming out from the Sun are directed towards polar regions by the Earth's magnetic field. When the particles hit atoms or molecules in the upper atmosphere, light is emitted.

...and beyond

Electric and magnetic forces are so closely related that they are classed as **electromagnetic forces**. The forces that hold atoms together are of this type. But three other kinds of force also operate in the natural world. Gravity is the most familiar. The others are the **weak nuclear force**, responsible for radioactivity, and the **strong nuclear force** which binds the particles in the nucleus of an atom. Today, physicists are developing models to link these forces, but gravity remains a problem, and the search is still on for a satisfactory unified-field theory that combines all the known forces of nature.

12.04 The Earth and beyond★

The centre of the Universe?

The ancient view of the Earth was that it was flat. However, by about 600 BCE, Greek mariners had observed how the positions of the stars altered as they sailed north or south, and realized that the Earth's surface must be curved. Around 350 BCE, Aristotle believed that the Earth was a stationary sphere at the centre of the Universe. The Sun, Moon, planets, and stars lay on transparent, crystal spheres which rotated about the Earth, so they moved in perfect circles.

The idea that the 'heavenly bodies' must move in perfect circles around the Earth was to cause difficulties for many centuries. When viewed from Earth, the planets do not move across the sky at a steady rate, and sometimes appear to move backwards and forwards. Around 150 CE, Ptolemy came up with an elaborate explanation for this. The planets *did* move in perfect circles, but sometimes they followed small circles superimposed on a larger one.

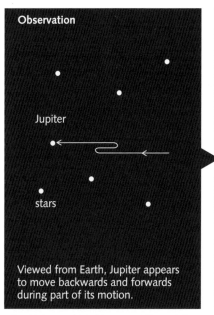

Observation

Jupiter

stars

Viewed from Earth, Jupiter appears to move backwards and forwards during part of its motion.

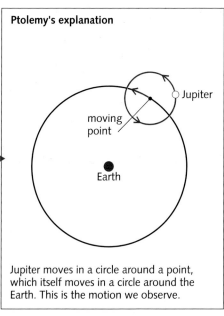

Ptolemy's explanation

Jupiter

moving point

Earth

Jupiter moves in a circle around a point, which itself moves in a circle around the Earth. This is the motion we observe.

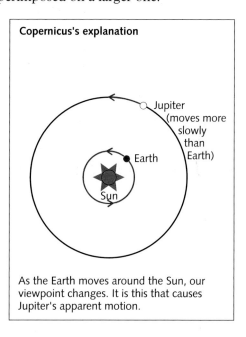

Copernicus's explanation

Jupiter (moves more slowly than Earth)

Earth

Sun

As the Earth moves around the Sun, our viewpoint changes. It is this that causes Jupiter's apparent motion.

▲ Galileo using a telescope he designed and built himself.

It was not until the 1500s that the views of Aristotle and Ptolemy were seriously questioned. The person responsible was Nicolaus Copernicus, who decided to take a fresh look at the problem of the observed motion of the planets. In 1543, he published his theory that the Sun must be at the centre of the Universe, with the Earth and planets moving around it. Over the following years, this idea was strongly opposed by the Church, which insisted that the Earth must be central. Later, Galileo supported Copernicus's ideas, but was forced to renounce them or risk torture and execution. In 1610, he had observed tiny moons moving around Jupiter – evidence that the Earth was not central to all objects in the heavens.

During the late 1500s, observations made by Tycho Brahe greatly increased the amount of accurate data on the positions of the planets. During the 1600s, the evidence for the Copernican model became overwhelming. Kepler established the laws of planetary orbits, Newton published his theory of gravitation, and put Kepler's laws on a firm mathematical basis.

++

◄ When William Herschel built this reflecting telescope in 1789, it was the largest in the world. It could be raised or lowered by pulleys, and there were rollers under the platform so that it could be turned.

Caroline Herschel, William's sister, was also an expert astronomer. She discovered comets and nebulae, and made a huge catalogue of her brother's observations.

Sun, stars, and galaxies

By the late 1600s, it was clear that the stars were similar to the Sun, but much further away. In the late 1700s, William Herschel used a large telescope to study how the stars were distributed. He concluded that the Sun was near the centre of a huge, lens-shaped system of stars, which he called the **Galaxy**.

By the early 1800s, astronomers were making increasingly accurate estimates of the distances to the stars. These were based on the following principle. During the course of a year, as the Earth moves round the Sun, our viewpoint in space changes, so nearby stars appear to move against the background of very distant stars. This apparent movement is called **parallax**. By measuring it, the distance to nearby stars can be calculated using trigonometry.

In 1918, Harlow Shapley mapped the relative distances of star clusters and found that the Sun was not at the centre of our Galaxy after all. And in the 1920s, Edwin Hubble discovered that our Galaxy was not alone. There were millions of other galaxies in the Universe. Hubble also made another significant discovery about galaxies. From the altered wavelengths of their light, he concluded that they must be rushing away from each other. This discovery led to the development of the **Big Bang theory** – the idea that, billions of years ago, the whole of space and everything in it started to expand from a single, atom-sized concentration of matter and energy.

The Hubble Space Telescope is in orbit around the Earth. It transmits pictures back to the ground which enable astronomers to see distant stars and galaxies without the distorting effects of the Earth's atmosphere.

281

Key developments in physics

c. 400 BCE	Democritus suggests that there might be a limit to the divisibility of matter. (*Atomos* is the Greek word for indivisible.)
c. 350 BCE	Aristotle suggests that the Earth is at the centre of the Universe, with the Sun, Moon, and planets on crystal spheres around it.
c. 240 BCE	Eratosthenes estimates the diameter of the Earth by comparing shadow angles in different places.
CE	
c. 60	Hero makes a small turbine driven by jets of steam.
c. 150	Ptolemy suggests that the Earth is at the centre of the Universe, and that the Sun, Moon, and planets are moving in perfect circles.
c. 1000	Magnetic compass used in China.
1543	Copernicus suggests that the Sun is at the centre of the Universe, with the Earth and planets moving around it.
1600	Gilbert suggests that the Earth acts like a giant bar magnet.
1604	Galileo shows that all falling objects should have the same, steady acceleration.
1621	Snell states his law of refraction.
1644	Torricelli makes the first mercury barometer.
1654	Guericke demonstrates atmospheric pressure.
1662	Boyle states his law for gases.
1678	Huyghens puts forward his wave theory of light.
1679	Hooke states his law for elastic materials.
1687	Newton publishes his theory of gravity and laws of motion.
1714	Fahrenheit makes the first mercury thermometer.
1752	Franklin performs a hazardous experiment with a kite to show that lightning is electricity.
c. 1790	Herschel discovers the shape of our galaxy.
1800	Volta makes the first battery.
1803	Dalton suggests that matter is made up of atoms. Young demonstrates the wave nature of light.
1821	Faraday makes a simple form of electric motor.
1825	Ampère works out a law for the force between current-carrying conductors.
1827	Ohm states his law for metal conductors.
1832	Faraday demonstrates electromagnetic induction.
1832	Sturgeon makes the first moving-coil meter.
1840	First use of the words 'physicist' and 'scientist'.
1849	Fizeau measures the speed of light. Joule establishes the link between heat and work.

1852	Kelvin states the law of conservation of energy.
1864	Maxwell predicts the existence of radio waves and other electromagnetic waves.
1877	Cailletet liquefies oxygen.
1879	Swan and Edison make the first electric light bulbs.
1888	Hertz demonstrates the existence of radio waves.
1894	Marconi transmits the first radio signals.
1895	Röntgen discovers X-rays.
1896	Becquerel discovers radioactivity.
1897	Thomson discovers the electron.
1898	M. Curie discovers radium and polonium.
1899	Rutherford identifies alpha and beta rays.
1900	Planck proposes the quantum theory.
1905	Einstein uses the quantum theory to explain the photoelectric effect, and publishes his special theory of relativity.
1911	Rutherford proposes a nuclear model of the atom.
1913	Bohr uses the quantum theory to modify Rutherford's model of the atom.
1916	Einstein publishes his general theory of relativity.
1919	Rutherford splits the atom and discovers the proton.
1924	De Broglie suggests that particles can behave as waves.
1925	Schrödinger develops a wave-mechanics model of the atom.
1927	Lemaitre suggests the possibility of the Big Bang.
1928	Geiger and Müller invent their radiation detector.
1929	Hubble discovers that the Universe is expanding.
1932	Chadwick discovers the neutron.
	Cockroft and Walton produce the first nuclear change using a particle accelerator.
1938	Hahn discovers nuclear fission.
1942	Fermi builds the first nuclear reactor.
1947	Bardeen, Brattain, and Shockley make the first transistor.
1957	First artificial satellite, *Sputnik I*, put into orbit.
1958	St Clair Kilby makes the first integrated circuit.
1960	Maiman builds the first laser.
1963	First geostationary communications satellite.
1969	First manned landing on the Moon.
1971	Intel Corporation makes the first microprocessor.
1977	First experimental evidence of quarks.
1990	Hubble Space Telescope launched.

Source: *the Biographical Encyclopedia of Scientists*, published by the Institute of Physics   *c. = circa* (about)

13

Experimental Physics

- SAFETY IN THE LABORATORY
- PLANNING AN EXPERIMENT
- OBTAINING EVIDENCE
- ANALYSING AND CONCLUDING
- EVALUATING EVIDENCE
- INVESTIGATIONS TO TRY
- PRACTICAL TESTS
- KEY EXPERIMENTAL SKILLS CHECK LIST

The worker inside the cage is quite safe, despite the 2.5 million volt sparks from the huge Van de Graaff generator. The electric discharges strike the metal bars, rather than pass between them, so the cage has a shielding effect. In fact, if safety procedures were ignored, some of the experiments done in a school laboratory would be much more dangerous than this one.

Safety in the laboratory

When doing any experimental work, the most important thing of all is safety. Here are some dos and don'ts:

General

● When moving around the laboratory, always walk. Don't run.

● Keep bags and coats away from the work area, so that they can't catch fire or be pushed against equipment.

● Keep bags and coats off the floor so that people can't trip over them.

● Never take food or drink into the laboratory.

Bunsens and tripods

● If a bunsen burner is alight, but not in use, always leave it on the yellow flame setting so that the flame can be seen.

● Make sure that bunsens and tripods have a heatproof mat underneath.

● Give a hot tripod plenty of time to cool down before attempting to move it.

● Don't attempt to move a tripod when there is a beaker resting on it.

Glass thermometers

● Don't put glass thermometers where they can roll off the bench.

● Keep glass thermometers away from bunsen flames.

● Support thermometers safely: see **Safe support** below.

● Mercury, used in some thermometers, is toxic. If a thermometer breaks and mercury runs out, don't handle it.

Glass tubing

● Never attempt to push glass tubing (or glass thermometers) through a hole in a bung. The laboratory technician has a special tool for doing this.

● Always handle hot glass tubing with tongs. Rest it in on a heatproof mat; don't put it straight on the bench.

● Hot glass tubing can stay hot for a long time. Give it plenty of time to cool down before you attempt to pick it up.

Safe support

● When clamping a test-tube, don't overtighten the clamp. And make sure that the clamp has soft pads to touch against the glass. This also applies when clamping a glass thermometer.

● In experiments where you have to suspend a load, make sure that the supporting clampstand is stable enough to take the heaviest load you will be using. You may need to weigh it down for this.

A yellow bunsen flame is easier to see than a blue one.

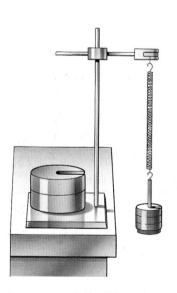

In experiments like this, make sure that the apparatus is stable enough to support the heaviest load.

Electricity

- Before making any changes to the wiring in your circuits, always switch off the power or disconnect the battery.

- Remember: low voltage circuits may not give you a shock, but they can cause burns if the current is too high and a wire overheats.

- Never make a direct connection across the terminals of a battery. Don't put wires or tools where they might connect across the terminals.

- If a mains appliance is faulty, switch off the power and pull out the plug. Don't change the fuse. Ask the laboratory technician to deal with the fault.

- Electrical fires: see **Fire** below.

In many modern laboratories, the mains circuits are protected by RCDs (residual current devices), so the risk of shocks is reduced. But...

- If someone has been electrocuted, and is still touching the faulty appliance, don't touch the person. Switch off the power and pull out the plug.

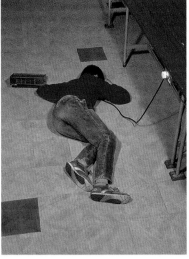

Emergency! But the first job is to switch off the power and pull out the plug.

Eye protection

- Always wear eye protection (e.g. safety goggles) when:
  - stretching metal wires or plastic cords
  - breaking or grinding solids (e.g. rock samples)
  - heating liquids
  - dealing with acids, alkalis, or any other liquid chemicals that might splash.

Light

- Don't look directly into a laser beam or other source of bright light. Don't stand where laser light might be reflected into your eyes.

- If you need to study the Sun's image, project it onto a card. Never look through a telescope or binoculars pointing straight at the Sun – even if there is a filter in front.

Radioactive sources

- The radioactive sources used in school laboratories should always be sealed.

- Radioactive sources should be kept well away from the body, and never placed where they are pointing at people.

Fire

- Don't heat flammable liquids (e.g. methylated spirits) over a bunsen. If heating is required, a water bath should be used – with hot water heated well away from the experiment.

- Don't throw water on burning liquids (e.g. methylated spirits). Smother the fire with a fire blanket or use a carbon dioxide extinguisher.

- Don't throw water on electrical fires. Switch off the supply and use a carbon dioxide extinguisher.

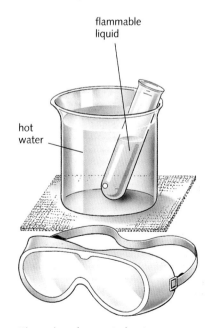

flammable liquid

hot water

The only safe way to heat a flammable liquid is to use a water bath.

13.02 Experimental procedures (1) – planning

This spread should help you plan an experimental procedure. The handwritten notes show part of one student's commentary on her procedure.

Presenting the problem
Start by describing the problem you are going to investigate, and the main features of the method you will use to tackle it.

I am going to investigate how the resistance of nichrome wire depends on its length.

I know that resistance can be calculated with this equation:

$$\text{resistance (in } \Omega) = \frac{\text{voltage (in V)}}{\text{current (in A)}}$$

So to find the resistance of a length of nichrome wire, I need to put the wire in a circuit, then measure the voltage across it and the current through it. I will do this for different lengths of nichrome.

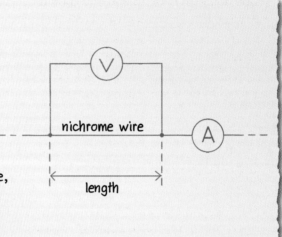

I think I can predict how the resistance will vary with length. If the length of wire is doubled, the current (flow of electrons) has to be pushed between twice as many atoms. So I would expect the resistance to double as well.

Making a prediction
You may have an idea of what you expect to happen in your experiment. This prediction is called your **hypothesis**. You should write it down. It may not be right! It is just an idea. The aim of your procedure is to test it.

In my experiment, three of the key variables are:

length of nichrome wire – to be measured with a ruler marked in mm

voltage – to be measured with a voltmeter

current – to be measured with an ammeter

I shall start with 50 cm of thin nichrome wire, put a voltage of 6 V across it, and measure the current through it. From the voltmeter and ammeter readings, I can calculate the resistance.

I will take more sets of readings, shortening the wire by 5 cm each time until it is only 10 cm long.

For convenience, I will probably keep the voltage fixed at 6 V throughout the experiment.

Dealing with variables
Quantities like length, current, and voltage are called **variables**. They can *change* from one situation to another.

Key variables These are the variables that can affect what happens in an experiment. You must decide what they are. For example, in the nichrome wire experiment, length is one of the key variables because changing the length of wire changes the resistance.

You must also decide how to measure the variables, and over what range. For example, in planning the nichrome wire experiment, you would have to:
- decide what the highest voltage and current values should be (safety must be considered here)
- decide what lengths of wire to use.

Controlling variables Some variables don't have to be measured, but they do need to be controlled. For example, in the nichrome wire experiment, you might want to keep the wire at a steady temperature, in case the temperature affects the resistance.

Some variables can be difficult to control. In your experiment, you may want to use the same thickness of nichrome wire each time, but this depends on how accurately the wire was manufactured. You must take factors like this into account when deciding how reliable your results are.

A fair test When doing an experiment, you should change just one variable at a time and find out how it affects one other. If lots of variables change at once, it will not be a fair test. For example, if you want to find out how the length of a wire affects its resistance, it wouldn't be fair to compare a long, thick wire with short, thin one.

Final preparations

Decide what equipment you need, how you will arrange it, and how you will use it.

To help your planning, you may need to carry out a trial run of the experiment. Before you do this, make sure that all your procedures are safe.

Prepare tables for your readings *before* you start your experiments. Look at the next spread on **obtaining the evidence** before doing this.

Equipment needed:

voltmeter (0–6V), ammeter (0–3A), 50cm of 0.28mm diameter nichrome wire, ...

---

nichrome: length cm	voltage V	current A	resistance Ω
50			
45			
40			
35			
30			
25			
20			

There are two more variables I need to control:

temperature – I know from reference books that the resistance of nichrome changes with temperature. So I will use a large beaker of cold water to keep the temperature of the nichrome steady.

diameter (thickness) of nichrome wire – this could affect the resistance. To make sure that I have the same diameter all the time, I will use lengths of wire taken from the same reel, and check each piece with a gauge before using it.

I will set up this circuit:

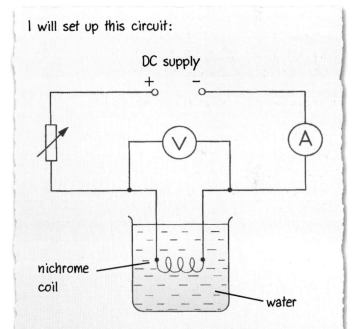

I am not sure how big the maximum current will be, so I will do a trial run of the experiment first. I will start with an ammeter that can measure several amperes, but may be able to change to a more sensitive meter for the main experiment.

Safety:
I must make sure that the power supply is switched off before I remove the nichrome wire to change its length.

287

Experimental procedures (2) – obtaining the evidence

voltage	
V	
2.3	
2.1	

Remember to include a unit in the heading at the top of each column.

You can only read this voltmeter to the nearest 0.1 V.

This spread should help you take and record measurements correctly.

Units

When you write down a measurement, remember to include the unit. For example:

voltage = 2.3 V

If you just write down '2.3', you may not be able to remember whether this was supposed to be a voltage of 2.3 V or 2.3 mV.

When writing measurements in a table, you don't need to put the unit after each number. But be sure to include the unit in the heading at the top of each column. You can see an example on the left.

Uncertainties

No measurement is exact. There is always some **uncertainty** about it. For example, you may only be able to read a voltmeter to the nearest 0.1 V.

Say that you measure a voltage of 2.3 V and a current of 1.2 A. To work out the resistance in ohms (Ω), you divide the voltage by the current on a calculator and get...

1.916 666 7

This should be recorded as 1.9 Ω. Uncertainties in your voltage and current readings mean that you cannot justify including any more figures. In this case, you are giving the result to two **significant figures**.

Taking enough readings

For a graph, you should have at least five sets of readings.

Not all experiments give you readings for a graph. Sometimes, you have to measure quantities that don't change – the diameter of a wire for example. In cases like this, you should repeat the measurement at least three times and find an average. Repeating a measurement helps you spot mistakes. It also gives you some idea of the uncertainty. Look at this example.

The diameter of a wire was measured four times:

1.41 mm   1.34 mm   1.19 mm   1.30 mm

You can work out the average like this:

$$\text{average} = \frac{(1.41 + 1.34 + 1.19 + 1.30)}{4} = 1.31 \text{ mm}$$

The original four numbers ranged from less than 1.2 to more than 1.4. So, the last figure, 1, in the average of 1.31, is completely uncertain. Therefore, you should write down the average diameter as 1.3 mm.

Reading scales

On many instruments, you have to judge the position of a pointer or level on a scale and work out the measurement from that. Here are some ways of making sure that you take the correct reading:

A Using a glass thermometer to measure the temperature of a liquid: keep the liquid well stirred, give the thermometer time to reach the temperature, and keep the bulb of the thermometer in the liquid while you take the reading.

B Using a ruler: be sure that the scale is right alongside the point you are trying to measure.

C Measuring a liquid level on a scale: look at the level of the liquid's flat surface, not its curved meniscus.

D Reading a meter: look at the pointer and scale 'square on'.
(The pointer may have a flat end like that shown here, so that you can look at it edge on.)

A

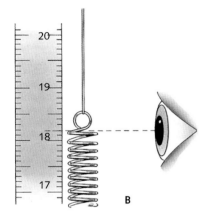

B

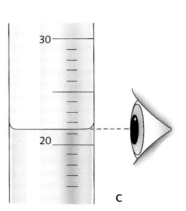

C

D

Can you read the instruments below correctly? The answers are on page 332.

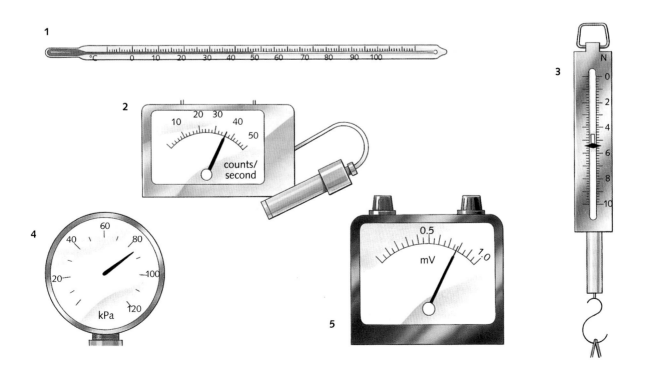

13.04 Experimental procedures (3) – analysing and concluding

I have used my voltage and current readings to calculate the resistance of each length of nichrome wire. Now I shall use these values to plot a graph of resistance against length.

Length is the <u>independent</u> variable (the one I chose to change), so it goes along the bottom axis. Resistance goes up the side.

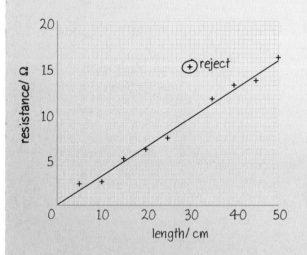

The points on my graph are a little scattered, but I think that the line of best fit is a straight line.

This line ought to go through the origin. If the wire has zero length, there is no metal to resist the current, so the resistance should also be zero.

I have rejected one point on my graph. In my table, the current reading for that point seems far too low. I probably misread the ammeter.

As the graph is a straight line through the origin, the resistance of the nichrome wire is in direct proportion to its length. This agrees with my original hypothesis that doubling the length of wire ought to make it twice as difficult to push electrons through.

This page should help you to analyse your data and draw conclusions from it. The handwritten notes show part of one student's commentary on her procedure.

Drawing a graph

A graph can help you see trends in your data.

Choosing axes Decide which variable to put along the bottom axis. Usually, it is the one you chose to vary by set amounts – the length of nichrome wire, for example. This is the **independent variable**. The resistance would be the **dependent variable** because its value depends on the length you chose. It goes up the side axis.

Choosing scales Check your highest readings, then choose the largest scales you can for your axes.

Labelling axes Along each axis, write in what is being measured and the units being used.

Drawing the best line Because of uncertainties, the points on a graph will be uneven. So don't join up the points! Instead, draw the straight line or smooth curve that goes closest to most of them. This is called a **line of best fit**. Before you draw it:

● Decide whether the line should go through the origin.
● Decide whether any readings should be rejected. Some may be so far out that they are probably due to mistakes rather than uncertainties. See if you can find out why they occurred.

From the way points scatter about a line of best fit, you can see how reliable your readings are. But for this, you need plenty of points.

Trends and conclusions

From the shape of your graph, you can draw conclusions about the data.

The simplest form of graph is a straight line through the origin. A graph of resistance against length of wire might be like this. If so, it means that if the length doubles, the resistance doubles... and so on. in this case, resistance and length are in **direct proportion**.

If you think that your graph supports your original prediction, then say so and explain your reasons.

Experimental procedures (4) – evaluating the evidence

This page should help you decide how reliable your conclusions are, and how your experiment could be improved or extended.

> The points on the graph are uneven. But as they zig–zag at random, I am fairly sure that, without uncertainties, they would lie on a straight line.
>
> There are several reasons why the points may have been so scattered...

> To get a more reliable graph, I need to find a more accurate method of measuring resistance...

> To extend my enquiry, I could find out how the resistance of the nichrome wire depends on the diameter...

Reliability

In reaching your conclusions, remember that there are uncertainties in your measurements, and variables that you may not have allowed for. So your results can never *prove* your original prediction. You must decide how far they *support* it.

If you think that your results are unreliable in any way, see if you can explain why.

You may have some results which do not agree with the others and look like mistakes. These are called **anomalous results**. Try to explain what caused them.

Suggesting improvements

Having completed your experiment, suggest ways of improving it so that your conclusions are more reliable.

Looking further

Suggest some further work which might produce extra evidence or take your investigation further.

Writing your report

The student's commentary was designed to help you understand the different stages of a procedure. It includes far more detail than you would normally put in a report. When producing your own report, these are the things you *should* include:

Planning **1**
- A description of what the procedure is about.
- A prediction of what you think will happen, and why.
- A list of key variables, and a description of how you will measure or control each one.
- A list of the equipment needed.
- Diagrams showing how the equipment will be set up.
- A description of what you plan to do.

Getting evidence **2**
- A description of what you did, including comments about any difficulties and how you overcame them.
- Tables showing all measurements, including units.

Analysing and concluding **3**
- Graphs and charts.
- Calculations based on your data.
- A conclusion, including details of:
  – what you found out.
  – whether your findings matched your prediction.

Evaluating **4**
- Comments about:
  – how reliable you think your results were
  – any anomalous results, and their possible causes
  – how your procedure could be improved
  – further work that could be done.

13.06 Some experimental investigations

Here are some suggestions for practical work. Some are full investigations. Others are shorter exercises to help you develop your experimental skills.

Measuring newspaper

Plan and carry out experiments to measure:
a) the thickness of one sheet of newspaper **b)** the mass of one sheet of newspaper **c)** the density of the paper used.

Start by thinking about the following:
If a single sheet is too thin to measure accurately, how can you improve the accuracy?

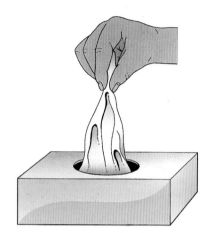

Wet or dry?

The makers of a well-known brand of soft tissue paper claim that their tissues are just as strong wet as dry. Are they right? Plan and carry out an enquiry to test their claim.

Start by thinking about the following:
What is meant by the 'strength' of a tissue? Do you need use a whole tissue? When comparing tissues, how can you make sure that your test is fair?

Fine or coarse?

Coarse glasspaper ('sandpaper') rubs through a wooden surface more quickly than fine glasspaper. But does it produce more friction? Plan and carry out experiments to find out.

Start by thinking about the following:
How can you measure the frictional force when glasspaper is rubbed on wood? How can you keep the glasspaper pressed against the wood? Will the force used to press the glasspaper against the wood affect the result? How can you make sure that your test is fair?

Pendulum

The time of one complete swing of a pendulum is called its period.
The period of swing *might* be affected by these factors: the mass of the bob, the amplitude (size) of the swing, the length of the pendulum.

Plan and carry out an enquiry to find out which factors affect the period.

Start by thinking about the following:
The period of your pendulum will probably be a couple of seconds at most. How are you going to find the time of one swing accurately? How are you going to measure the size of the swing?

Note: make sure that the top of the pendulum string is firmly held so that there is no movement at that point.

Further work:
Find out how the period of one pendulum compares with another of four times the length. Is there a simple connection between the length and the period? Does the connection work for other lengths as well?

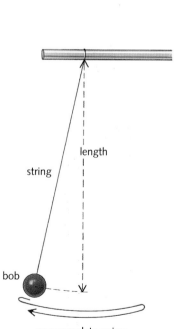

length
string
bob
one complete swing

Stretching rubber

A company wants to market a cheap spring balance for weighing letters. Their designer suggests that, to save money, they could use a rubber band instead of a spring. Their technician says that this would be unsatisfactory because rubber bands change length and 'springiness' once they have been stretched. Who is correct? Plan and carry out an enquiry to find out.

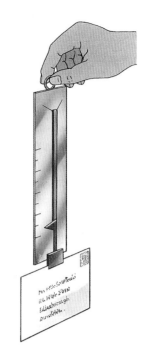

Find the mass

Plan and carry out an experiment to find the mass of a lump of Plasticine (or some other solid). You are not allowed to use a balance with a mass scale already marked on it. And you are not allowed to use slotted masses of less than 50 g.

Further work:
Take the problem a stage further. Plan and carry out experiments to measure a much smaller mass – such as the mass of a pen or pencil.
This time, you can use a selection of standard masses down to 5 g.

Start by thinking about the following:
Your original design will probably not be sensitive enough to measure a small mass. Can it be modified in some way to make it more sensitive?

Bouncing ball

Some table tennis balls have more 'bounce' than others. Plan and carry out an enquiry to compare the bounce of two table tennis balls.

Start by thinking about the following:
What is meant by 'bounce'? What do you need to measure? When comparing the balls, how can you make sure that your test is fair?

Parachute design

The diagram on the right shows a simple model parachute. Plan and carry out an enquiry to find out if there is a link between the design of the parachute and the speed at which it falls.

Start by thinking about the following:
Shape and area are two possible features of the design. Will you investigate both? How will you make sure that your tests are fair? How will you work out the speed of fall?

Double-glazing

In cooler countries, people fit double-glazing in their houses because two layers of glass, with air between, are supposed to lose thermal energy (heat) more slowly than a single layer. But does double-glazing cut down thermal energy loss? Plan and carry out an enquiry to find out.

Start by thinking about the following:
How are you going to set up a double layer of glass with air between? What will you use as a source of thermal energy? How will you tell whether the flow of thermal energy is reduced when the extra layer of glass is added? Will your test be fair?

glass

glass

air

Salt on ice

During winter, salt is often sprayed on the roads to melt the ice. Pure ice has a melting point of 0°C. Adding salt to ice affects the melting point.

Plan and carry out experiments to find out how the melting point of ice changes when salt is mixed in. Find out if there is a connection between the melting point and the concentration of salt in the ice. (The concentration can be measured in grams of salt per cm³ of ice.)

Start by thinking about the following:
How will you make sure that the salt and ice are properly mixed? How are you going to measure the melting point?

The speed of sound

length of air column

In the diagram on the left, someone is holding a vibrating tuning fork above a measuring cylinder. Sound waves travel down the cylinder and back, and make the air inside vibrate. If the length of the air column is exactly a quarter of the wavelength of the sound, the air vibrations are strongest and the air gives out its loudest note. The effect is called resonance.

The speed of sound is linked to its frequency and wavelength by this equation:

$$\text{speed (m/s)} \;=\; \text{frequency (Hz)} \times \text{wavelength (m)}$$

Using the information above, plan and carry out an enquiry to find the speed of sound in air.

Start by thinking about the following:
As a measuring cylinder has a fixed length, how will you vary the length of the air column inside?

Apparent depth

The person in the diagram on the left is looking at a pin on the bottom of a beaker of water. Light from the pin is refracted (bent) when it leaves the water. As a result, the water looks less deep than it really is and the pin appears closer to the surface than it really is.

Plan and carry out an experiment to find the apparent depth of some water in a beaker.

Start by thinking about the following:
If you look at a pin in some water, it is an image of the pin which you are seeing. How can you locate the position of this image? Could you use a similar method to that used to find the position of an image in a mirror?

Two pairs or one?

People claim that two pairs of socks are warmer than one. But does an extra pair cut down the loss of thermal energy (heat)? Plan and carry out an enquiry to find out. (You do not have to use warm feet as your source of thermal energy!)

Start by thinking about the following:
How are you going to tell that one object is losing heat more rapidly than another? How will you make sure that your test is fair?

Image size and distance

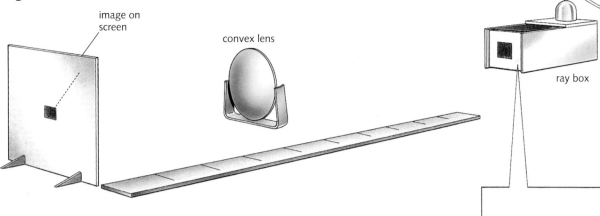

image on screen

convex lens

ray box

tissue paper

card with square hole in it

Place a bright object well away from a convex lens as in the diagram, and you can get a clear image on a screen. If you move the object closer, the size and the position of the image both change, and you need to move the screen to get a clear image again.

Is there a connection between the size of the image and its distance from the lens? Plan and carry out an enquiry to find out.

Current–voltage investigations

Plan and carry out experiments to find out how the current through each of the following depends on the voltage across it:

1) nichrome wire, kept at constant temperature

2) the filament of a bulb

3) a semiconductor diode.

Start by thinking about the following:
How will you vary the voltage across each component and measure the current through it? What checks must you do to make sure that the current through each component is safe, and does not cause damage?

Making a resistor

Resistors are used for keeping voltages and currents at correct levels in electronic circuits.

Using nichrome wire, make a resistor with a resistance of $5\,\Omega$.

Start by thinking about the following:
How does the length of wire affect its resistance? How is resistance calculated? What circuit will you use to test the nichrome? From your measurements, how can you work out how much wire you need?

Thermistor investigation

thermistor

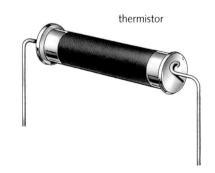

Thermistors have a resistance that varies considerably with temperature. They can be used as temperature sensors. Plan and carry out an experiment to find out how the resistance of a thermistor varies between $0\,°C$ and $100\,°C$.

Start by thinking about the following:
How will you change and control the temperature of the thermistor? How will you measure the resistance of the thermistor? How will you make sure that your circuit doesn't heat up the thermistor?

295

Taking a practical test

This spread should help you if you have to take a practical test in physics. The skills that might be tested are listed below. There are two typical questions on the opposite page.

If you need to sit an alternative-to-practical paper instead, there are some sample questions in Section 15 (IGCSE practice questions). However, you may be assessed on coursework, rather than on the results of a practical test or its alternative. Your teacher will be able to tell you which form of assessment applies to you.

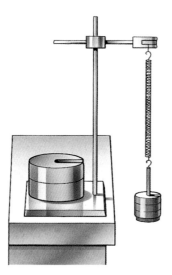

Skills assessed in a practical test

In a typical practical test, you will be assessed on your ability to do the following:

● Follow written instructions for the assembly and use of the apparatus provided. For example, wire up a simple electrical circuit.

● Select, from the items given you, the measuring device that is most suitable for the task. For example, choose an ammeter with a suitable scale.

● Carry out the specified manipulation of the apparatus. For example, in a spring-stretching experiment, add loads to the spring in stages, and follow any other instructions.

● Take readings from a measuring device:
   – read the scale with appropriate accuracy
   – use an appropriate number of significant figures
   – estimate the reading between any scale divisions
   – allow for zero errors, where appropriate
   – take repeated measurements to obtain an average value.

● Record your observations systematically, with appropriate units

● Process your data as required.

● Present your data graphically, using suitable axes, labelled scales, and points that have been accurately plotted.

● Decide on the best straight line or curve through the points.

● Take readings from the graph.

● Work out the gradient on a graph.

● Reach a conclusion and report the result clearly.

● Indicate how you carried out a required instruction.

● Describe any precautions taken in carrying out a procedure.

● Give reasons for choosing particular items of apparatus.

● Comment on a procedure used in an experiment and suggest an improvement.

1 *In this experiment you are to investigate the effect of insulation on the rate of cooling of hot water.*

Record your results in tables like those at the bottom of the page. Carry out the following instructions, referring to the diagram below.

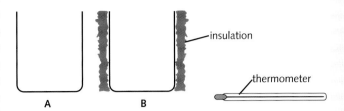

You are provided with two beakers labelled **A** and **B**. Beaker **B** is insulated. Do **not** remove this insulation. You also have a supply of hot water.

a) Pour hot water into beaker **A** until it is approximately two thirds full.
b) Measure the temperature θ of the hot water. Record this temperature in the table for time $t = 0$ s.
c) Start the stopwatch and then record the temperature of the water at 30 s intervals for a total of 4 minutes.
d) Complete the column headings in the table.
e) Use the data in the table to plot a graph of θ (*y*-axis) against *t* (*x*-axis). Draw the best fit curve.
f) Repeat steps **a)** – **d)** using beaker **B**.
g) Use the data obtained from part **f)** to plot another curve on the same graph axes that you used for part **e)**.
h) The experiment you have just done was designed to investigate the effect of insulation on the rate of cooling. Suggest two improvements that could be made to the design of the experiment.

Beaker **A**		Beaker **B**	
t/	θ/	*t/*	θ/
0		0	
30		30	
60		60	
90		90	
120		120	
150		150	
180		180	
210		210	
240		240	

CIE (0625/05 N '05 Q3)

2 *In this experiment you are to investigate the period of oscillation of a mass attached between two springs.µ*

Record your results in a table like the one at the bottom of the page. Carry out the following instructions, referring to the diagram below.

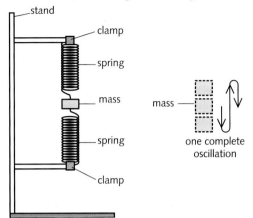

You are provided with a mass *m* of 400 g attached between two springs.

a) Displace the mass a *small* distance downwards and release it so that it oscillates. Record the time t_1 taken for 10 complete oscillations of the mass. Also record the time t_2 taken for another 10 complete oscillations of the mass.
b) Calculate *t*, the average value of t_1 and t_2.
c) Calculate the period *T* of the oscillations. *T* is the time for one complete oscillation.
d) Calculate the value of $\frac{T}{M}$
e) Repeat steps **a)** – **d)** using values for *m* of 300 g and 200 g.
f) Complete the final column heading in the table.
g) A student suggests that *T* should be directly proportional to *m*. State with a reason whether or not your results support this suggestion.
h) In the experiment you have just done, the mass oscillates rapidly so that it is difficult to take the times accurately. The instructions in the question included methods of improving the accuracy of the value obtained for the period *T*. Describe briefly one of these methods and any calculation involved to obtain the *T* value.

m/g	t_1/s	t_2/s	*t/s*	*T/s*	$\frac{T}{m}/$
400					
300					
200					

CIE (0625/05 J '05 Q3)

Key experimental skills checklist

Planning

In planning an experimental procedure, you should be able to:

a turn an idea into a form that can be investigated. . . . . ☐

b carry out trial runs if necessary. . . . . ☐

c make predictions, where appropriate. . . . . ☐

d decide what the key factors (e.g. key variables) are. . . . . ☐

e plan how to vary or control the key variables. . . . . ☐

f decide on the range and number of measurements needed. . . . . ☐

g decide how much evidence you will need, and recognize which
variables may be difficult to control. . . . . ☐

h select equipment, and plan how to use it safely. . . . . ☐

Obtaining evidence

In obtaining your evidence, you should be able to:

a use the equipment safely, and with skill. ☐

b make accurate measurements. . . . . ☐

c make enough measurements to produce reliable evidence. . . . . ☐

d consider what uncertainties there are. . . . . ☐

e repeat measurements as necessary. . . . . ☐

f record measurements clearly. . . . . ☐

Analysing and concluding

In analysing your evidence and reaching conclusions, you should be able to:

a present data clearly. . . . . ☐

b present data in the form of a graph using a line of best fit. . . . . ☐

c identify trends or patterns in your data. . . . . ☐

d use a graph to identify how variables are related. . . . . ☐

e present numerical results to an appropriate level of accuracy. . . . . . ☐

f check that your conclusions match the evidence. . . . . ☐

g explain whether or not your results match your original prediction. . . ☐

h explain your conclusions, if you can, using your knowledge of physics. ☐

Evaluating

In evaluating your procedure, you should be able to:

a decide whether you have enough evidence to reach a firm conclusion. ☐

b decide whether any results should be rejected, and if so, why. . . . . . ☐

c decide, from the uncertainties, how reliable your results are. . . . . . . ☐

d suggest improvements to the methods you have used. . . . . ☐

e suggest further work that could produce extra evidence or take
your investigation further. . . . . ☐

14

Mathematics for physics

● THE ESSENTIAL MATHEMATICS

A summary of the mathematical concepts and skills required for IGCSE examinations.

The essential mathematics

For IGCSE examinations, you will need some basic skills in mathematics. The following are typical of what is required.

Adding, subtracting, multiplying, and dividing

You should be familiar with the symbols $+$, $-$, $\times$, and $\div$ and the processes they represent. This may sound obvious, but it includes understanding the link between division and fractions, as described next.

Using fractions and decimals

A half can of course be written as $\frac{1}{2}$ (sometimes printed as ½ or 1/2).

However, you should also understand that it means $1 \div 2$

Improper fractions are those with a bigger number on the top than the bottom: for example, $\frac{48}{12}$, which means $48 \div 12$.

You should be able to write fractions using decimals. So, one half is 0.5. Decimals may have several numbers after the point. However, you should understand that 0.489, for example, is smaller than 0.5.

Using percentages

You should understand that percentages are fractions of one hundred. So, for example, one half is $\frac{50}{100}$, which is 50%. The percent symbol % really means 'divided by 100'.

Using ratios

Ratios are another way of expressing fractions. If some apples are being shared between two people in the ratio 2 : 3, there are 5 'parts' to divide up, so one person gets $\frac{2}{5}$ of the total, and the other gets $\frac{3}{5}$ of the total.

If there were 10 apples to share:

one person would get $\frac{2}{5} \times 10$, which is 4 apples;

the other person would get $\frac{3}{5} \times 10$, which is 6 apples.

Working out reciprocals

$\frac{1}{\text{number}}$ is called the **reciprocal** of the number. For example:

The reciprocal of 2 is $\frac{2}{5}$, or 0.5.

The reciprocal of 10 is $\frac{2}{5}$, or 0.1

You may have to work out reciprocals when preparing readings for a graph. Many calculators have a special key for doing this.

Drawing and interpreting graphs and charts

Graphs are a way of presenting sets of scientific data so that trends or laws can more easily be seen. For more information on how to prepare a graph and draw a line of best fit, see Spread 13.04.

You should be able to read off new values from a graph line. Finding values between existing points is called **interpolation**. Extending a graph line to estimate new values beyond the measured points is called **extrapolation**.

Data can also be presented in charts. A **table** is one form of chart. A **pie chart** is another: it shows the different proportions or percentages making up the whole. From the example on the right, you should be able to deduce that 25% of the people at a football match supported the away team.

Proportions of home and away supporters attending a football match.

Understanding direct and inverse proportions

Look at the sets of values for X and Y in the table on the right. If X doubles, Y doubles; if X triples, Y triples. Also, dividing Y by X always gives the same number (4), and a graph of Y against X is a straight line through the origin. These all indicate that X and Y are in **direct proportion**. This can be expressed in the form $Y \propto X$, where $\propto$ is the symbol for 'directly proportional to'.

Now look at the sets of values for X and Z on the right. In this case, if X doubles, Z halves, and so on. And *multiplying* Z by X always gives the same number (12). Here, Z and X are in **inverse proportion**.

The table also includes a column for the reciprocals of X. Note that Z and $1/X$ are in *direct* proportion. So, another way of expressing the inverse proportion between Z and X is to write: $Z \propto 1/X$

X	Y
1	4
2	8
3	12
4	16

X	Z	$1/X$
1	12	1
2	6	0.5
3	4	0.33
4	3	0.25

Understanding indices

You should know that 2^3 means $2 \times 2 \times 2$, and that 10^4 means $10 \times 10 \times 10 \times 10$. The tiny numbers ^{3} and ^{4} are called **indices**.

For more advanced work, it is useful to know about negative indices.

For example, 10^{-4} means $\dfrac{1}{10^4}$

Understanding numbers in standard notation

In standard notation (also called standard form, or scientific notation), numbers are expressed using powers of 10. For example, 1500 is written as 1.5×10^3. Using standard notation, you can indicate how accurately a value is known. This is explained in Spread 1.01.

Using a calculator

You need to able to use a calculator correctly. For example, to work out a value for $\dfrac{6 \times 7}{11 \times 3}$, you would key in this: $6 \times 7 \div 11 \div 3 =$

As a result, the calculator display will show this: 1.2727273

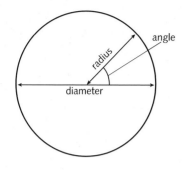
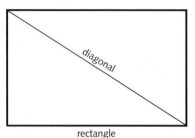
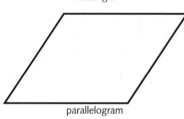

rectangle

parallelogram

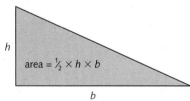

area = ½ × h × b

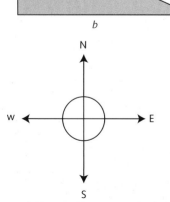

If a calculator display reads 1.2727273, you must be able to interpret this correctly. If the original numbers came from experimental data, you could not justify giving the result so accurately. 1.27, or 1.3 if you round it up, would be more appropriate.

You should also be able to interpret high numbers on a calculator. For example, 2.5×10^9 will probably be displayed as 2.5 E 09, or just 2.5   09

Making approximations and estimations

You should be able to check whether a result is reasonable by doing a rough estimate without a calculator. For example, if you divide 12 by 2.95, you should realise that the answer will be just over 4. So, if the calculator displays 40.67, you have made a mistake in keying in the numbers.

Understanding units

Most measurements have units as well as numbers: for example, a speed of 10 m/s. When giving a result, you must always include the unit. For more about units, see Spread 1.01.

Understanding number accuracy

You should understand the significance of whole numbers. You may count 12 students in a room as an exact number, but if a length measurement is given as 12 mm, this only indicates that the value lies somewhere between 11.5 mm and 12.5 mm.

Manipulating equations

If you are given an equation like this: $Z = \dfrac{Y}{X}$

you should be able to rearrange it to give $Y = Z \times X$, and $X = \dfrac{Y}{Z}$

Understanding the terms for shapes, lines, and angles

As well as the circle, sphere, triangle, square, and cube, you need to know the terms shown in the first three diagrams on the left.

Using the links between length, area, and volume

You should be able to calculate the area of a rectangle, the area of a right-angled triangle (see left), and the volume of a rectangular block (see Spread 1.05).

Using mathematical instruments

The basic instruments are a ruler for measuring length, a protractor for measuring angles in degrees (°), compasses for drawing circles, and a set square for use in drawing right angles.

Knowing the points of the compass

The directions north, south, east, and west are called the points of the compass. To make sure you know which is which, look at the diagram on the left.

15

IGCSE Practice Questions

MULTICHOICE QUESTIONS • IGCSE THEORY QUESTIONS

ALTERNATIVE-TO-PRACTICAL QUESTIONS

A selection of questions from IGCSE examination papers and other sources.

1 The diagram below shows the mass of a measuring cylinder before some liquid is poured into it and then after.

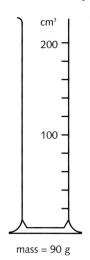 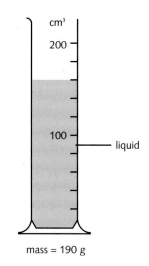

mass = 90 g   mass = 190 g

What is the density of the liquid?

A $\dfrac{100}{160}$ g/cm³   **B** $\dfrac{100}{130}$ g/cm³

C $\dfrac{190}{160}$ g/cm³   **D** $\dfrac{100}{130}$ g/cm³

2 A motorcycle accelerates from rest. The graph shows how its speed changes with time.

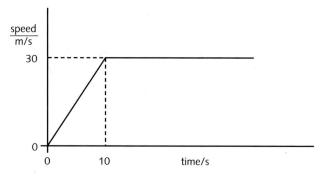

What distance does the motorcycle travel before it reaches a steady speed?

A 3 m   **B** 30 m
C 150 m   **D** 300 m

3 Which type of power station does *not* use steam from boiling water to turn the generators?

A coal-fired
B hydroelectric
C geothermal
D nuclear

4 A large electric motor is used to lift a load off a lorry. Which of the following values would be enough for you to calculate the power of the motor?

A the current used and the work done
B the force used and the distance moved
C the mass lifted and the distance moved
D the work done and the time taken

5 The diagram shows how much a spring stretches when objects with different masses are hung from it.

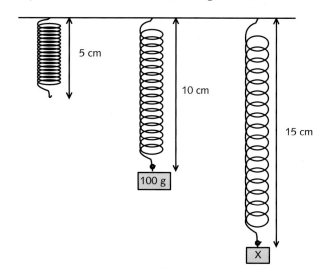

The extension of the spring is directly proportional to the mass hung on it. What is the mass of object X?

A 125 g   **B** 150 g   **C** 200 g   **D** 300 g

6 A manometer, containing water, is used to measure the pressure of a gas supply in a school laboratory. Its reading is *h* cm of water.

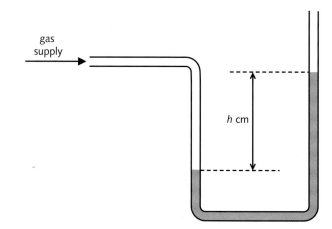

Why is it better to use water in the manometer, rather than mercury?

A With mercury, a narrower tube would be needed.
B With mercury, a wider tube would be needed.
C With mercury, *h* would be too large.
D With mercury, *h* would be too small.

7 When a vehicle is travelling along, the temperature of its tyres increases. This causes the air pressure in the tyres to rise. Why is this?

A The molecules in the air increase in number.
B The molecules in the air move at a higher speed.
C Molecules in the air expand with the rise in temperature.
D There is more force between the molecules in the air.

8 A vacuum flask has double walls of glass or steel with a vacuum between them. Which kinds of heat transfer are reduced by the vacuum?

A convection and radiation

B conduction and convection

C conduction and radiation

D conduction, convection, and radiation

9 An alarm is too quiet, so a technician adjusts it to produce a louder note of the same pitch. What effect does this have on the amplitude and on the frequency of the sound?

	amplitude	frequency
A	larger	same
B	same	larger
C	larger	same
D	same	same

10 Radio waves are detected by a receiver which is in a house on the other side of a hill.

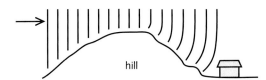

hill

The radio waves reach the receiver because the hill has caused them to be

A diffracted   **B** reflected   **C** refracted   **D** radiated.

11 The voltage and current ratings of four electric heaters are shown in the table below. Which of the heaters has the highest resistance?

	voltage/V	current/A
A	110	4.0
B	110	8.0
C	230	4.0
D	230	8.0

12 Which component is used in a time delay circuit to store energy?

A capacitor

B diode

C thermistor

D resistor

13 In the circuit below, the 12 V lamp glows when switch S is open.

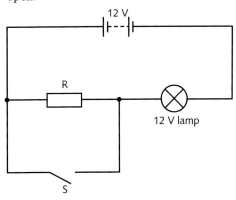

If switch S is then closed, what happens to the brightness of the lamp?

A It stays the same.   **B** It goes off.

C It becomes dimmer.   **D** It becomes brighter.

14 A transformer has 200 turns on its primary coil and 400 turns on its secondary coil. An AC voltage of 50 V is applied to the primary coil.

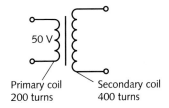

Primary coil   Secondary coil
200 turns   400 turns

What is the voltage across the secondary coil?

A 25 V   **B** 50 V   **C** 100 V   **D** 200 V

15 A sample contains 800 mg of a radioactive material, which emits α-particles. The material has a half-life of 6 days. What mass of material is still radioactive after 18 days?

A 0 mg   **B** 100 mg   **C** 200 mg   **D** 400 mg

16 An unstable nucleus has 137 neutrons and 88 protons. It decays by emitting a β-particle. How many neutrons and protons does the nucleus have after emitting the β-particle?

	neutrons	protons
A	136	88
B	136	89
C	137	87
D	137	89

Questions from IGCSE theory papers
including core and extended* theory papers

Assume $g = 10$ m/s^2

1 The clock on a public building has a bell that strikes each hour so that people who cannot see the clock can know what hour of the day it is. At precisely 6 o'clock, the clock starts to strike. It strikes 6 times.

A

B

At the first strike of the bell, a man's wrist-watch is as shown in diagram A. When the bell strikes for the sixth time, the wrist-watch is as shown in diagram B.
a) Calculate the time interval (in s) between the 1st strike and the 6th strike. [1]
b) Calculate the time interval (in s) between one strike and the next. [2]
c) At precisely 11 o'clock, the clock starts to strike. Calculate the time interval (in s) between the 1st strike and the 11th strike. [2]
CIE (0625/02 J'04 Q1)

2 Some fat purchased from a shop is supplied as the block shown below.

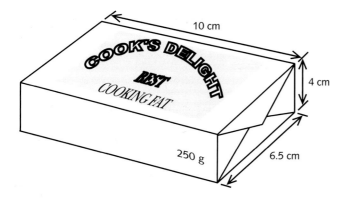

Use the information in the diagram to calculate
a) the volume of the block (in cm^3). [2]
b) the density of the fat. Give your answer to 2 significant figures. [5]
CIE (0625/02 N'05 Q10)

3 a) State what is meant by the terms
　(i) *weight* [1]
　(ii) *density* [1]
b) A student is given a spring balance that has a scale in newtons. The student is told that the acceleration of free-fall is 10 m/s^2.
　(i) Describe how the student could find the mass of an irregular solid object. [2]
　(ii) Describe how the student could go on to find the density of the object. [2]
c) The diagram below shows three forces acting on an object of mass 0.5 kg. All three forces act through the centre of mass of the object.

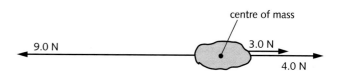

Calculate
　(i) the magnitude and direction of the resultant force on the object. [2]
　(ii) the magnitude of the acceleration of the object. [2]
CIE (0625/03 N'05 Q1)*

4 The diagram below shows the speed–time graph of part of a short journey made by a cyclist.

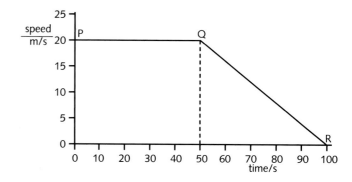

a) Which part of the graph shows when the cyclist is travelling at constant speed? [1]
b) State what is happening during the rest of the journey shown in the graph. [1]
c) (i) Calculate the distance travelled during the first 50 s.
　(ii) Calculate the distance travelled between 50 s and 100 s.
　(iii) Calculate the total distance travelled.
　(iv) Calculate the average speed during the 100 s. [8]
CIE (0625/02 J'04 Q3)

5 A wheel is rotating at approximately 2 revolutions per second. Describe how you would use a stopwatch to measure as accurately as possible the time for one revolution of the wheel. Make sure you include all the relevant information. [5]

CIE (0625/02 N'03 Q2)

6 The speed of a cyclist reduces uniformly from 2.5 m/s to 1.0 m/s in 12 s.
a) Calculate the deceleration of the cyclist. [3]
b) Calculate the distance travelled by the cyclist in this time. [2]

CIE (0625/03 J'03 Q2)★

7 The diagram below shows the speed–time graph for a bus during tests. At time $t = 0$, the driver starts to brake.

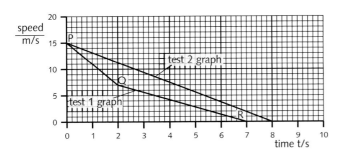

a) For test 1:
  (i) determine how long the bus takes to stop.
  (ii) state which part of the graph shows the greatest deceleration.
  (iii) use the graph to determine how far the bus travels in the first 2 seconds. [4]
b) For test 2, a device was fitted to the bus. The device changed the deceleration.
  (i) State two ways in which the deceleration during test 2 is different from that during test 1.
  (ii) Calculate the value of the deceleration in test 2. [4]
c) The diagram below shows a sketch graph of the magnitude of the acceleration for the bus when it is travelling around a circular track at constant speed.

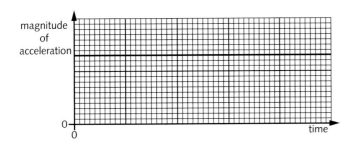

  (i) Use the graph to show that there is a force of constant magnitude acting on the bus.
  (ii) State the direction of this force. [3]

CIE (0625/03 N'03 Q1)★

8 A solid plastic sphere falls towards the Earth. The diagram below shows the speed–time graph of the fall up to the point where the sphere hits the Earth's surface.

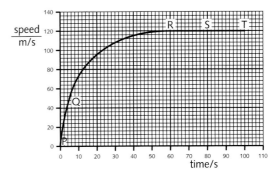

a) Describe in detail the motion of the sphere shown by the graph. [3]
b) Draw a circle to represent the sphere, and add arrows to show the directions of the forces acting on the sphere when it is at the position shown by point S on the graph. Label your arrows with the names of the forces. [2]
c) Explain why the sphere is moving with constant speed at S. [2]
d) Use the graph to calculate the approximate distance that the sphere falls
  (i) between R and T [2]
  (ii) between P and Q. [2]

CIE (0625/03 J'05 Q1)★

9 The diagram below shows apparatus for investigating moments of forces.

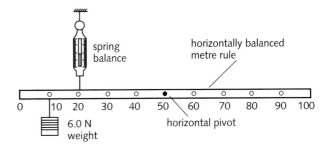

The uniform metre rule shown in the diagram above is in equilibrium.
a) Write down two conditions for the metre rule to be in equilibrium. [2]
b) Show that the value of the reading on the spring balance is 8.0 N. [2]
c) The weight of the uniform metre rule is 1.5 N. Calculate the force exerted by the pivot on the metre rule and give its direction. [2]

CIE (0625/03 N'05 Q2)★

10 A mass of 3.0 kg accelerates at 2.0 m/s^2 in a straight line.
a) State why the velocity and the acceleration are both described as vector quantities. [1]
b) Calculate the force required to accelerate the mass. [2]
c) The mass hits a wall. The average force exerted on the wall during the impact is 120 N. The area of the mass in contact with the wall at impact is 0.050 m^2. Calculate the average pressure that the mass exerts on the wall during the impact. [2]
CIE (0625/03 J'05 Q3)★

11 The diagram below shows apparatus that may be used to compare the strengths of two springs of the same size, but made from different materials.

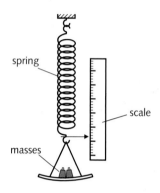

a) (i) Explain how the masses produce a force to stretch the spring.
(ii) Explain why this force, like all forces, is a vector quantity. [2]
b) The diagram below shows the graphs obtained when the two springs are stretched.

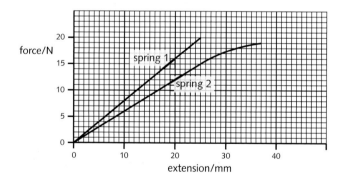

(i) State which spring is more difficult to extend. Quote values from the graphs to support your answer.
(ii) Make a sketch of the graph of spring 2. On it, mark a point P at the limit of proportionality. Explain your choice of point P.
(iii) Use the graphs to find the difference in the extensions of the two springs when a force of 15 N is applied to each one. [6]
CIE (0625/03 J'03 Q1)★

12 The diagram below shows a diver 50 m below the surface of the water.

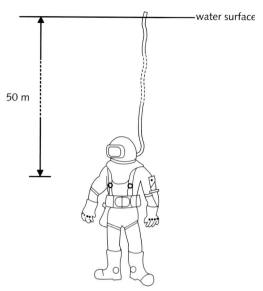

a) The density of water is 1000 kg/m^3 and the acceleration of free fall is 10 m/s^2. Calculate the pressure that the water exerts on the diver. [3]
b) The window in the diver's helmet is 150 mm wide and 70 mm from top to bottom. Calculate the force that the water exerts on this window. [3]
CIE (0625/03 N'03 Q2)★

13 The diagram below shows a simple pendulum that swings backwards and forwards between P and Q.

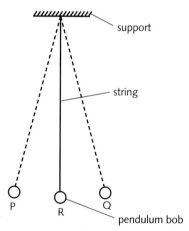

a) The time taken for the pendulum to swing from P to Q is approximately 0.5 s. Describe how you would determine this time as accurately as possible. [2]
b) (i) State the two vertical forces acting on the pendulum bob when it is at position R. [1]
(ii) The pendulum bob moves along the arc of a circle. State the direction of the resultant of the two forces in (i). [1]
c) The mass of the bob is 0.2 kg. During the swing it moves so that P is 0.05 m higher than R. Calculate the increase in potential energy of the pendulum bob between R and P. [2]
CIE (0625/03 J'05 Q2)★

14 The diagram below shows the arm of a crane when it is lifting a heavy box.

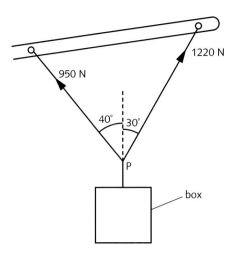

a) By the use of a scale diagram (not calculation) of the forces acting at P, find the weight of the box.   [5]

b) Another box of weight 1500 N is raised vertically by 3.0 m.

(i) Calculate the work done on the box.

(ii) The crane takes 2.5 s to raise this box 3.0 m. Calculate the power output of the crane.   [4]

CIE (0625/03 J'03 Q3)★

15 A rock climber climbs up a rock face, as shown in the diagram below.

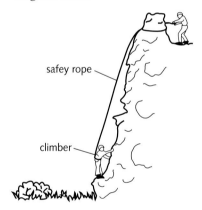

a) To climb the rock face, the climber must do work. Which one of the following forces must the climber work against as he climbs?   [1]

air resistance; friction on the rock;
his weight; tension in the safety rope

b) What other quantity, as well as the force ticked in **a)**, must be known in order to find the work done by the climber?   [1]

c) One climber weighs 1000 N and another weighs 800 N. They both take the same time to climb the cliff.

(i) Which one has done the most work?

(ii) Which one has the greater power rating?   [2]

d) When the first climber reaches the top, he has more gravitational potential energy than he had at the bottom.

(i) What form of energy, stored in his body, was used to give him this extra gravitational potential energy?

(ii) Where did he get this energy from?

(iii) Other than increasing gravitational potential energy on the way up, how else was energy in his body used? State one way.   [3]

CIE (0625/02 J'03 Q12)

16 A man is delivering a cupboard to a house.

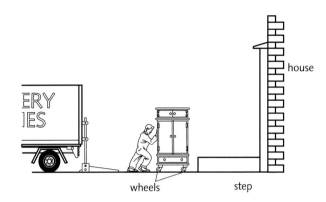

a) The man rolls the cupboard at a steady speed from the lorry to the house. The friction force in the wheels is 40 N. State the force with which the man has to push.   [1]

b) The cupboard weighs 720 N. State the smallest force needed to lift the cupboard.   [1]

c) The step is 0.20 m high. Calculate the work required to lift the cupboard onto the step.   [4]

d) The man has to ask his assistant to help him lift the cupboard onto the step. Together, they lift it onto the step in 1.2 s. The men work equally hard. Calculate the power developed by each man.   [4]

CIE (0625/02 N'05 Q4)

17 The diagram below shows a sealed glass syringe that contains air and many very tiny suspended dust particles.

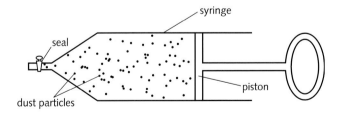

a) Explain why the dust particles are suspended in the air and do not settle to the bottom.   [3]

b) The air in the syringe is at a pressure of 2.0×10^5 Pa. The piston is slowly moved into the syringe, keeping the temperature constant, until the volume of the air is reduced from 80 cm³ to 25 cm³. Calculate the final pressure of the air.   [3]

CIE (0625/03 J'03 Q4)★

18 a) The diagram below shows the paths of a few air molecules and a single dust particle. The actual air molecules are too small to show on the diagram.

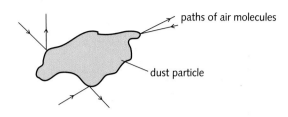

paths of air molecules

dust particle

Explain why the dust particle undergoes small random movements. [4]

b) The diagram below shows the paths of a few molecules leaving the surface of a liquid. The liquid is below its boiling point.

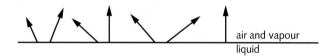

air and vapour

liquid

(i) State which liquid molecules are most likely to leave the surface. [1]

(ii) Explain your answer to (i). [2]

CIE (0625/03 J'05 Q5)★

19 The apparatus shown in the diagram below is set up in a laboratory during a morning science lesson.

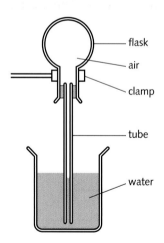

flask

air

clamp

tube

water

Later in the day, the room temperature is higher than in the morning.

a) What change is observed in the apparatus? [1]

b) Explain why this change happens. [1]

c) Suggest one disadvantage of using this apparatus to measure temperature. [1]

CIE (0625/03 N'03 Q5)

20 a) Equal volumes of nitrogen, water and copper at 20 °C are heated to 50 °C.

(i) Which one of the three will have a much greater expansion than the other two?

(ii) Explain your answer in terms of the way the molecules are arranged in the three substances. [3]

b) The diagram below shows a thermometer with a range of −10 °C to 50 °C.

−10 °C   50 °C

Explain what is meant by

(i) the sensitivity of a thermometer

(ii) the linearity of a thermometer. [2]

CIE (0625/02 N'05 Q5)★

21 The diagram below shows a way of indicating the positions and direction of movement of some molecules in a gas at one instant.

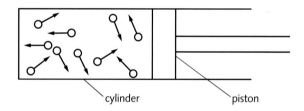

cylinder   piston

a) (i) Describe the movement of the molecules. [1]

(ii) Explain how the molecules exert a pressure on the container walls. [1]

b) When the gas in the cylinder is heated, it pushes the piston further out of the cylinder.

State what happens to

(i) the average spacing of the molecules [1]

(ii) the average speed of the molecules. [1]

c) The gas shown in the diagram above is changed into a liquid and then into a solid by cooling.

Compare the gaseous and solid states in terms of

(i) the movement of the molecules [1]

(ii) the average separation of the molecules. [1]

CIE (0625/03 N'05 Q5)★

22 The diagram below shows a thermocouple set up to measure the temperature at a point on a solar panel.

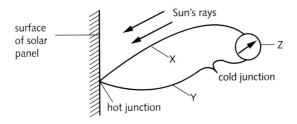

a) X is a copper wire.
  (i) Suggest a material for Y
  (ii) Name the component Z. [2]
b) Explain how a thermocouple is used to measure temperature. [3]
c) Experiment shows that the temperature of the surface depends upon the type of surface used. Describe the nature of the surface that will cause the temperature to rise most. [1]

CIE (0625/03 J'03 Q5)★

23 The diagram below shows apparatus that a student uses to make an estimate of the specific heat capacity of iron.

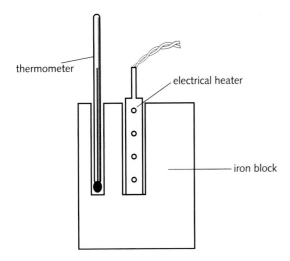

a) The power of the heater is known. State the four readings the student must take to find the specific heat capacity of iron. [3]
b) Write down an equation, in words or in symbols, that could be used to work out the specific heat capacity of iron from the readings in **a)**.
c) (i) Explain why the value obtained with this apparatus is higher than the actual value. [1]
  (ii) State one addition to the apparatus that would help to improve the accuracy of the value obtained. [1]

CIE (0625/03 J'05 Q4)★

24 The diagram below shows water being heated by an electrical heater. The water in the can is not boiling, but some is evaporating.

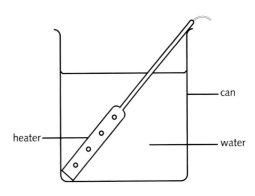

a) Describe, in terms of the movement and energies of the water molecules, how evaporation takes place. [2]
b) State two differences between evaporation and boiling. [2]
c) After the water has reached its boiling point, the mass of water in the can is reduced by 3.2 g in 120 s. The heater supplies energy to the water at a rate of 60 W. Use this information to calculate the specific latent heat of vaporization of water. [3]

CIE (0625/03 N'03 Q4)★

25 The diagram below shows the diffraction of waves by a narrow gap. P is a wavefront that has passed through the gap.

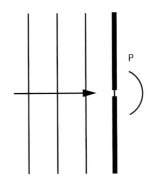

a) Copy the diagram and draw three more wavefronts to the right of the gap. [3]
b) The waves travel towards the gap at a speed of 3×10^8 m/s and have a frequency of 5×10^{14} Hz. Calculate the wavelength of these waves. [3]

CIE (0625/03 N'03 Q6)★

26 The diagram below shows the path of a sound wave from a source X.

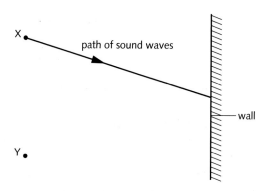

a) State why a person standing at point Y hears an echo. [1]
b) The frequency of the sound wave leaving X is 400 Hz. State the frequency of the sound wave reaching Y. [1]
c) The speed of the sound wave leaving X is 330 m/s. Calculate the wavelength of these sound waves. [2]
d) Sound waves are longitudinal waves. State what is meant by the term *longitudinal*. [1]

CIE (0625/03 N'05 Q6)★

27 The diagram below shows a ray of light OPQ passing through a semi-circular glass block.

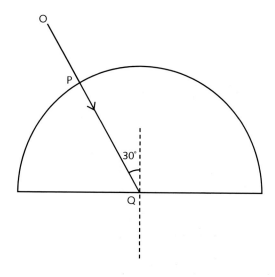

a) Explain why there is no change in the direction of the ray at P. [1]
b) State the changes, if any, that occur to the speed, wavelength and frequency of the light as it enters the glass block. [2]
c) At Q some of the light in ray OPQ is reflected and some is refracted. Copy the diagram and draw in the approximate positions of the reflected ray and the refracted ray. Label these rays. [2]

d) The refractive index for light passing from glass to air is 0.67. Calculate the angle of refraction of the ray that is refracted at Q into air. [3]

CIE (0625/03 J'05 Q6)★

28 In a thunderstorm, both light and sound waves are generated at the same time.
a) How fast does the light travel towards an observer? [1]
b) Explain why the sound waves always reach the observer after the light waves. [1]
c) The speed of sound waves in air may be determined by experiment using a source that generates light waves and sound waves at the same time.
  (i) Draw a labelled diagram of the arrangement of suitable apparatus for the experiment.
  (ii) State the readings you would take.
  (iii) Explain how you would calculate the speed of sound in air from your readings. [4]

CIE (0625/03 J'03 Q7)★

29 The speed of sound in air is 332 m/s. A man stands 249 m from a large flat wall, as shown in the diagram below, and claps his hands once.

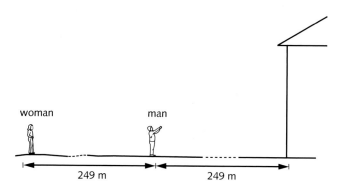

a) Calculate the time interval (in s) between the time when the man claps his hands and the time when he hears the echo from the wall. [3]
b) A woman is standing 249 m further away from the wall than the man. She hears the clap twice, once directly and once after reflection from the wall. How long after the man claps does she hear these two sounds? Choose your two answers from the following:
  0.75 s
  1.50 s
  2.25 s
  3.00 s
[2]

CIE (0625/02 J'05 Q9)

30 a) The diagram below shows two rays of light from a point O on an object. These rays are incident on a plane mirror.

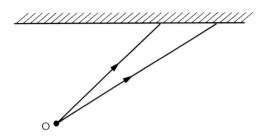

(i) Copy the diagram and continue the paths of the two rays after they reach the mirror. Hence locate the image of the object O. Label the image I. [2]

(ii) Describe the nature of the image I. [2]

b) The diagram below is drawn to scale. It shows an object PQ and a convex lens.

(i) Copy the diagram on graph paper and draw two rays from the top of the object P that pass through the lens. Use these rays to locate the top of the image. Label this point T. [3]

(ii) On your diagram, draw an eye symbol to show the position from which the image T should be viewed. [1]

CIE (0625/03 N'05 Q7)★

31 The diagram below shows the parts of the electromagnetic spectrum.

γ-rays and X-rays	ultra-voilet	v i s i b l e	infra-red	radio waves

a) Name one type of radiation that has
    (i) a higher frequency than ultraviolet. [1]
    (ii) a longer wavelength than visible light. [1]
b) Some γ-rays emitted from a radioactive source have a speed in air of 3.0×10^8 m/s and a wavelength of 1.0×10^{-12} m. Calculate the frequency of the γ-rays. [2]
c) State the approximate speed of infrared waves in air. [1]

CIE (0625/03 J'05 Q7)★

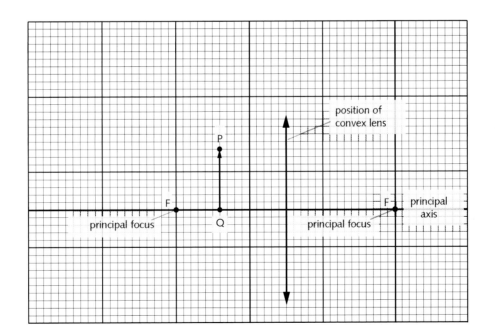

32 a) The diagram below shows a circuit containing a lamp and a variable resistor.

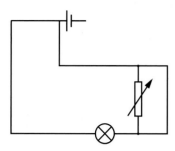

The circuit does not work. The lamp does not light and altering the setting on the variable resistor makes no difference. Re-draw the diagram, showing a circuit in which the variable resistor may be used to change the brightness of the lamp. [2]

b) The diagram below shows two resistors and an ammeter connected in series to a 6 V DC supply. The resistance of the ammeter is so small that it can be ignored.

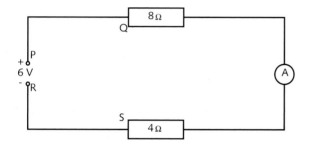

(i) Calculate the combined resistance (in Ω) of the 8 Ω and 4 Ω resistors in series. [2]

(ii) 1. Calculate the current supplied by the 6 V DC supply.
2. State the value of the current..
..in section PQ of the circuit.
..recorded by the ammeter
..in section SR of the circuit. [5]

(iii) Copy the diagram above, and show a voltmeter connected to measure the potential difference across the 4 Ω resistor. [1]

CIE (0625/02 J'03 Q11)

33 A student has a power supply, a resistor, a voltmeter, an ammeter and a variable resistor.

a) The student obtains five sets of readings from which he determines an average value for the resistance of the resistor. Draw a labelled diagram of a circuit that he could use. [3]

b) Describe how the circuit should be used to obtain the five sets of readings. [2]

c) The diagram below shows another circuit.

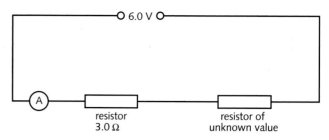

When the circuit is switched on, the ammeter reads 0.50 A.

(i) Calculate the value of the unknown resistor. [2]

(ii) Calculate the charge passing through the 3.0 Ω resistor in 120 s. [1]

(iii) Calculate the power dissipated in the 3.0 Ω resistor. [2]

CIE (0625/03 J'05 Q8)★

34 The diagram below shows a battery with an EMF of 12 V supplying power to two lamps. The total power supplied is 150 W when both lamps are on.

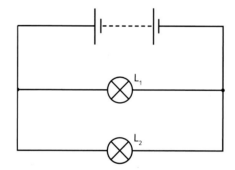

a) Calculate the current supplied by the battery when both lamps are on. [2]

b) The current in lamp L_2 is 5.0 A. Calculate

(i) the current in lamp L_1.

(ii) the power of lamp L_1.

(iii) the resistance of lamp L_1. [6]

CIE (0625/03 N'03 Q10)★

35 The diagram below shows a high-voltage supply connected across two metal plates.

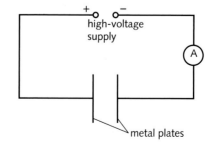

When the supply is switched on, an electric field is present between the plates.

a) Explain what is meant by an electric field.   [2]

b) Copy the diagram above. Draw the electric field lines between the plates and indicate their direction by arrows.[2]

c) The metal plates are now joined by a high-resistance wire. A charge of 0.060 C passes along the wire in 30 s. Calculate the reading on the ammeter.   [2]

d) The potential difference of the supply is re-set to 1500 V and the ammeter reading changes to 0.0080 A. Calculate the energy supplied in 10 s. Show your working.   [3]

CIE (0625/03 N'05 Q8)★

36 The diagrams show two views of a vertical wire carrying a current up through a horizontal card. Points P and Q are marked on the card.

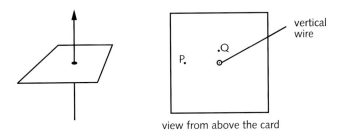

view from above the card

a) Copy the first diagram. On your copy
  (i) draw a complete magnetic field line (line of force) through P and indicate its direction with an arrow,
  (ii) draw an arrow through Q to indicate the direction in which a compass placed at Q would point.   [3]

b) State the effect on the direction in which compass Q points of
  (i) increasing the current in the wire.
  (ii) reversing the direction of the current in the wire.   [2]

c) The diagram below shows the view from above of another vertical wire carrying a current up through a horizontal card. A cm grid is marked on the card. Point W is 1 cm vertically above the top surface of the card.

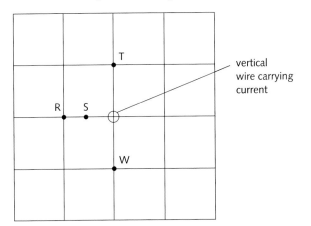

vertical wire carrying current

State the magnetic field strength at S, T and W in terms of the magnetic field strength at R. Use one of the alternatives, *weaker*, *same strength* or *stronger* for each answer.   [3]

CIE (0625/03 J'03 Q10)

37 The diagram below shows the outline of an AC generator. The peak output voltage of the generator is 6.0 V and the output has a frequency of 10 Hz.

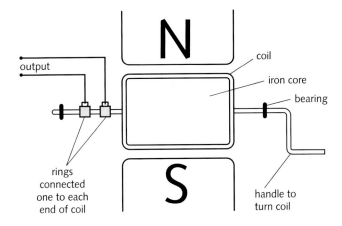

a) The diagram below shows the axes of a voltage–time graph for the generator output.

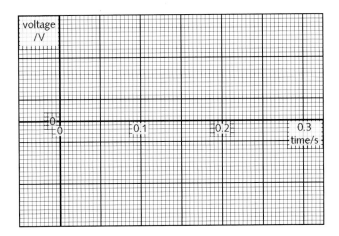

Copy the axes onto graph paper. On your copy
  (i) mark suitable voltage values on the voltage axis.
  (ii) draw a graph of the generator output.   [3]

b) The generator shown above works by electromagnetic induction. Explain how this effect produces the output voltage.   [3]

c) State the energy changes that occur in the generator when it is producing output.   [2]

CIE (0625/03 N'03 Q8)★

38 A transformer has an output of 24 V when supplying a current of 2.0 A. The current in the primary coil is 0.40 A and the transformer is 100% efficient.

a) Calculate
  (i) the power output of the transformer.
  (ii) the voltage applied across the primary coil.   [4]

b) Explain
  (i) what is meant by the statement that the transformer is 100% efficient.
  (ii) how the transformer changes an input voltage into a different output voltage.   [4]

CIE★

39 a) The diagram below shows an AC supply connected to a resistor and a diode.

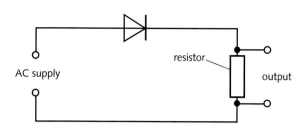

(i) State the effect of fitting the diode in the circuit. [1]

(ii) Copy the axes below and sketch graphs to show the variation of the AC supply voltage and the output voltage with time. [2]

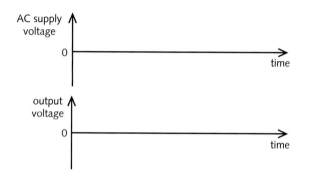

b) (i) Draw the symbol for a NOT gate. [1]

(ii) State the action of a NOT gate. [2]

CIE (0625/03 J'05 Q9)

40 a) Draw the symbol for a NOR gate. Label the inputs and the output. [2]

b) State whether the output of a NOR gate will be high (ON) or low (OFF) when

(i) one input is high and one input is low

(ii) both inputs are high. [1]

c) The diagram below shows a digital circuit made from three NOT gates and one NAND gate.

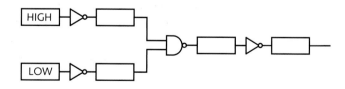

(i) Copy the diagram and write HIGH or LOW in each of the boxes. [2]

(ii) State the effect on the output of changing both of the inputs. [1]

CIE (0625/03 N'05 Q9)★

41 a) The diagram below shows the screen of a CRO (cathode ray oscilloscope). The CRO is being used to display the output from a microphone. The vertical scale on the screen is in volts.

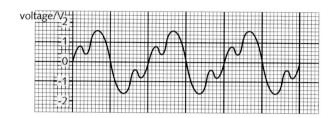

(i) Describe the output from the microphone.

(ii) Use the graph to determine the peak voltage of the output.

(iii) Describe how you could check that the voltage calibration on the screen is correct. [4]

b) The diagram below shows the screen of the CRO when it is being used to measure a small time interval between two voltage pulses.

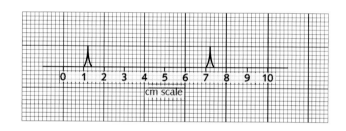

(i) What is the distance on the screen between the two voltage pulses?

(ii) The time-base control of the CRO is set at 5.0 ms/cm. Calculate the time interval between the voltage pulses.

(iii) Suggest one example where a CRO can be used to measure a small time interval. [4]

CIE (0625/03 N'03 Q9)★

42 a) When a nucleus decays by emitting an α-particle, what happens to

(i) the number of neutrons in the nucleus

(ii) the number of protons in the nucleus

(iii) the charge on the nucleus? [5]

b) On 1st January 1900, a sample of a particular radioactive nuclide had an activity of 3200 count/min. The nuclide has a half-life of 22 years. Calculate the activity of the nuclide remaining in the sample on 1st January 1966. [4]

CIE (0625/02 N'03 Q9)

43 a) In an electronic circuit, what is a capacitor designed to store?   [1]

b) The circuit below contains a large-value resistor and a capacitor.

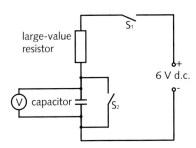

large-value resistor

V capacitor

6 V d.c.

S₁

S₂

(i) Switch S₁ is open. Switch S₂ is closed and then opened again. What reading (in V) now shows on the voltmeter?

(ii) S₂ is left open and S₁ is closed and left closed. Describe what happens to the reading on the voltmeter.

(iii) The circuit in the diagram above is an example of a simple time-delay circuit. State one use of a time-delay circuit.   [4]

CIE (0625/02 J'03 Q8)

44 a) The graph below is the decay curve for a radioactive isotope that emits only β-particles.

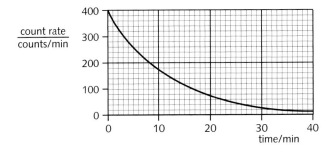

count rate
counts/min

Use the graph to find the value of the half-life of the isotope. Indicate, on the graph, how you arrived at your value.   [2]

b) A student determines the percentage of β-particles absorbed by a thick aluminium sheet. He uses a source that is emitting only β-particles and that has a long half-life.

(i) Draw a labelled diagram of the apparatus required, set up to make the determination.   [2]

(ii) List the readings that the student needs to take.   [3]

CIE (0625/03 J'05 Q10)*

45 a) A radioactive isotope emits only α-particles.

(i) Draw a labelled diagram of the apparatus you would use to prove that no β-particles or γ-radiation are emitted from the isotope.

(ii) Describe the test you would carry out.

(iii) Explain how your results would show that only α-particles are emitted.   [6]

b) The diagram below shows a stream of α-particles about to enter the space between the poles of a very strong magnet.

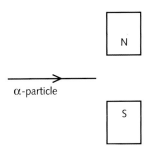

α-particle

Describe the path of the α-particles in the space between the magnetic poles.   [3]

CIE (0625/03 J'03 Q11)*

46 a) A sodium nucleus decays by the emission of a β-particle to form magnesium.

(i) Copy and complete the decay equation below.

$$^{24}_{11}Na \rightarrow Mg +$$

(ii) The diagram below shows β-particles from sodium nuclei moving into the space between the poles of a magnet.

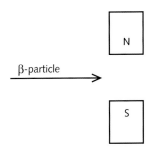

β-particle

Describe the path of the β-particles between the magnetic poles.   [5]

b) Very small quantities of a radioactive isotope are used to check the circulation of blood by injecting the isotope into the bloodstream.

(i) Describe how the results are obtained.

(ii) Explain why a γ-emitting isotope is used for this purpose rather than one that emits either α-particles or β-particles.   [4]

CIE (0625/03 N'03 Q11)*

If you do not take a practical examination, you may need to sit an alternative-to-practical paper instead. Here are some typical questions. For some of them, you will require graph paper.

1 A student carried out an experiment to find the spring constant of a steel spring. The apparatus is shown in the diagram below.

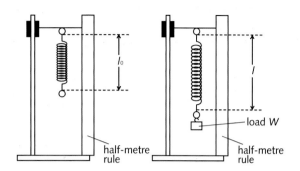

The student recorded the unstretched length l0 of the spring. Then she added loads W to the spring, recording the new length l each time. The readings are shown in the table below.

W/N	l/mm	e/mm	$l_0 = 30$ mm
0	30		
1	32		
2	33		
3	36		
4	39		
5	40		
6	42		

a) Calculate the extension e of the spring produced by each load, using the equation
$$e = (l - l_0)$$
Record the values of e in the table. [2]
b) Plot the graph of e/mm (y-axis) against W/N (x-axis). [4]
c) Draw the best-fit straight line for the points you have plotted. Calculate the gradient of the line. Show clearly on the graph how you obtained the necessary information. [4]
CIE (0625/06 J'03 Q1)

2 A student is investigating the oscillation of a metre rule that has one end resting on the laboratory bench. The other end is held above the level of the bench by a spring attached at the 90.0 cm mark. The arrangement is shown in the diagram below.

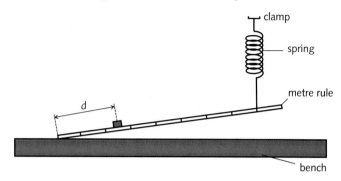

The period of oscillation is changed by moving a 200 g mass to different positions along the rule. The student records the time t taken for 10 oscillations of the end of the rule for each position of the mass. He measures the distance from the end of the rule to the mark under the centre of the mass. The readings are shown in the table below.

d/cm	t/s	T/s
20.0	3.4	
40.0	4.4	
50.0	4.9	
60.0	5.3	
70.0	6.0	
80.0	6.3	

a) Copy the table. Calculate the period T for each set of readings and enter the values in your table. [2]
b) Plot a graph of d/cm (x-axis) against T/s (y-axis). [5]
c) Using the graph, determine the period T when the distance d is 55.0 cm. [2]
d) The student suggests that T should be proportional to d. State with a reason whether your results support this suggestion. [2]
CIE (0625/06 J'04 Q2)

3 A student investigates the resistance of wire in different circuit arrangements. The circuit shown in the diagram below is used.

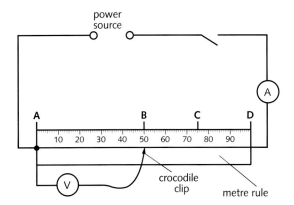

The student measures the current I in the wire. She then measures the PD V across **AB**, **AC**, and **AD**.
The student's readings are shown in the table below.

section of wire	I/cm	I/A	V/V	R
AB		0.375	0.95	
AC		0.375	1.50	
AD		0.375	1.95	

a) Copy the table. Then, using the diagram, record in your table the length l of each section of wire. [1]
b) Copy the diagrams below and show the positions of the pointers of the ammeter reading 0.375 A, and the voltmeter reading 1.50 V. [2]

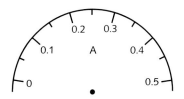

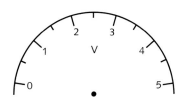

c) Calculate the resistance R of the sections of wire **AB**, **AC**, and **AD** using the equation

$$R = \frac{V}{I}$$

Record these values of R, to a suitable number of significant figures, in the table. [2]
d) Complete the column heading for the R column of the table. [1]
e) Use your results to predict the resistance of a 1.50 m length of the same wire. Show your working. [2]

CIE (0625/06 J'05 Q)

4 The diagram below shows the circuit that a student uses to find the resistance of a combination of three lamps.

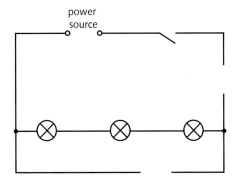

The voltmeter and the ammeter have not been drawn in.
a) Copy the diagram and complete it by drawing in the voltmeter and the ammeter, using conventional symbols. [2]
b) The student obtains these readings.
current $I = 0.54$ A
potential difference $V = 1.8$ V
Calculate the resistance R using the equation $R = \dfrac{V}{I}$ [2]

c) The three lamps are now connected in parallel with one another. Draw a circuit diagram of the three lamps connected to the power supply. Include in your circuit diagram
(i) an ammeter to record the total current through the lamps.
(ii) a variable resistor to vary the brightness of all three lamp.
(iii) a voltmeter to record the potential difference across the lamps. [3]

CIE (0625/06 J'04 Q3)

5 The IGCSE class carries out an experiment using a convex lens, an illuminated object and a screen. The diagram at the bottom of the page shows the apparatus. A sharp image is obtained on the screen.

a) (i) Use a rule to measure, on the diagram, the distance x from the illuminated object to the centre of the lens.

(ii) Use a rule to measure, on the diagram, the distance y from the centre of the lens to the screen.

(iii) The diagram shows the apparatus drawn to 1/5th of actual size. Calculate the actual distance u between the object and the lens, and the actual distance v between the lens and the screen.

(iv) Calculate the magnification m using the equation

$$m = \frac{v}{u}$$ [5]

b) The illuminated object is triangular in shape, as shown in the diagram below.

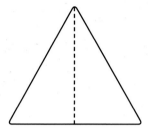

Draw a diagram of the image as it would appear on the screen. [1]

c) State two precautions that the IGCSE class should take to obtain experimental readings that are as accurate as possible. [2]

CIE (0625/06 J'04 Q4)

6 a) The IGCSE class carries out an experiment to investigate the rate of cooling from 100 °C of a range of hot liquids. Copy out any of the following variables that are likely to have a significant effect on the temperature readings. (You may copy one, two, or all three of the suggested variables.)

type and size of container
volume of liquid
temperature of the surroundings [2]

b) In an experiment to find the resistance of a wire, the students record the current in the wire and the potential difference across it. They then calculate the resistance. Copy out any of the following variables that are likely to have a significant effect on the current and/or potential difference readings. (You may copy one, two, or all three of the suggested variables.)

atmospheric pressure
temperature of the wire
length of wire [2]

c) In an experiment, a short pendulum oscillates rapidly. A student is asked to find the period of oscillation T of the pendulum using a stopwatch. The student sets the pendulum swinging and records the time for one oscillation. A technique for improving the accuracy of the value obtained for the period T should be used in this experiment. State, briefly, what this technique is and any calculation involved to obtain the value of T. [2]

CIE (0625/06 N'05 Q5)

Reference

Useful equations

In most cases, the equations below are given in both word and symbol form.

$g = 10$ N/kg (Earth's gravitational field strength)

 $= 10$ m/s² (acceleration of free fall)

Density, mass, and volume

$$\text{density} = \frac{\text{mass}}{\text{volume}}$$

$$\rho = \frac{m}{V}$$

Speed

$$\text{average speed} = \frac{\text{distance moved}}{\text{time taken}}$$

Acceleration

$$\text{average acceleration} = \frac{\text{change in velocity}}{\text{time taken}}$$

$$a = \frac{v - u}{t}$$

Force, mass, and acceleration

force = mass × acceleration

$$F = ma$$

Weight

weight = mass × g

$$W = mg$$

Moment of a force

$$\begin{matrix}\text{moment of force} \\ \text{about a point}\end{matrix} = \text{force} \times \begin{matrix}\text{perpendicular} \\ \text{distance from point}\end{matrix}$$

Stretched spring

load = spring constant × extension

$$F = kx$$

Pressure and force

$$\text{pressure} = \frac{\text{force}}{\text{area}}$$

$$p = \frac{F}{A}$$

Pressure in a liquid

pressure = density × g × depth

$$p = \rho g h$$

Temperature

Kelvin temperature = temperature in °C + 273

Compressing gases

For a fixed mass of gas at constant temperature:

pressure₁ × volume₁ = pressure₂ × volume₂

$$p_1 V_1 = p_2 V_2$$

(Boyle's law)

Heating gases

For a fixed mass of gas at constant volume:

$$\frac{\text{pressure}_1}{\text{Kelvin temperature}_1} = \frac{\text{pressure}_2}{\text{Kelvin temperature}_2}$$

$$\frac{p_1}{T_1} = \frac{p_2}{T_2}$$

Work

$$\text{work done} = \text{force} \times \begin{matrix}\text{distance moved} \\ \text{in direction of force}\end{matrix}$$

$$W = Fd$$

Gravitational potential energy

gravitational potential energy = mass × g × height

$$PE = mgh$$

Kinetic energy

kinetic energy = ½ × mass × velocity²

$$KE = \tfrac{1}{2}mv^2$$

Energy and temperature change

$$\text{energy transferred} = \text{mass} \times \text{specific heat capacity} \times \text{temperature change}$$

$$E = mc\Delta T$$

Energy and state change

energy transferred = mass × specific latent heat

$$E = mL$$

Power

$$\text{power} = \frac{\text{work done}}{\text{time taken}} = \frac{\text{energy transformed}}{\text{time taken}}$$

Efficiency

$$\text{efficiency} = \frac{\text{useful work done}}{\text{total energy input}}$$

$$= \frac{\text{useful energy output}}{\text{total energy input}}$$

$$= \frac{\text{useful power output}}{\text{total power input}}$$

Waves

speed = frequency × wavelength

$$v = f\lambda$$

Frequency and period

$$\text{frequency} = \frac{1}{\text{period}}$$

$$f = \frac{1}{T}$$

Refraction of light

$$\text{refractive index} = \frac{\text{sine of angle of incidence}}{\text{sine of angle of refraction}}$$

$$n = \frac{\sin i}{\sin r}$$

Total internal reflection

$$\text{sine of critical angle} = \frac{1}{\text{reflective index}}$$

$$\sin c = \frac{1}{n}$$

Charge and current

charge = current × time

$$Q = It$$

Resistance, PD (voltage), and current

$$\text{resistance} = \frac{\text{PD}}{\text{current}}$$

$$R = \frac{V}{I}$$

Resistors in series....

total resistance $R = R_1 + R_2$

...and in parallel

$$\frac{1}{R} = \frac{1}{R_1} + \frac{1}{R_2}$$

Electrical power

power = PD × current

$$P = VI$$

Electrical energy

energy transformed = power × time

$$= \text{PD} \times \text{current} \times \text{time}$$

$$E = VIt$$

Transformers

$$\frac{\text{output voltage}}{\text{input voltage}} = \frac{\text{output turns}}{\text{input turns}}$$

$$\frac{V_2}{V_1} = \frac{n_2}{n_1}$$

For 100% efficient transformer:

power input = power output

$$V_1 I_1 = V_2 I_2$$

SI units and prefixes

quantity	unit	symbol
mass	kilogram	kg
length	metre	m
time	second	s
area	square metre	m^2
volume	cubic metre	m^3
force	newton	N
weight	newton	N
pressure	pascal	Pa
energy	joule	J
work	joule	J
power	watt	W
frequency	hertz	Hz
PD, EMF (voltage)	volt	V
current	ampere	A
resistance	ohm	Ω
charge	coulomb	C
capacitance	farad	F
temperature	kelvin	K
	degree Celsius	°C

prefix	meaning	
G (giga)	1 000 000 000	(10^9)
M (mega)	1 000 000	(10^6)
k (kilo)	1000	(10^3)
d (deci)	$\dfrac{1}{10}$	(10^{-1})
c (centi)	$\dfrac{1}{100}$	(10^{-2})
m (milli)	$\dfrac{1}{1000}$	(10^{-3})
μ (micro)	$\dfrac{1}{1\,000\,000}$	(10^{-6})
n (nano)	$\dfrac{1}{1\,000\,000\,000}$	(10^{-9})
p (pico)	$\dfrac{1}{1\,000\,000\,000\,000}$	(10^{-12})

Examples

1 μF (microfarad) = 10^{-6} F

1 ms (millisecond) = 10^{-3} s

1 km (kilometre) = 10^3 m

1 MW (megawatt) = 10^6 W

Note: 'micro' means 'millionth'; 'milli' means 'thousandth'.

Elements

For simplicity, many of the rarer elements have been omitted from the table below.

atomic number (proton number)	element	chemical symbol
1	hydrogen	H
2	helium	He
3	lithium	Li
4	beryllium	Be
5	boron	B
6	carbon	C
7	nitrogen	N
8	oxygen	O
9	fluorine	F
10	neon	N
11	sodium	Na
12	magnesium	Mg
13	aluminium	Al
14	silicon	Si
15	phosphorus	P
16	sulphur	S
17	chlorine	Cl
18	argon	Ar
19	potassium	K
20	calcium	Ca
22	titanium	Ti
25	manganese	Mn
26	iron	Fe
27	cobalt	Co
28	nickel	Ni
29	copper	Cu
30	zinc	Zn
35	bromine	Br
38	strontium	Sr
47	silver	Ag
48	cadmium	Cd
50	tin	Sn
53	iodine	I
55	caesium	Cs
74	tungsten	W
78	platinum	Pt
79	gold	Au
80	mercury	Hg
82	lead	Pb
86	radon	Rn
88	radium	Ra
90	thorium	Th
92	uranium	U
94	plutonium	Pu

Electrical symbols and codes

Electrical symbols

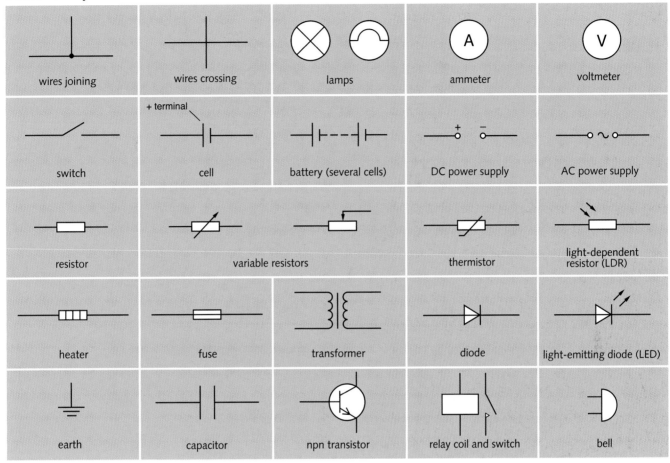

wires joining	wires crossing	lamps	ammeter	voltmeter
switch	cell	battery (several cells)	DC power supply	AC power supply
resistor	variable resistors		thermistor	light-dependent resistor (LDR)
heater	fuse	transformer	diode	light-emitting diode (LED)
earth	capacitor	npn transistor	relay coil and switch	bell

Resistor codes

The resistance of a resistor in ohms (Ω) is normally marked on it using one of these codes:

The resistor is marked with coloured rings. Each colour stands for a number:

black	0
brown	1
red	2
orange	3
yellow	4
green	5
blue	6
violet	7
grey	8
white	9

You 'read' the first three rings like this:

first figure   second figure   number of noughts

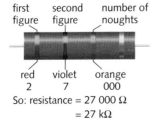

red   violet   orange
2   7   000

So: resistance = 27 000 Ω
= 27 kΩ

The fourth ring gives the **tolerance**. This tells you by how much the resistance may differ from the marked value:

gold ±5%   silver ±10%   no colour ±20%

The resistance is printed on the resistor:

R27 means 0.27 Ω
2R7 means 2.7 Ω
3K0 means 3000 Ω
5K6 means 5600 Ω
47K means 47 kΩ
2M2 means 2.2 MΩ

So: resistance = 8.2 kΩ

The extra letter at the end gives the tolerance:

F ±1%   G ±2%   J ±5%   K ±10%   M ±20%

Answers

The following have been omitted: some short answers, where the information is readily available from the text; answers requiring longer descriptions, explanations, or diagrams.

1.01 (page 11)
1 1000 g **2** 1000 mm **3** 1×10^6 μs **4** 6 m² **5** 2 km, 0.2 km, 20 km **6** 5 s, 50 s **7** 1.5×10^3 m, 1.5×10^6 m, 1.5×10^{-1} m, 1.5×10^{-2} m

1.02 (page 13)
5 a) 1.564 m **b)** 1.750 kg **c)** 26 000 kg (2.6×10^4 t) **d)** 6.2×10^{-5} s (0.000 062 s) **e)** 36.5 kg **f)** 6.16×10^{-10} m **6 a)** 5×10^{-3} kg **b)** 5000 mg

1.03 (page 15)
1 a) 2.3 s **b)** Time more swings **3 a)** 0.03 mm **b)** 6.31 mm

1.04 (page 17)
1 10^6 cm³ **2** 10^3 cm³ **3** 10^6 ml **4 a)** 200 l **b)** 2×10^5 cm³ **c)** 2×10^5 ml **5 a)** 2.7 g/cm³ **b)** 54 g **c)** 10 cm³ **6** Steel (stainless) **7** 39 kg **8** 4 m³ **9** 22.8×10^3 kg

1.05 (page 19)
1 Crowns: A silver, B gold, C mixture **2 a)** 80 g, 100 cm³, 0.8 g/cm³ **b)** 120 g, 48 cm³, 2.5 g/cm³

1.06 (page 20)
1 a) Yes **b)** No **2 a)** 2600 kg **b)** 2200 kg **c)** 400 kg

Examination questions (pages 22–23)
2 a, b) 1000 **c)** 1 000 000 **d)** 4 000 000 **e)** 500 000
3 0.3 m **b)** 0.5 kg **c)** 1.5 km **d)** 0.25 s **e)** 500 ms **f)** 750 m **g)** 2500 g **h)** 800 mm
4 24 cm³, 4 cm, 10 cm, 0.5 cm
5 a) 2500 m **b)** 2 m **c)** 3000 kg **d)** 2 litres
6 B and D
8 D
9 B
10 1.25 kg/m³
11 a) 0.1 m³, 0.05 m³ **b)** 800 kg **c)** 800 kg/m³ **d)** 1000 kg/m³ **12** A only is true
13 a) No; too many significant figures **b)** Time more swings **c)** 0.928 s

2.01 (page 27)
1 20 m/s **3 a)** 64 m **b)** 20 s **4** Runner 6.67 m/s, Grand Prix car 100 m/s, jet car 333 m/s, Concorde 667 m/s, Space Shuttle 10 000 m/s **6** 2.5 m/s² **7** 4 m/s² **8 a)** 12 m/s **b)** 44 m/s **9** 17 m/s

2.02 (page 29)
1 b) A and B **c)** B and C **d)** 4 m/s **e)** 60 m **f)** 3 m/s
2 a) 30 m/s **b)** 3 m/s² **c)** 6 m/s² **d)** 150 m **e)** 525 m **f)** 25 s **g)** 21 m/s

2.03 (page 31)
2 a) 0.1 s **b)** 200 mm/s **c)** 800 mm/s **d)** 600 mm/s²
3 a) 10 mm **b)** 100 mm/s **c)** 50 mm **d)** 500 mm/s **e)** 400 mm/s **f)** 1000 mm/s²

2.04 (page 33)
1 a) 10 m/s **b)** 20 m/s **c)** 50 m/s **2 a)** 30 m/s **b)** 40 m/s **c)** 70 m/s
3 a) 10 m/s **b)** 0 m/s **c)** 30 m/s **4 a)** Downwards **b)** B **c)** B **d)** 10 m/s² **e)** 10 m/s² **f)** 10 m/s² **g)** C

2.05 (page 35)
1 a) CD **b)** AB **c)** DE **d)** AB **e)** BC, DE **f)** DE
2) Line is a curve, rising and levelling off.

2.06 (page 37)
3 c) Equal **d)** Less

2.07 (page 39)
1 b) 10 N, 20 N **2 a)** 1000 N **b)** 1.25 m/s² **c)** Acceleration = 0

2.08 (page 41)
1 a) Brakes, tyres on road **b)** Air resistance, engine parts **2** Lower fuel consumption **3 a)** 18 m **b)** 87 m **c)** 105 m **4** Changed into thermal energy (heat) **5** Kinetic energy to be lost is more than doubled

2.09 (page 43)
1 a) 50 N, 100 N **b)** 10 m/s² (both) **c)** 10 N/kg
2 a) 1000 N **b)** 100 kg **c)** 100 kg **d)** 37 N **e)** 3.7 m/s²

2.10 (page 45)
1 a) 500 N **b)** 500 N

2.11 (page 47)
2 a) 17 N **b)** 7 N **c)** 13 N at 23° to 12 N force
3 a) Horizontal component 87 N, vertical component 50 N **b)** 350 N **c)** Force reduced (to 250 N)

2.12 (page 49)

1 Path is at a tangent to circle **2** Friction between tyres and road **3** Centripetal force **a)** less **b)** less **c)** more
4 b) Speed unchanged **c)** Speed less **d)** Centripetal force less

Examination questions (pages 50–51)

1 b) 100 m
2 a) (i) 8 m (ii) 2.0 s **b)** 4.0 m/s **c)** Increasing distance between positions (ii) Weight has a component down slope, force causes acceleration
3 a) (iii) 2.0 kg
4 a) 25 s **b)** 1080 N **c)** Resisting force (air resistance) increases with speed, so resultant force less
5 a) 2000 N **b)** (i) 1200 N (iii) 1.5 m/s^2
6 a) 5 N
7 a) 5 km **b)** (i) 10 m/s (ii) 8 min 20 s **c)** 2 m/s^2
8 a) 4 s **b)** Friction **c)** On tyres, from road **d)** 3000 N
9 b) 20 m/s **c)** 4 s; reduction in speed **d)** (i) Weight (gravity) (ii) Air resistance (iii) Weight; air resistance; equal **e)** Straight and level **f)** (i) No change (ii) Greater loss of speed
10 b) (i) 1.33 m/s^2 (ii) 225 m
11 a) Two of driver's reaction time, mass, braking force **c)** 16 m
12 b) Ball travels in straight line at tangent to circle **c)** Gravity

3.01 (page 55)

3 a) 16 N m **b)** 12 N m **c)** No; clockwise **d)** 1 N
e) Downwards **4 a)** 21 N **b)** 10 N, 8 N, and 3 N forces; 84 N m **c)** 21 N force; 84 N m **d)** Yes

3.02 (page 57)

1 b) Shorter legs, wider apart **2 b)** 1 N

3.03 (page 59)

1 b) 720 N m **c)** 180 N **d)** 600 N **e)** 420 N **f)** 0
g) 0.5 m

3.04 (page 61)

3 No, not a straight line **4 a)** 40 mm **e)** Up to 76 mm length **f)** 3.89 N **g)** 2.78 N

3.05 (page 63)

1 a) 50 Pa **b)** 100 Pa **2 a)** 200 N **b)** 400 N
4 a) 300 N **b)** Face measuring 0.1 m × 0.4 m is on ground, 7500 Pa **c)** Face measuring 0.4 m × 1.5 m is on ground, 500 Pa

3.06 (page 65)

1 a) Less **b)** Same **c)** Same **d)** Less **2 a)** 24 m^3
b) 19 200 kg **c)** 192 000 N **d)** 16 000 Pa **3** 20 000 Pa

3.07 (page 67)

1 a) 200 Pa **b)** 200 Pa **c)** 100 N **2 a)** Increased output force **b)** Increased output force

3.08 (page 69)

4 a) 100 mm of mercury **b)** 860 mm of mercury
c) 113 000 Pa
5 a) 96 000 Pa **b)** 0.96 atm **c)** 960 mb **6 a)** 9810 Pa
b) 10.3 m

3.09 (page 71)

2 a) 12 m^3 **b)** 15 m^3 **3 a)** Pressure × volume is constant **b)** Straight line

3.10 (page 73)

1 a) 20 000 Pa **b)** 100 000 Pa **c)** a – same, b – halved
2 100 cm^3

Examination questions (pages 74–75)

1 b) Larger force, further from pivot
2 a) 0.4 N m **b)** 1.6 N
3 a) 100 kPa, 200 kPa, 300 kPa, 400 kPa **b)** 2 m^3
4 b) (i) 30 N m (ii) 0.38 m
5 b) (i) 10.0 cm (ii) 14.0 cm (iii) 2.5 N
6 a) 100 000 Pa
7 b) Force spread over greater area, so less pressure on heel **c)** 200 N
8 a) the same as; more than; less than **c)** 2.5 N/cm^2
9 a) (i) 0.5 m^2 (ii) 2.0 m^2 **b)** 50 000 N/m^2
10 a) 12 m^3 **b)** 12 000 kg **c)** 120 000 N **d)** 20 000 Pa

4.01 (page 79)

1 60 J **2** 0.5 m **3 a)** 10 000 J **b)** 35 000 000 J
c) 500 000 J **d)** 200 J **4** 18 **5 a)** kinetic, gravitational potential, chemical **b)** chemical

4.02 (page 81)

1 a) 50 J **b)** 50 J **c)** Changed to thermal energy (heat)

4.03 (page 83)

1 a) 240 J **b)** 360 J **2 a)** 75 J **b)** 300 J **3** 20 m/s
4 a) 25 J **b)** 25 J **c)** 5 m **d)** 5 m

4.04 (page 85)

1 a) 30% **2** 500 W **3 a)** 3000 W **b)** 3000 W
c) 60 000 J **d)** 75% **4 a)** 6000 J **b)** 300 J **c)** 300 W
5 a) 6000 N **b)** 4000 W **6** 50 000 W

4.05 (page 87)

2 a) Turning turbines **b)** Condense steam (turn it back
to liquid) **3 a)** Turbines **b)** Thermal energy (heat)
c) X: 2000 MW, Y: 500 MW **e)** X: 36%, Y: 27%

4.06 (page 89)

2 b) B **c)** E **d)** A **e)** A **f)** No fuel required or burned

4.07 (page 91)

2 a) Wind, hydroelectric **8** Hydroelectric, tidal, wave

Examination questions (pages 94–95)

1 a) Wound-up spring, stretched rubber bands
2 a) (i) PE (ii) PE + KE (iii) PE + KE **b)** Changed
into thermal energy (heat)
3 a) (i) Elastic potential energy (strain energy)
(ii) Changed to KE + gravitational PE **b)** 0.75 J
5 a) 225 000 J **b)** 225 000 J **c)** 1.25 m
6 a) 3000 N **b)** 180 000 J **c)** 4000 W
7 a) kettle 2 kW; food mixer 600 W **b)** television
c) food mixer
10 a) wood – yes, no; uranium – no, no **b)** (ii) Not
renewable, use causing global warming
11 b) energy, burned
12 a) (i) heat (ii) kinetic, heat (iii) light, heat
(iv) sound, heat **b)** elastic; gravitational; chemical

5.01 (page 99)

1 a, e, g) Gas **b, c)** Solid **d, f)** Liquid **3** Move faster
on average

5.02 (page 101)

1 a) 100 °C **b)** 373 K **c)** −273 °C **d)** 0 K **e)** 0 °C
f) 273 K **2 a)** Volume increase with temperature
b) Change of conducting ability (resistance) with
temperature **3 a)** Slower on average in A **b)** B to A
c) When temperatures are the same

5.03 (page 103)

1 a) (i) 0 °C (ii) 100 °C **b)** (i) 75 °C (ii) 25 °C
(iii) −50 °C **c)** Mercury freezes above this
temperature **2 a)** C **b)** A

5.04 (page 105)

2 a) Bimetal strip bends, so contacts separate **b)** Right

5.05 (page 107)

2 Increases **3** 6 atm **4 a)** −273 °C **b)** Yes; graph of
pressure against Kelvin temperature is straight line
through origin

5.06 (page 109)

3 Thicker lagging, keeping water at a lower average
temperature **4 a)** Copper **b)** Length, diameter,
temperature difference same for all the metals **5** Free
electrons present

5.07 (page 111)

3 B; hot water rises, so collects from top down

5.08 (page 113)

4 a) Temperature, detector distance and area same for
all the surfaces **b)** Plate area, distance, and radiation
source same for both surfaces

5.10 (page 117)

2 a) 400 J **b)** 200 000 J **c)** 2 100 000 J **3 a)** 8400 J
b) 42 000 J **c)** 5 °C

5.11 (page 119)

1 b) 68 °C **3 a)** 3 300 000 J **b)** 23 000 000 J **4** 0.12 kg

Examination questions (pages 120–121)

2 B
3 d) 2 kW **e)** 5 m²
4 b) (i) 2 mm (ii) 200 mm
5 a) (i) liquid (ii) liquid **b)** (i) 440 °C
6 a) Evaporation **b, c, e)** Convection **d)** Conduction
7 b) Hot water rises by convection, so collects from the
top down **c)** (i) kilo (× 1000) (ii) 3000 J
(iii) 1 260 000 J **d)** (i) 4200 J (ii) 420 000 (iii) 3 °C
8 a) Larger surface area gives increased heat transfer
rate **b)** 12.6 MJ (12 600 000 J)
9 a) 80 °C **b)** None

6.01 (page 125)

1 a) Transverse **b)** 2 m **c)** 0.5 m **d)** (i) 2 Hz (ii) 0.5 s
e) 4 m/s **f)** 4 m **g)** 1 Hz

6.02 (page 127)

1 b) Refraction **c, d, e)** Diffraction **2 a)** Reflect
b) Refract (bend) **c)** Diffract (spread out) **d)** Less diffraction (less spreading)

6.03 (page 129)

2 a) No medium to carry vibrations **b)** Sound waves diffract

6.04 (page 131)

1 b) 1320 m **2 a)** Warm air **b)** gas **4 a)** 440 m
b) 1.33 s **c)** 82.5 m

6.05 (page 133)

1 a) C **b)** A and D **c)** B **3 a)** Peaks closer together
b) Peaks higher (greater amplitude) **4 a)** 20 kHz
b) 16.5 m **c)** 0.016 5 m

6.06 (page 135)

4 a) 40 000 Hz **b)** 21 m **c)** 0.035 m

Examination questions (pages 136–137)

1 a) Circular, transverse **b)** Transverse waves produce only up and down motion **c)** (i) 2 Hz (ii) 0.25 m
2 a) 15 s **b)** (ii) 264 m **c)** Sound waves are longitudinal, and much faster
 3 a) Waves have same spacing but higher peaks
b) (ii) 3.3 m
4 a) (i) A (ii) B **b)** (i) From greater than average to less than average (ii) Repeatedly backwards and forwards
5 a) (ii) 3 (iii) 0.05 s (iv) 20 Hz **b)** (ii) Louder
6 b) (i) Greater amplitude (greater forwards-and-backwards motion) (ii) More vibrations per second
c) 680 Hz
7 a) (i) 0.48 ms (ii) 5000 m/s
8 a) B louder than A **b)** C higher pitch than A
c) B **d)** C **e)** 1.5 m **f)** 440 Hz
9 b) Safer, can distinguish between tissue layers
c) Cleaning or metal testing

7.01 (page 141)

3 a) Reflected **b)** Absorbed **4 a)** 1.28 s **b)** 500 s
5 Shorter wavelength **6** Single wavelength (and colour)

7.02 (page 143)

1 c) Virtual **d)** No **2** 7.5 m

7.03 (page 145)

1 b) A 26.6°, B 63.4° **e)** 63.4°

7.04 (page 147)

1 b) Larger angle of refraction **2 b)** Violet **c)** Red
3 226 000 km/s

7.05 (page 149)

2 b) No; angle of incidence less than critical angle

7.06 (page 151)

1 a) 17.8° **b)** 36.9° **2 a)** 30° **b)** greater
3 a) 124 000 km/s **b)** 24.4°

7.07 (page 153)

1 a) A **b)** A **2 a)** At principal focus **b)** Further from lens, larger

7.08 (page 155)

1 a) 12 cm from lens, height 2 cm, real and inverted
b) 15 cm from lens, height 3 cm, real and inverted
c) 1.5 **2 a)** Closer than principal focus **b)** At twice focal length **c)** Closer than in b), but no closer than principal focus

7.09 (page 157)

1 Aperture, shutter speed **2** Away from film
5 a) Smaller **b)** Towards screen **6 a)** increase
b) decrease

7.10 (page 159)

1 a) Retina **b)** Cornea and liquid behind it **c)** Iris
2 a) Accommodation **5 a)** Convex **b)** Concave

7.11 (page 161)

3 a) Light **b)** Infrared **c)** Radio waves **d)** Ultraviolet
e) Microwaves **f)** X-rays or gamma rays
4 a) 100 000 000 Hz **b)** 3 m **c)** 1500 m

Examination questions (pages 166–167)

1 a) Speed, direction **b)** Totally internally reflected
2 a) 1st – smaller, closer etc; 2nd line – real, further etc; 3rd line – virtual, magnified **b)** (ii) Virtual, upright, magnified
3 c) Waves change speed
4 c) 1.5 **6 c)** 3
5 Larger, further from lens
6 b) Two of virtual, magnified, upright **c)** 3
7 b) B
8 a) No change in direction **b)** 28° **c)** 2×10^8 m/s
9 b) Two of frequency, wavelength, penetrating power
c) 5×10^{14}
10 a) (i) X-rays (ii) infrared **b)** (ii) 3×10^{11} Hz
(iii) microwaves

11 a) 35° **b)** 42° **c)** Strikes KL at more than critical angle but, after reflection, strikes LM at less than critical angle

12 a) (i) greater than (ii) the same as (iii) greater than **b)** (i) Microwaves (ii) Ultraviolet or gamma

8.02 (page 173)

2 1 000 000 **3 c)** Away from the can **d)** Positive

8.03 (page 175)

1 a) Arrows point away from sphere **b)** Away from the sphere **c)** Towards the sphere **d)** Become less (because of flow through point)

8.04 (page 177)

1 a) 0.5 A **b)** 2.5 A **2 a)** 2000 mA **b)** 100 mA **3 c)** 0.5 A **d)** A and B **4 a)** 50 C **b)** 10 C

8.05 (page 179)

1 a) volt **b)** volt **c)** coulomb **d)** ampere **e)** joule **2 a)** Ammeter **b)** Voltmeter **c)** 8 V **d)** 12 J **e)** 4 J **f)** 2 C **g)** 8 J

8.06 (page 181)

1 a) 23 Ω **b)** Would not heat up without resistance **2** Bulb gets brighter; less resistance in circuit, so more current **3 a)** LDR **b)** thermistor **c)** diode

8.07 (page 183)

1 B **2 a)** 2 Ω **b)** 4 Ω **3** reverse **4 a)** 16 V **b)** 32 V **c)** 0.75 A

8.08 (page 185)

1 2 Ω **2 a)** 720 mm **b)** 250 Ω **c)** 200 mm

8.09 (page 187)

1 In series **4 a)** X: 2 A, Y: 2 A **b)** 6 V

8.10 (page 189)

1 a) 1.5 A **b)** 6 V (both) **2 a)** 3 A **b)** 3 A (both) **c)** 6 A **d)** 2 Ω **3** D (9.9 Ω)

8.11 (page 191)

1 a) 2000 W **b)** 2 kW **2** 920 W **3** 3 A **4 a)** R_1: 4 A, R_2: 2 A **b)** R_1: 48 W, R_2: 24 W **c)** 72 W **d)** 288 W

8.12 (page 193)

1 a) live **b)** earth **c)** live **4** Double insulation instead **5** Hairdryer: 13 A, drill: 3 A, iron: 13 A, table lamp: 3 A

8.14 (page 197)

1 a) 60 J **b)** 3600 J **2 a)** 36 W **b)** 21 600 J **3 a)** 8 kWh **b)** 28 800 000 J **4 a)** 450cu **b)** 36cu **c)** 3cu **5** Approx 3900cu

Examination questions (pages 198–199)

2 a) (i) Like charges repel (ii) Negative

3 b) Brighter, more current

4 a) 1 A **b)** 3 A

5 a) 4.5 J **b)** 4.5 W

6 a) 6 Ω **b)** 0.2 A **c)** 0.2 A **d)** 0.8 V **e)** 0.16 W

7 c) 3.26 A **d)** (ii) 5 A **e)** Much less energy (heat) per second delivered

8 a) 21 A **b)** (i) 2.1 kW **d)** 960 Ω **e)** (i) 50 Hz

9 c) 0.4 A **d)** 5 Ω **e)** 7.5 Ω **f)** Increases

10 a) 5 A **b)** 2.4 Ω **c)** 100 C **d)** 1200 J

9.01 (page 203)

5 Bars 1 and 3 are permanent magnets, bar 2 is not

9.02 (page 205)

1 a) N is at top end **b)** N pole at right-hand end of magnet **c)** X

9.03 (page 207)

1 c) Arrowhead points away from + end of battery; N pole is at left end of coil **2** Needles form part of a circle with black ends pointing clockwise

9.04 (page 209)

3 b) Trips (cuts off) at lower current

9.05 (page 211)

1 b) Upwards

9.06 (page 213)

2 a) Coil horizontal **b)** Coil vertical **c)** Anticlockwise

9.07 (page 215)

1 a) Current direction reversed **b, c)** No current **2 a, c)** Greater EMF **b)** Current direction reversed

9.08 (page 217)

1 Unchanged **2 a)** S pole **b)** AB

9.09 (page 219)
1 a) AC **c)** Horizontal **d)** Vertical

9.10 (page 221)
1 a) Galvanometer needle flicks **b)** ...stays at zero
c) ...flicks opposite way
2 a) Needle deflection much more **b)** AC induced in coil (so average deflection of needle is zero) **3 a)** 3 V
b) 3

9.11 (page 223)
3 a) 10 V **b)** 23 W **c)** 23 W **d)** 2.3 A

9.12 (page 225)
5 2640 MW **6** 0.02 W **7 a)** 2 kW **b)** 0.002 W

Examination questions (pages 226–227)
2 a) a magnetic material **b)** a magnet **c)** a magnetic material **d)** a non-magnetic material
4 a) F is to the right **b)** (i) Stronger (ii) Weaker
(ii) Opposite direction
5 a) (i) Needle deflects to right (ii) Needle deflects to right (ii) Larger deflection to left
6 c) 32 000
7 c) 115 V
8 a) 90 C **b)** (i) Magnetic field (ii) Become magnetized, so will repel
9 b) (i) 2 (ii) 4 **c)** Stronger magnet or faster rotation

10.01 (page 231)
1 a) capacitor **b)** LED **c)** IC **d)** relay **3 a)** bell push
b) bell **4 a)** analogue **b)** digital

10.02 (page 233)
2 a) Y **b)** X **4** Reduced to 2 V

10.03 (page 235)
1 c) ON **d)** Bulb OFF **2 a)** 5 mA **b)** 99

10.04 (page 237)
3 Bulb comes on when it is light, goes off when it is dark **4 a)** Higher temperature needed to make bell ring
b) Replace 10 kΩ resistor with variable one

10.05 (page 239)
1 AND gate **2** OR gate **3 b)** Both LOW (0)

10.06 (page 241)
1 b) HIGH (1) **2 b)** (i) NOT gate (ii) AND gate
10.07 (page 243)

1 d) No **2 a)** 8.0×10^{-17} J **b)** 0.1 C **c)** 6.3×10^{17}
10.08 (page 245)
3 a) 1.5 cm **c)** 15 V **d)** 20 ms **e)** 50 Hz

Examination questions (pages 246–247)
1 b) (i) 0.95 mA (iii) 2400 Ω
2 a) Light-dependent resistor **b)** Loudspeaker
c) AND gate
3 b) (i) X = OR, Y = AND (ii) D (iii) Motion sensor, or reed switch linked to door or window
4 a) (i) A = AND (ii) B = NOT (iii) C = AND
5 c) Bell keeps ringing
6 a) LED **b)** Relay **c)** Thermistor **d)** LDR **e)** Diode
7 a) Falls **b)** Rises
8 a) Resistance falls **b)** LDR
10 a) Heat, thermionic emission **b)** Electric field, or attracted to positive anode **c)** Apply PD so that Y_1 is + and Y_2 is −

11.01 (page 251)
2 a) 13 **b)** 13 **c)** 14 **4 a)** **b)** **c)** **5** X is carbon, Y is carbon, Z is nitrogen

11.02 (page 253)
1 a, d, f, h, i) gamma **b, e, g)** alpha **c)** beta

11.03 (page 255)
1 Radon gas from ground **3 a)** Gamma **b)** Alpha
4 a) 2 counts per second **b)** 26 counts per second
c) Gamma

11.04 (page 257)
1 a) Alpha **b)** $A = 228$, $Z = 88$ **c)** Radium
e) radium-228, alpha **2 a)** 0 **b)** −1
c) Beta particle (electron)

11.05 (page 259)
1 Strontium-90 **2 a)** 400 Bq **b)** 200 Bq **c)** 50 Bq
3 b) 1.5 hours

11.06 (page 261)
5 a) 391.924×10^{-27} kg **b)** 391.614×10^{-27} kg
c) 2.8×10^{-11} J

11.07 (page 263)
4 a) Hydrogen **b)** Fusion **c)** Helium

11.08 (page 265)
3 b) Reading goes down

Examination questions (pages 270–271)
1 a) (i) nuclei (ii) electrons (iii) waves **b)** Alpha particles
2 a) 17 electrons, 17 protons, 18 neutrons
3 a) (i) 33 (ii) 52 **c)** Use fact that lead will stop alpha and beta particles but not gamma rays
4 a) (i) Nucleus of phosphorus-32 has extra neutron (ii) Same electron arrangement **b)** (ii) 16, 32
5 a) Too easily absorbed by tissue **b)** (i) 12 hours (ii) $\frac{1}{16}$ × original value
6 b) Total of protons and neutrons in nucleus
c) (ii) 8 days **d)** Much longer half-life
7 b) (i) 48 counts/minute (ii) 40 hours approx
c) Gamma not very ionizing **d)** (i) beta (ii) I 22 II 13
8 c) (i) 86 (ii) 86 (iii) 136
9 a) Too easily absorbed by tissue
10 a) 146

13.03 (page 289)
1 47 °C **2** 36 counts/second **3** 5.4 N **4** 86 kPa
5 0.79 mV

Multichoice questions (pages 304–305)
1 A **2** C **3** B **4** D **5** C **6** D **7** B **8** B **9** A **10** A **11** C
12 A **13** D **14** C
15 B **16** B

Further examination questions (pages 306–317)
1 a) 10 s **b)** 2 s **c)** 20 s
2 a) 260 cm³ (ii) **b)** 0.96 g/cm³
3 b) (i) Weigh object, in N; multiply result by 10 m/s² to give mass in kg (ii) Find volume by displacement of liquid; divide mass by volume to find density
c) (i) 2.0 N to left (ii) 4.0 m/s²
4 a) PQ **b)** Cyclist decelerates (uniformly) until stationary **c)** (i) 1000 m (ii) 500 m (iii) 1500 m
(iv) 15 m/s
5 Measure time for 20 revolutions (for example); divide result by 20
6 a) 0.125 m/s² **b)** 21 m
7 a) (i) 7 s (ii) PQ (iii) 22 m **b)** (i) uniform, between values for PQ and QR (ii) 1.875 m/s² **c)** (i) Acceleration is constant so, from $F = ma$, force must be constant (ii) Towards centre of circle

8 a) Speed increases to a terminal value, while acceleration decreases to zero **b)** Weight downwards, equal air resistance upwards **c)** Resultant force zero, so acceleration zero **d)** (i) 4800 m (ii) 150 m
9 a) Moments about any point (e.g. pivot) equal, resultant force in any direction zero **b)** Taking moments about pivot, 6.0 × 40 = 8.0 × 30 c) 0.5 N downwards
10 a) Have direction as well as magnitude **b)** 6.0 N
c) 2400 Pa
11 a) (i) Masses have weight because of Earth's gravitational pull (ii) Has magnitude and direction
b) (i) spring 1 (ii) P is at end of straight section
(iii) 6.5 mm
12 a) 500 000 Pa **b)** 5250 N
13 a) Measure (for example) 50 swings, divide total time by 50 **b)** (i) weight, tension in string
(ii) Upwards, towards centre of arc **c)** 0.1 J
14 a) 1784 N **b)** (i) 4500 J (ii) 1800 W
15 a) (i) his weight **b)** distance moved **c)** (i) and
(ii) 1000 N climber **d)** (i) chemical energy (ii) food (respiration) (iii) producing heat
16 a) 40 N b) 720 N **c)** 144 J **d)** 60 W
17 a) Bombarded by molecules in air (ii) **b)** 6.4 × 105
18 a) Random bombardment by molecules in air
(ii) **b)** (i) Faster (more energetic) ones (ii) Attractions not strong enough to keep them in liquid
19 a) Water level in tube is lower **b)** Air has expanded because of temperature rise **c)** Results only consistent if outside pressure is constant
20 a) (i) Nitrogen (assuming pressure is constant)
(ii) In gas, molecules free to move and not bound together **b)** (i) Distance moved along scale per degree change in temperature (ii) How close thermometer is to having the same scale distance for every degree
21 a) (i) Fast, in random directions (ii) Exert force when they bounce off walls **b)** (i) Decreases
(ii) Increases **c)** (i) Molecules in solid vibrate
(ii) Separation less in solid
22 a) (i) iron (ii) milliammeter (or millivoltmeter)
b) Greater the temperature difference between junctions, greater the voltage (and current) produced, so higher the meter reading **c)** matt black
23 a) mass of block, initial and final temperatures, time heater is switched on **b)** $Pt = mc\Delta T$ **c)** (i) Extra energy supplied to make up for heat losses
(ii) Insulation round block

24 a) Faster (more energetic) water molecules overcome attractions in liquid and escape from surface **b)** boiling is very rapid, it occurs when vapour pressure matches atmospheric pressure **c)** 2 250 000 J/kg

25 b) 6×10^{-7} m (0.000 000 7 m)

26 a) Sound from X reaches Y first, then reflection from wall arrives after **b)** 400 Hz **c)** 0.825 m **d)** Oscillations backwards and forwards

27 a) Angle of incidence is zero, so angle of refraction is zero **b)** Speed and wavelength decrease, frequency stays the same d) 48°

28 a) 3×10^8 m/s **b)** Sound much slower (330 m/s) **c)** (ii) distance to sound detector, time for sound to reach detector (ii) Assume time for light is zero, divide distance by time for sound

29 a) 1.50 s b) 0.75 s, 2.25 s

30 a) (i) For similar image position, see p142 (ii) virtual, laterally inverted **b)** (i) and (ii) For similar ray diagram, see p154 (top)

31 a) (i) X-rays (or γ-rays) (ii) infrared (or radio waves) **b)** 3.0×1020 Hz **c)** 3×10^8 m/s

32 a) Cell, lamp, and variable resistor should all be in series, with no breaks in circuit and no wire across resistor **b)** (i) 12 Ω (ii) 1 is 0.5 A, 2 is 0.5 A for all three (iii) Voltmeter should be in parallel with 4 Ω resistor

33 a) and **b)** Diagram and description as on p182 **c)** (i) 9 Ω (ii) 60 C (iii) 0.75 W

34 a) 12.5 A **b)** (i) 7.5 A (ii) 90 W (iii) 1.6 Ω

35 a) Region in which electric charge feels a force **b)** Field lines as on p174 (bottom right) **c)** 0.002 A **d)** 120 J

36 a) (i) Field line is anticlockwise circle with wire at centre, as on p206 (top) (ii) Needle would point to left **b)** (i) No change (ii) Points in opposite direction **c)** S is stronger, T is same strength, W is same strength

37 a) (i) and (ii) Graph is three waves, each similar to that on p218, each lasting 0.1 s **b)** Voltage induced when coil cuts through field lines; rate of cutting is maximum when plane of coil is parallel to field, as in diagram, but zero when plane is at right-angles to field **c)** kinetic energy from handle → electrical (+ heat) energy in coil

38 a) (i) 48 W (ii) 48 W **b)** (i) No energy (or power) wasted (ii) Current in primary coil causes changing magnetic field which generates changing voltage in each turn of secondary coil; more turns gives greater voltage

39 a) (i) Allows current in one direction only (ii) Graphs as on p232 (bottom) **b)** (i) See p239 (ii) Output is high if input is not high, and vice versa

40 a) See p240 **b)** (i) low (ii) low **c)** (i) Sequence from top to bottom then left to right: LOW, HIGH, HIGH, LOW (ii) No effect

41 a) (i) Alternating voltage with higher frequencies superimposed on basic frequency (ii) 1.6 V (iii) Connect CRO's input terminals to AC source of standard voltage **b)** (i) 6.1 cm (ii) 30.5 ms (ii) In finding speed of sound

42 a) (i) Decreases by 2 (ii) Decreases by 2 (iii) Decreases by +2 **b)** 400 count/min

43 a) Charge (or energy) (ii) **b)** (i) Zero (ii) Gradually rises to 6 V (iii) Allowing time between setting alarm and exiting house

44 a) 8 minutes **b)** (i) Arrangement similar to that on p255 (Q4) but with aluminium, not lead (ii) Total count without sheet present, time taken, total count with sheet present, time taken

45 a) (i) Arrangement similar to that on p255 (Q4) but with thick card, not lead (ii) Detect radiation without card present, then with (iii) α-particles stopped by card, so only background radiation detected when card present **b)** Beam follows circular path (out of paper) at right-angles to field

46 a) (i) $^{24}_{11}\text{Na} \rightarrow\ ^{24}_{12}\text{Mg} +\ ^{0}_{-1}\beta$ (ii) Beam follows circular path (into paper) at right-angles to field **b)** (i) GM tube or similar can detect arrival of isotope at different times in different parts of body (ii) γ-radiation not absorbed by body, so safer and can be detected outside

Alternative-to-practical questions (pages 318–320)

1 a) Values of e (in mm): 2, 3, 6, 9, 10, 12 **c)** gradient = 2 N/mm

2 a) Values of T (in s): 0.34, 0.44, 0.49, 0.53, 0.60, 0.63 **c)** 0.51 s **d)** No; line may be straight but does not pass through origin

3 a) Values of l (cm): 50, 75, 100 **b)** Ammeter needle three-quarters of way between 0.3 and 0.4; voltmeter needle half way between 1 and 2 **c)** Resistance values: 2.5, 4.00, 5.20 **d)** R/Ω **e)** 7.8 Ω

4 a) Ammeter is in right gap, voltmeter in bottom gap **b)** 3.4 Ω **c)** Ammeter and variable resistor in series with power source; three lamps and voltmeter all in parallel with these

5 a) (i) 15 mm (ii) 77.5 mm (iii) $u = 75$ mm, $v = 388$ mm (iv) 5.2 **b)** larger (5×), inverted **c)** Check focus for sharpness; make sure that distances are measured from centre of lens, and not from supporting block

6 a) All three **b)** temperature, length **c)** Measure time for 50 oscillations (for example); divide result by 50 to find T

Index

If a page number is given in **bold**, you should look this up first.